Science and Sustainable Food Security

Selected Papers of M S Swaminathan

IISc Centenary Lecture Series

Science and Sustainable Food Security

Selected Papers of M S Swaminathan

M S Swaminathan

M S Swaminathan Research Foundation, India

IISc Press

World Scientific

NEW JERSEY • LONDON • SINGAPORE • BEIJING • SHANGHAI • HONG KONG • TAIPEI • CHENNAI

Published by

World Scientific Publishing Co. Pte. Ltd.
5 Toh Tuck Link, Singapore 596224
USA office: 27 Warren Street, Suite 401-402, Hackensack, NJ 07601
UK office: 57 Shelton Street, Covent Garden, London WC2H 9HE

British Library Cataloguing-in-Publication Data
A catalogue record for this book is available from the British Library.

IISc Centenary Lecture Series — Vol. 3
SCIENCE AND SUSTAINABLE FOOD SECURITY
Selected Papers of M S Swaminathan

ISBN-13 978-981-4282-10-9
ISBN-10 981-4282-10-3

Editor: Tjan Kwang Wei

Printed in Singapore by B & Jo Enterprise Pte Ltd

Norman Borlaug
1914–2009

Dedicated to
Dr Norman E Borlaug for his unparalleled contributions to shaping our agricultural destiny and to saving millions of lives from hunger and famine.

ACKNOWLEDGEMENTS

I express deep sense of gratitude to the various publishers for giving copyright permission to reproduce the following articles in the book:

Taylor & Francis Group: Articles (1) Swaminathan, M.S. (1983). Agricultural Progress-key to third world prosperity. Third World Quarterly 5(3):553- 566. and (2) Swaminathan, M. S. 2003. Sustainable Food Security in Africa: Lessons from India's Green Revolution. South African Journal of International Affairs. 10(1): 11-26

Nature: Articles (1) Swaminathan, M.S. (1955). Overcoming cross-incompatibility among some Mexican diploid species of ***Solanum***. Nature 176:887-888 (2) Swaminathan, M.S. and A.T. Natarajan (1957). Polyploidy and radiosensitivity. Nature 179:479-480 and (3) Swaminathan, M.S. (1956). Disomic and tetrosomic inheritance in a ***Solanum*** hybrid. Nature 178:599-600.

Elsevier: Sanchez Pedro, A and Swaminathan M S (2005). Hunger in Africa : the link between unhealthy people and unhealthy soils. The Lancet – Vol. 265, Issue 9457, 29 January 2005, 442-444 Pp.

In addition, there are a number of articles in this book that do not require formal copyright permissions but at the same time I am thankful to these Publishers and the Journals for the rewarding association that I had with them over several decades. They are: The Little Magazine, Consultative Group on International Agricultural Research, TWAS Newsletter, Science, Field Crops Research, Cambridge University Press, Ankuram, Jawaharlal Nehru Memorial Lecture Organization, Current Science, Wheat Information Service, Springer, Wiley, Genetics, Australian National University Press, United Nations, Institute of Biology, World Meteorological Organisation. The Royal Society, Oxford & IBH Publishing Company, and The Hindu.

I also wish to thank Dr N Parasuraman for his painstaking efforts in getting all the papers assembled for this publication.

M S Swaminathan

Foreword

The Indian Institute of Science is one of India's oldest research institutions. In commemorating a century of existence a program of Centenary Lectures was initiated in 2008. These lectures were intended to bring scientists of the highest accomplishment to the Institute to inform and inspire a new generation of researchers. For Indians who have lived through the decades of the 1960s and 1970s two images are etched in memory. The first is the spectre of food shortages and rationing, which impacted almost all sections of the population. The second is the remarkable transformation that took place in the 1970s, when the Green Revolution took root, quickly consigning the era of large scale food imports to the realm of history. Prof. M.S. Swaminathan is one of the acknowledged architects of this agricultural revolution; a transformation that required the happy confluence of political will, administrative efficiency, scientific input and the enthusiastic acceptance by farmers of new crop varieties. Over three decades later India stands on the threshold of a new and looming crisis in agriculture which Prof. Swaminathan articulates so clearly: "We should look upon agriculture not just as a food-producing machine for the urban population, but as the major source of skilled and remunerative employment and the backbone of the rural livelihood system". This volume is intended to focus attention on these important issues, conveying a sense of urgency on the need to catalyze a second transformation in Indian agriculture in this new century.

P. Balaram
Director
Indian Institute of Science Bangalore
India

About Prof M.S. Swaminathan

Professor M S Swaminathan has been acclaimed by the TIME magazine as one of the twenty most influential Asians of the 20th century and one of the only three from India, the other two being Mahatma Gandhi and Rabindranath Tagore. He has been described by the United Nations Environment Programme as "the Father of Economic Ecology" because of his leadership of the ever-green revolution movement in agriculture and by Javier Perez de Cuellar, Secretary General of the United Nations, as "a living legend who will go into the annals of history as a world scientist of rare distinction". He was Chairman of the UN Science Advisory Committee set up in 1980 to take follow-up action on the Vienna Plan of Action. He has also served as Independent Chairman of the FAO Council (1981–85) and President of the International Union for the Conservation of Nature and Natural Resources (1984–90). He was President of the World Wide Fund for Nature (India) from 1989–96. He also served as President of the Pugwash Conferences on Science and World Affairs (2002–07), President of the National Academy of Agricultural Sciences (1991–96 and 2005–07) and Chairman, National Commission on Farmers (2004–06).

He served as Director of the Indian Agricultural Research Institute (1961–72), Director General of Indian Council of Agricultural Research and Secretary to the Government of India, Department of Agricultural Research and Education (1972–79), Principal Secretary, Ministry of Agriculture (1979–80), Acting Deputy Chairman and later Member (Science and Agriculture), Planning Commission (1980–82) and Director General, International Rice Research Institute, the Philippines (1982–88).

A plant geneticist by training, Professor Swaminathan's contributions to the agricultural renaissance of India have led to his being widely referred to as the scientific leader of the green revolution movement. His advocacy of sustainable agriculture leading to an ever-green revolution makes him an acknowledged world leader in the field of sustainable food security. The International Association of Women and Development conferred on him the first international award for significant contributions to promoting the knowledge, skill, and technological empowerment of women in agriculture and for his pioneering role in mainstreaming gender considerations in agriculture and rural development. Professor Swaminathan was awarded the Ramon Magsaysay Award for Community Leadership in 1971, the Albert Einstein World Science Award in 1986, the first World Food Prize in 1987, and Volvo and Tyler Prize for Environment, the Indira Gandhi Prize for Peace, Disarmament and Development in 2000 and the Franklin D Roosevelt Four Freedoms Medal, the Mahatma Gandhi Prize of UNESCO in 2000 and the Lal Bahadur Sastri National Award (2007).

Professor Swaminathan is a Fellow of many of the leading scientific academies of India and the world, including the Royal Society of London and the U S National Academy of Sciences. He has received 59 honorary doctorate degrees from universities around the world. He currently holds the UNESCO Chair in Ecotechnology at the M S Swaminathan Research Foundation in Chennai (Madras), India. He is a Member of the Parliament of India (Rajya Sabha), to which position he was nominated by the Government of India in May 2007 in recognition of his contributions in the field of agricultural research and development.

Contents

Dedication v

Acknowledgment vii

Foreword ix

About the Editor xi

Introduction 1

Section I: Food Security and Economic Development — *How science is applied to solve problems of poverty, drought and famine*

Commentary

1. Swaminathan, M.S. (1983). Agricultural Progress — Key to Third World Prosperity. *Third World Quarterly* **5**(3): 553–566. 27

2. Swaminathan, M.S. (1990). Changing Nature of the Food Security Challenge: Implications for Agricultural Research and Policy. *Sir John Crawford Memorial Lecture (Consultative Group on International Agricultural Research (CGIAR)*, Washington DC. 41

3. Swaminathan, M.S. (2001). Bridging the Nutritional Divide — Building Community Centred Nutrition Security Systems. *The Little Magazine*, Year end, 2001, pp. 15–25. 67

4. Swaminathan, M.S. (2004). Africa's Rainbow Revolution. *TWAS Newsletter 3/4*, **10**, 72–76. 79

5. Sanchez Pedro, A. and Swaminathan, M.S. (2005). Hunger in Africa: The Link Between Unhealthy People and Unhealthy Soils. *The Lancet* — Vol. **265**, Issue 9457, 29 January 2005, 442–444. 85

6. Sanchez Pedro, A. and Swaminathan, M.S. (2005). Cutting World Hunger in Half. *Science* **307**(5708): 357–359. 89

7. Swaminathan, M.S. (2007). Can Science and Technology Feed the World in 2025? *Field Crops Research* **104**: 3–9. 93

8. Parry, Martin L. and Swaminathan, M.S. (1992). Effects of Climate Change on Food Production. *Confronting Climate Change: Risks, Implications and Responses*. Irwing, M, Montzer (ed.) Cambridge University Press, pp. 113–115. 107

9. Swaminathan, M.S. (2003). Sustainable Food Security in Africa: Lessons from India's Green Revolution. *South African Journal of International Affairs* **10**(1): 11–26. 133

10. Swaminathan, M.S. (2003). Sustainable Food and Water Security. *Ankuram – Environment Quarterly* **1**(1): 23–33. 147

Section II: Science and Food Security — *How science is used to generate efficient and optimal agricultural outputs*

Commentary

11. Swaminathan, M.S. (1999). Science and Sustainable Food Security. *31st Jawaharlal Nehru Memorial Lecture*, New Delhi, 39 pp. 185

12. Swaminathan, M.S. (1981). Indian Agriculture at the Crossroads. *Current Science* **51**: 13–24. 225

13. Bui Ba Bong and Swaminathan, M.S. (1995). Magnitude of Hybrid Vigor Retained in Double Haploid Lines of Some Heterotic Rice Hybrids. *Theor. Appl. Genet.* **90**: 253–257. 237

14. Swaminathan, M.S., Chopra, V.L., Joshi B.C. and Singh, D. (1968). Development of Monosomic Series in an Indian Wheat and Isolation of a Nullisomic Lines. *Wheat Information Service* **27**: 19–20. 243

15. Kumar, S., Pai, R.A. and Swaminathan, M.S. (1967). Consanguineous Marriages and the Genetic Load Due to Lethal Genes in Kerala. *Ann. Hum. Genet. Lond.* **131**: 141–147. 245

16. Swaminathan, M.S. (1965). The Experimental Manipulation of Genes. *Curr. Sci.* **34**: 108–111. 253

17. Swaminathan, M.S. (1954). Nature of Polyploidy in Some 48-Chromosome Species of the Section ***Tuberarium*** Genus ***Solanum***. *Genetics* **39**: 56–76. 257

18. Swaminathan, M.S. (1955). Overcoming Cross-Incompatibility Among some Mexican Diploid Species of ***Solanum***. *Nature* **176**: 887–888. 275

19. Swaminathan, M.S. and Natarajan, A.T. (1957). Polyploidy and Radiosensitivity. *Nature* **179**: 479–480. 277

20. Swaminathan, M.S. (1956). Disomic and Tetrosomic Inheritance in a ***Solanum*** Hybrid. *Nature* **178**: 599–600. 279

21. Swaminathan, M.S. (1992). The Green Revolution in Indian Agriculture From an Environmentally Sound Technology Point of View. *Environmentally sound technology for sustainable development. Advanced Technology Assesment System Bulletin (7).* Dirk Pilari and David Philips Eade (eds.). United Nations Publication, New York, pp. 143–148. 281

22. Swaminathan, M.S. (2006). Science and Shaping Our Agricultural Future. *The First Ten K. R. Narayanan Orations: Essays by Eminent Persons on the Rapidly Transforming Indian Economy.* Raghbendra Jha (ed.). The Australian National University Press, Canberra, pp. 155–183. 293

Section III: Food Security and Ecological Balance — *How the gains of green revolution are impacted by climate change, how science will be helpful in ensuring sustainable food security, green revolution to ever-green revolution — a roadmap*

Commentary

23. Swaminathan, M.S. (2000). An Evergreen Revolution, *Biologist* **47**(2): 85–89. 325

24. Swaminathan, M.S. (1990). Agriculture and Food Systems. *Proceedings of the Second World Climate Conference, Geneva, World Meteorological Organisation.* 331

25. Kesavan, P.C. and Swaminathan M.S. (2006). Managing Extreme Natural Disasters in Coastal Areas. *Philosophical Transactions of The Royal Society*, pp. 2191–2216. 345

26. Swaminathan, M.S. (1999). Ecological Security — A Prerequisite for Food and Livelihood Security. Ahmad K. Hegazy (ed.). *Environment 2000 and Beyond.* International Center for Environment and Development, Cairo, Egypt, pp. 203–222. 367

27. Swaminathan, M.S. (1984). Genetic Conservation: Microbes to Man. Presidential Address, *XV International Congress on Genetics. Genetics: New Frontiers*, Vol. 1. Oxford & IBH Publishing Co., New Delhi, pp. 29–56. 387

28. Monsoon Management in an Era of Climate Change. 417

From Green to an Ever-green Revolution

M.S. Swaminathan

The rediscovery of Mendel's laws of inheritance in 1900 helped plant breeders to initiate scientifically designed efforts to develop new varieties of crops based on the principles of genetic segregation, recombination and mutation. Genetic selection has however been in progress almost from the beginning of agriculture or settled cultivation over 10,000 years ago. This explains the basis for the occurrence of a rich diversity of land races or locally adapted strains of both plants and farm animals. Genetics however helped to undertake carefully planned hybridization and selection experiments. From 1930 onwards, mutation breeding using ionizing radiations also became feasible, thanks to the discovery by H J Muller and L J Stadler that exposure of animals and plants to X-rays and other radiations can result in the induction of mutations.

From the early years of agriculture, increasing crop productivity and resistance to pests and diseases have been major goals of farm families. The discovery of mineral fertilizers in the later part of the 19th century by Liepzig and other German scientists helped to halt the law of diminishing return of the soil. Genetic resistance, crop rotations and chemical and biological pesticides helped to minimize crop losses due to pest epidemics. Inspite of such progress, famines were frequent in the past due to crop failure caused by both pest epidemics and adverse meteorological conditions. Even today, the saying that "Indian agriculture is a gamble in the monsoon" has some validity.

The rediscovery of Mendel's laws of inheritance almost coincided with the establishment in 1905 of the Indian (then Imperial) Agricultural Research Institute at Pusa, Bihar, and the Agricultural Colleges at Coimbatore, Nagpur, Pune, Kanpur and Lyalpur (now Faizalabad in Pakistan). At all these centres, research, education and extension education were integrated in order to help achieve a technological upgrading of traditional farming practices. The early breeding method was selection among naturally occurring variability. This led to wheat varieties like NP4 developed by Sir Albert and Gabrielle Howard at Pusa and rice varieties like GEB 24 bred by Dr K Ramaiah at Coimbatore.

The early breeding work centred around the development of crop varieties with resistance to pests and diseases, adaptation to local growing conditions and good culinary and organoleptic properties. Often dual purpose varieties were selected – good grain quality for human consumption and good straw quality for the consumption of ruminating farm animals. In wheat, varieties like NP809 possessing resistance to stem, leaf and stripe rusts were developed at IARI by a team led by Dr B P Pal.

After independence in 1947, increase in human population occurred at an accelerated pace, thanks to advances in preventive and curative medicine. Also, India's independence was born in the backdrop of the Bengal Famine of 1942-43, which resulted in the death of over 2 million children, women and men. Thus led Jawaharlal Nehru, the first Prime Minister of India, to observe in 1947, "everything else can wait but not agriculture". The Nehru era (1947-1964) was marked by infrastructure development for scientific agriculture, such as irrigation projects, fertilizer factories, Agricultural Universities and research institutes and extension agencies. Inspite of all these efforts, food production could not keep pace with population increase, with the result that the country started depending upon imported food, particularly wheat under the PL480 programme of the United States of America. As much as 10 million tonnes of wheat were imported in 1966. Indians were thus reduced to a "ship to mouth" existence.

It is in this background that the Government of India introduced in 1961 an Intensive Agricultural District Programme (IADP) in order to maximize the benefits of assured irrigation facilities, by providing farmers a package of inputs like seeds, fertilizers and extension advice. Early results from IADP were disappointing in relation to the enhancement of productivity per hectare. Advances in production were largely due to an increase in area rather than in yield, ie, a horizontal expansion in cultivated area and not a vertical expansion in productivity. It was observed that traditional wheat and rice varieties, if fed with mineral fertilizers, tended to lodge because of their tall height and thin straw. Since wheat and rice plants need about 25 and 20 kgs of Nitrogen respectively to yield one tonne of grain, it became clear that unless the architecture of the plant can be redesigned in order to enable them to utilize higher quantities of nutrients and irrigation water, low yields will continue to prevail.

My first assignment on my return from foreign studies early in 1954 was in the *Indica-Japonica* hybridization programme in rice at the Central Rice Research Institute at Cuttack. The aim of this programme which was a brain child of Dr K Ramaiah was to transfer genes for fertilizer response from *japonica* to *indica* rice varieties. The *indica* varieties cultivated then had tall straw and will lodge if the quantity of mineral fertilizers essential for raising the yield level from 1 to 4 tonnes of rice, was applied. The *indica-japonica* hybridization programme resulted in a few good varieties like ADT-27 in Tamil Nadu and Mashuri in Malaysia. However, real progress in yield improvement had to await the introduction of semi-dwarf *indica* varieties, based on the Dee-gee-woo-gen dwarfing gene identified in China and used in Taiwan to develop varieties like Taichung Native 1. Later, this gene was used at the International Rice Research Institute (IRRI) at Los Banos, the Phillipines, leading to the breeding of the high yielding variety, IR8 in 1968.

I joined the Indian Agricultural Research Institute (IARI), New Delhi in late 1954. IARI, which was originally located at Pusa in Bihar, was shifted to New Delhi in 1936, following the damage caused to its original laboratories at Pusa by the Bihar earthquake of 1934. My first aim, on joining IARI, was to repeat the efforts in rice, namely the breeding of high yielding varieties which can take advantage of good soil fertility management and irrigation water availability. A three pronged strategy was developed for this purpose.

- First, efforts were made to develop **erectoides** mutants on the lines of the work done in barley in Sweden by a team led by Prof Ake Gustafson. X-rays, gamma rays and fast and thermal neutrons, as well as chemical mutagens like ethyl-methane-sulphonate were used for this purpose. A Gamma Garden using a 200 Curie Cobalt 60 source was also set up at IARI for subjecting plants to chronic doses of gamma radiation.

- Secondly, a search was made for naturally occurring dwarf wheat species and varieties like *Triticun aestivum* sub.sp *sphaerococcum* (popularly known as Mohanjadaro wheat, since its Kernels were identified in the Mohanjadaro excavation). *T. aestivum sub.sp compactum* and Tom Thumb. They were crossed with wheat varieties from IARI as well as Punjab (eg. C591, C306 etc bred by Chaudhary Ram Dhan Singh).
- Thirdly, attempts were made to induce chemical dwarfing by spraying maleic hydrazine.

Unfortunately, all these pathways of breeding fertilizer responsive, semi-dwarf wheat varieties did not succeed, since the dwarf height was associated with a short panicle with a low grain number and grain weight. Thus the dwarf plant habit had a pleiotropic effect on the panicle and thereby on yield.

It is in this background, I was attracted to the work of Dr Orville Vogel of the Washington State University, Pulman, USA when he announced the release of the winter wheat variety Gaines, capable of yielding over 10 tonnes per ha, based on the Norin dwarfing gene identified by Dr Gonziro Inazuka at the Norin Experiment Station in Japan (Fig. 1). I wrote to Dr Vogel in 1960 requesting seeds of Gaines. While sending the seeds, he also mentioned that winter wheats, which required long days and mild temperature to flower and set grain, will not do well under the short day conditions prevailing during the *rabi* season (November to April) in North India. He suggested that I should obtain seeds of spring wheat varieties with the same dwarfing gene from Dr Norman E Borlaug, who was developing semi-dwarf varieties under a Rockefeller Foundation – Mexican Government programme in Mexico using the Norin dwarfing genes made available to him by Dr Vogel. Meanwhile, I and my colleague, Dr M V Rao, observed during 1960-61 semi-dwarf wheat varieties with long panicles and high yield potential in the International Wheat Rust Nursery sent by the US Department of Agriculture as part of an international programme to map the wheat rust profile in different countries and to identify genes for resistance to the dominant races of leaf, stripe and stem rusts. On enquiring from USDA, the semi-dwarf, long panicle entries were traced to the wheat breeding programme of Dr Norman Borlaug in Mexico. I then wrote to Dr Borlaug requesting seeds of a wide range of semi-dwarf material. He promptly replied saying that he will be happy to provide the material but it will help him to select material for being sent to India, if he visits the wheat growing areas of India. Early in 1961, I wrote to Dr B P Pal then Director of IARI, requesting him to extend an invitation to Dr Borlaug to visit India. It took over a year for the Ministry of Agriculture to send a letter of invitation to Dr Borlaug through the Rockefeller Foundation (then headed in India by Dr Ralph W Cummings). Finally Dr Borlaug visited us in March 1963.

Figure 1

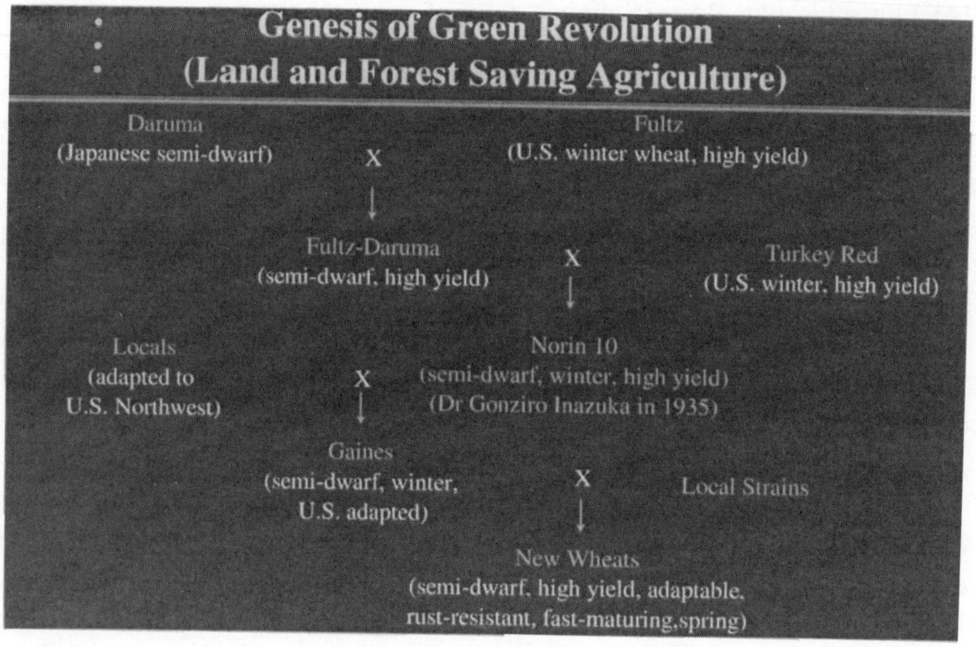

During March 1963, I and some of my colleagues took Dr Borlaug to the major wheat growing regions of the country. It was a wonderful experience traveling with him since I found him to be not only a brilliant scientist, but a humanitarian to the core. In an article in Yojana (published by the Planning Commission in 1965) I referred to Dr Borlaug as the Albert Schweitzer of Agriculture and I was happy that like Schweitzer, he also received the Nobel Peace Prize in 1970.

On the conclusion of his visit towards the end of March 1963, Dr Borlaug mentioned that since most of his semi-dwarf wheat material was growing in Pakistan – Punjab, he would like to observe their behaviour before selecting strains to be sent to India. He therefore went in late March 1963, from Delhi to Lahore and spent a week in Pakistan studying the performance of the Mexican wheat material he had sent there under a FAO supported Cereal Improvement Programme. In May, 1963, he wrote to me an ecstatic letter saying that several of the Mexican semi-dwarf strains are performing very well in Pakistan. I requested him to send, in addition to segregating populations, atleast 100 kgs of seeds of 4 semi-dwarf varieties released in Mexico, namely Sonora 63, Sonora 64, Majo 64, and Lerma Rojo 64-A. He was good enough to send in September 1963, bulk quantities of seeds of these varieties, in addition to a wide range of segregating material. A five year plan (1963-68) was developed to prepare a road map for the Wheat Revolution.

To purchase time, I decided to divide the seeds received into several lots, so that we can get data simultaneously from Delhi, Ludhiana, Pant Nagar, Kanpur and Pusa during the *rabi* season of 1963-64. This helped us to study the genotype X environment interaction. I summarized this data at the All India Wheat Workshop held at IARI in August 1964 (Swaminathan MS 1965,

"The Impact of Dwarfing Genes on Wheat Production" Journal of the IARI Post-graduate School, 1965, Vol.2, pp 57-62). It was clear that with the semi-dwarf varieties we can atleast treble the yield immediately (the average yield of wheat at that time was less than 1 tonne per ha). In addition, the semi-dwarf wheat strains from Mexico showed wide adaptation to diverse growing conditions, since Dr Borlaug had introduced the character of photo-insensitivity through shuttle breeding. I therefore proposed that a National Demonstration Programme may be initiated in small farmers' fields during *rabi* 1964-65, in order to serve as windows into the world of high wheat yield awaiting our farmers (Swaminathan, M.S. 1966. National Demonstrations in Rice. Indian Farming 16(6):67-70). *I had stipulated in my proposal that the demonstrations should be laid in the fields of small farmers since yield results from demonstrations in rich farmers' fields will be attributed to affluence and not to technology*. There were objections to this approach from the Ministry of Agriculture, since according to them, demonstrations should be laid out in the fields of "progressive farmers", an euphemism for resource–rich farmers. Fortunately, Bharat Ratna C Subramaniam, who had then joined as Union Minister of Food and Agriculture, overruled this objection and approved my proposal for organizing National Demonstrations in the fields of resource poor farmers. Seeds and fertilizers were provided to farmers under this programme but all the cultivation work was done by the farm family. The small farmers chosen under this programme adopted the right agronomic practices and harvested on an average over 4 tonnes per ha during April 1965. This revolutionary change in yield stirred up great enthusiasm among wheat farmers and there was a huge demand for seeds of the semi-dwarf wheat varieties. The Government of India approved in April 1966, our proposal to import 18,000 tonnes of seeds of Lerma Rojo 64A and Sonora 64 from Mexico. These seeds arrived on time for *rabi* sowing in September 1966 and were distributed among all major wheat growing states. At the same time, a Seed Village programme was started in the Jounti Village in Delhi State, where all farmers agreed to multiply the seeds of the semi-dwarf wheat varieties (Swaminathan, M S, 1968. The Evolution and Significance of Jounti Seed Village. Indian Farming. Jan 1-4). As a result of this *"purchase time operation"* nearly a million ha came under semi-dwarf varieties in the rabi season of 1967-68, resulting in an increase of nearly 7 million tonnes in wheat production in April 1968. Meanwhile, we also developed semi-dwarf varieties with amber grains and good chapathi making qualities from the segregating material received from Dr Borlaug, such as Kalyan Sona and Sonalika.

In February, 1968, Smt Indira Gandhi came to IARI for delivering the Convocation Address. I then mentioned in my welcome address that the wheat harvest of 1968 will mark the beginning of a new era in Indian agriculture. I requested her to release a special stamp to bring this significant event in India's agricultural history to the attention of the public. She approved this proposal and a special stamp titled "the Wheat Revolution" was released by her in July 1968 at the IARI Auditorium. The details of the scientific and developmental efforts which led to the Wheat Revolution are chronicled in a Monograph titled "Wheat Revolution: a Dialogue" (Swaminathan, M.S. (Ed.) 1993. *Wheat Revolution: A dialogue*. Macmillan India Ltd., Madras 164pp).

During my visits in 1967 to villages in the Punjab, Haryana and Western UP region, which later became the heartland of the Green Revolution, I observed that farmers were already wanting what they called "triple dwarfs" so that they can apply more fertilizer and irrigation water and thereby go up to about ten tonnes per hectare in yield. I warned them not to apply too much fertilizer and also not to pump too much ground water, in order to prevent the green revolution from becoming a "greed revolution". I summarized my views in an address to the Agricultural Sciences Section of the Indian Science Congress held at Varanasi from January 3-7, 1968 in the following words:

> "*Exploitive agriculture offers great dangers if carried out with only an immediate profit or production motive. The emerging exploitive farming community in India should become aware of this. Intensive cultivation of land without conservation of soil fertility and soil structure would lead, ultimately, to the springing up of deserts. Irrigation without arrangements for drainage would result in soils getting alkaline or saline. Indiscriminate use of pesticides, fungicides and herbicides could cause adverse changes in biological balance as well as lead to an increase in the incidence of cancer and other diseases, through the toxic residues present in the grains or other edible parts. Unscientific tapping of underground water will lead to the rapid exhaustion of this wonderful capital resource left to us through ages of natural farming. The rapid replacement of numerous locally adapted varieties with one or two high-yielding strains in large contiguous areas would result in the spread of serious diseases capable of wiping out entire crops, as happened prior to the Irish potato famine of 1854 and the Bengal rice famine in 1942. Therefore the initiation of exploitive agriculture without a proper understanding of the various consequences of every one of the changes introduced into traditional agriculture, and without first building up a proper scientific and training base to sustain it, may only lead us, in the long run, into an era of agricultural disaster rather than one of agricultural prosperity.*"

This line of thinking led me later to coin the term Ever-green Revolution to emphasise the need for improving productivity in perpetuity without associated ecological harm (Swaminathan, M.S. 1996. Sustainable Agriculture: Towards an Evergreen Revolution. Konark Publishers Pvt Ltd, Delhi. pp 232). The ever-green revolution concept is based on a paradigm shift from a commodity centred to a farming system centred approach in technology development and dissemination. It involves attention to Integrated Natural Resources Management.

The pathways to ever-green revolution could be either organic farming or green agriculture. Organic farming precludes the use of mineral fertilizers, chemical pesticides and genetically modified crops. Green agriculture, on the other hand involves the adoption of integrated pest management, integrated nutrient supply, scientific water management and use of appropriate crop varieties whether bred by Mendelian or Molecular techniques. Green agriculture also involves improved post-harvest technology.

In India, nearly eighty percent of farm families own one hectare or less of arable land. Therefore, there is need for higher productivity in small farms in order to ensure adequate marketable surplus to earn cash income. Organic farming will be more easy to adopt if the farmer has several farm animals. The greatest scientific challenges in organic farming are soil fertility restoration and plant protection. Organic farming is particularly important in vegetable crops and medicinal plants, in order to avoid pesticide residues in the edible product. The following paradigm shift in research and development strategies will be needed to provide the technological underpinning of an ever-green farm revolution.

- Research Strategies to meet the challenge of Diversity in Farming Systems and Agro-Ecological, Socio-Economic and Socio-Cultural Conditions
- Participatory Breeding (combining genetic efficiency with genetic diversity)
- Participatory Knowledge Management
- Proactive advice on land use based on home needs, nutritional deficiencies and market opportunities
- From cropping system to farming system research based on integrated natural resources and pest management, crop-livestock-fish integration, concurrent attention to production and post-harvest technologies and to on-farm and non-farm employment
- Proactive research on the mitigation of the adverse impact of climate change
- Improving Soil Fertility and Fostering a Soil Health Care Movement and giving every farmer a Soil Health Monitoring Card
- Adoption of agro-forestry systems involving nitrogen-fixing shrubs and trees like *Leucina leucocephala* (Subabul) and *Faidherbia albida*
- Rain Water Harvesting and Watershed Management – Concurrent attention to Supply Augmentation and Demand Management – More income per unit of water

Ever-green agriculture will lead to a small farm management revolution because of its emphasis on the scientific management of the ecological foundations essential for sustainable agriculture. It will thus uphold the following principle enunciated by Indira Gandhi in her address to the UN Conference on Human Environment, Stockholm, 1972:

> *"Modern man must reestablish an unbroken link with nature and with life. He must again learn to invoke the energy of growing things and to recognize, as did the ancients in India centuries ago, that* ***one can take from the Earth and the atmosphere only so much as one puts back into them.,***

Fortunately, frontier technologies like space, information and biotechnology offer opportunities for fostering sustainable land and water use and also for developing novel genetic combinations to meet challenges arising from expanding biotic and abiotic stresses. The concept of ever-green revolution is now widely accepted as the road to sustainable food security. For eg., E O Wilson in his book "Future of Life", made the following remarks (Wilson E O, 2002. The Future of Life, Alfred A Knopf, pp. 229):

The problem before us is how to feed billions of new mouths over the next several decades and save the rest of life at the same time without being trapped in a Faustian bargain that threatens freedom from security. ***The benefits must come from an evergreen revolution.*** *The aim of this new thrust is to lift production well above the levels attained by the Green Revolution of the 1960s, using technology and regulatory policy more advanced and even safer than now in existence.*

Combining Yield and Quality in Rice

In 1965, semi-dwarf varieties of rice from Taiwan like Taichung Native One, Tainan Three, etc., became available through the International Rice Research Institute (IRRI) in the Philippines. I then felt that it will be useful if we can transfer the gene(s) for dwarfing from TN1 to our Basmati rice varieties in order to combine the semi-dwarf nature with the excellent grain quality and aroma of Basmati 370 and other Basmati strains. I therefore initiated a back-crossing programme between semi-dwarf rice strains and Basmati 370 with the help of Post Graduate scholars like Dr E A Siddiq. In 1967, the precursor lines of IR8 also became available. The expected genetic combination of reduced plant height and basmati grain characters however was very difficult to achieve. A few better quality rices like *Sabarmati* were developed, but none of the back-cross derivatives had the basmati kind of grain quality. However IARI continued this work under the guidance of Drs Siddiq, V P Singh, A K Singh, Zaman and many other dedicated young scientists which finally resulted in strains known as Pusa Basmati. The latest in this series, Pusa 1121 has outstanding basmati kind of grain quality and has recently been approved for release as a basmati strain. In view of the importance of this scientific development in rice, I would like to give brief information on Pusa 1121, now named Pusa Basmati 4.

Pusa Basmati 4 (Pusa 1121) combines unique Basmati quality characters. It possesses extra long slender (7.71 mm), highly aromatic grains with 52.9% HRR and very occasionally chalky grains. It has the highest KLAC of 17.8 mm and exceptionally high cooked kernel elongation with an elongation ratio of 2.5 and volume expansion nearly four times. It has intermediate amylase content, appealing taste, good mouth feel and easy digestibility. In the Panel test conducted by the Directorate of Rice Research, the proposed basmati variety Pusa 1121 featuring highly elongated long slender grains on cooking, achieved high scores for taste, aroma and attractive appearance during all three years of testing (2002-04).

Economics of Pusa Basmati 4 cultivation:

Because of its superior grain and cooking quality attributes, Pusa 1121 is being offered approximately $200/ tonne more than Traditional Basmati (Taraori Basmati) in the international market (Personal communication with trade). Since its release in 2003, the area under Pusa 1121 has been growing very fast. Encouraged with response in the international market, the traders/ millers in India were able to pay good price for Pusa 1121 to farmers. During *Kharif* 2006, the average yield of Pusa 1121 on farmers' fields was recorded as 5 tons / ha and the average paddy price of this variety in the Mandi was Rs. 1800/q bringing a gross return of Rs. 90,000/ ha to farmers as against Rs. 45,000/ha from cultivation of Taraori Basmati and Rs. 60,000/ ha from Pusa Basmati 1.

Area under Pusa 1121 cultivation during Kharif 2007:

The high profitability from Pusa 1121 resulted in major shift in cultivation of Basmati rice towards Pusa 1121, even the area under Permal rice (Pusa 44 and PR series non-Basmati rices) in Punjab and Haryana has been diverted to a large extent to Pusa 1121. According to the estimates provided by the President, All India Rice Exporters' Association (AIREA), during the *Kharif* 2007, nearly 50-60% of 2 million ha area under Basmati rice has been planted to Pusa 1121. The Breeder seed indent for this variety for production during *Kharif* 2007 was 23 q, which is a clear reflection its popularity with farmers. As per the seed production plan prepared by the Haryana State for 2008-10, 1500 q of certified seed of Pusa 1121 is to be produced each year, which is three times more than CSR30 and Taraori Basmati. This renamed variety Pusa Basmati 4, matures a week earlier than Taraori Basmati allowing timely sowing of wheat.

Release of Pusa 1121 in Punjab:

During *Kharif* 2007, Pusa 1121 was grown in substantially large area in the State of Punjab. Having seen its unprecedented success, the Govt. of Punjab has released Pusa 1121 for cultivation in the state, as a replacement to Traditional Basmati rice variety Basmati 386. Pusa 1121 is likely to replace partly some of the widely grown non-Basmati rice varieties such as Pusa 44, PR 106, PR 113, PR 116 and PR 118 during the *Kharif* 2008 crop season.

After the official recognition of Pusa 1121 as a Basmati variety, the price of Pusa Basmati 4 went up to nearly Rs. 3200 per quintal. Farmers in Haryana alone have derived a profit of over Rs. 2000 crores from this variety. This is a good illustration of the power of modern science to improve the economic well being of farmers.

State of Indian Agriculture Today and the Way Ahead

Indian agriculture is characterized by the following distinctive features as per the data collected under the 59th Round of NSSO (Report No. 497) as well as other studies:

- Over 60% of the population depends on crop and animal husbandry, inland and marine fisheries, forestry and agro-processing for their livelihood
- Over 60% of the consumers are farm families and the prevailing widespread malnutrition can be eliminated only by improving their income and purchasing power
- Farming is the largest private sector enterprise in the country; State-wise information shows that the lowest percentage of farm households, out of total rural households is 35% in Tamil Nadu. The highest at 91% is in Jammu and Kashmir

- Out of 115 million holdings, 90 million are 1 hectare or less in size. Tiny holdings limit the capacity of farmers to take to the technological upgrading of farm enterprises. They also limit their coping capacity to withstand economic shocks caused by meteorological and / or marketing policies
- Out of 178.25 million farmers, 136.85 million (76.7%) are foodgrain farmers. 13 million farmers produce cotton and sugarcane
- Rural poverty and under-nutrition are higher than urban poverty and hunger
- There is increasing feminisation of agriculture. Women constitute over 50% of the agricultural work force. Yet, women have no land rights and are push-outs of the credit system
- Rural infrastructure is poor, particularly in the area of post harvest technology. There is a mismatch between production and post-harvest technologies in most farming systems
- Capital formation in agriculture is going down and the credit-deposit ratio is adverse in rural branches
- The principle of financial inclusion is yet to become a reality in rural India
- Prime farmland is being diverted for non-farm purposes, due to the lack of an agricultural land conservation policy; consequently more food and other agricultural products will have to be produced under conditions of diminishing per capita availability of arable land and irrigation water
- The rate of growth of employment in rural areas is poor, leading to the population supporting capacity of the ecosystems being exceeded in most parts of the country and to the unplanned migration of the rural poor to urban slums
- The average monthly income per farmer household is Rs. 2115 and does not cover the average per capita monthly consumption expenditure of Rs. 2770. Only households with landholding of 4 ha and above have a surplus of income over expenditure
- The cost-risk-return structure of farming is getting adverse leading to a sense of despair in the rural areas. Consequently, the youth in villages are reluctant to take to farming as a career
- The ecological foundations essential for sustainable agriculture like land, water, forests and biodiversity are in varying degrees of decay and adverse changes in climate appear a distinct possibility as a result of global warming
- In the prevailing atmosphere of gloom and doom, **there are many bright spots** – farmers with access to adequate land, water, capital and technology are doing very well in areas like horticulture and animal husbandry and are demonstrating the strengths of Indian agriculture, arising from abundant sunshine and a rich diversity of agro-ecological and agro-climatic conditions.

Keeping in view the above scenario, the National Commission on Farmers (NCF) recommended a Five Point Action Plan to rejuvenate our agriculture. These are:

- Conservation and enhancement of soil health and the issue of Soil Health Cards to every farm family
- Harvesting, storage and efficient use of rain water, as well as the conjunctive use of different water sources

- Pro-small farmer credit and insurance system
- Environmentally friendly technologies supported by appropriate inputs
- Assured and remunerative marketing
- The food security basket should be widened to include the crops of the dry farming areas like **bajra, jowar, ragi, minor millets** and **pulses**. The Public Distribution System (PDS) should include **these nutritious cereals and pulses purchased at a remunerative Minimum Support Price (MSP)**. This will be a win-win situation both for the dryland farmer and the consumer. **We will witness neither a second green revolution, nor much progress in dryland farming, unless farmers get assured and remunerative prices for their produce.** A nutritional literacy and food safety programme also needs to be launched through Gyan Chaupals in conjunction with PDS.
- Food security with home grown foodgrains can alone eradicate widespread rural poverty and malnutrition, since farming is the backbone of the livelihood security system in rural India. This will enable **Government to remain at the commanding height of the national food security system**. Building a food security system and containing price rise with imported foodgrains may sometimes be a short term necessity, but will be a long term disaster to our farmers and farming. **A well-defined, pro-farmer and pro-resource poor consumer Food Security Policy is an urgent necessity. Also, horticultural remedies should be promoted for nutritional maladies, by mainstreaming nutrition in the National Horticulture Mission.**
- Both universal PDS and enforcing MSP throughout the country for the selected crops are essential for imparting dynamism to agriculture. Higher consumption enhances demand and gives a stimulus to increased production.

Just as IT industries have specialized in handling outsourcing assignments efficiently, we must enable our farm graduates and farmers to take up outsourcing jobs, in areas where we have a comparative advantage. Some examples are hybrid seed production, tissue culture propagated plants, organic farm products, biological software for sustainable agriculture like biopesticides, biofertilizers, pheremones as well as herbal products, fruits, flowers and vegetables, vaccines and sero-diagnostics and veterinary pharmaceuticals based on medicinal plants. There is also scope for becoming a global outsourcing hub in the areas of plant and animal genomics and ICT for rural poor. Farm, Veterinary, Fisheries and Home Science Graduates should be trained to become genome and digital entrepreneurs. To start with, a few of our Agricultural, Animal Sciences and Fisheries Universities could setup **Bureaus for Outsourcing Business in Agriculture** to facilitate contacts between farmers' organizations as well as agri-business centers operated by farm and home science graduates and external agri-business enterprises. Outsourcing should not only be from other countries to urban India, but also from urban to rural India, so that educated youth continue to live in villages.

We need a new vision for agriculture. That vision should aim to spread happiness among farm and rural families. Bio-happiness through the conversion of our bio-resources into wealth meaningful to our rural families should be the goal of our national policy for farmers. The hidden and unrecognized opportunities for creating more skilled jobs and income in the farm and non-farm sectors need to be tapped through appropriate public policies and programmes. Technology Missions should be revamped and revitalized on the lines envisaged by Bharat Ratna Rajiv Gandhi when he first proposed them as organizational and management tools to help the nation leapfrog in the production of essential commodities like oilseeds and pulses.

A structurally progressive economy should reduce the share of people dependent on a sector as the share of that sector falls in the GDP. As the share of agriculture in the GDP falls, the share of people dependent on agriculture is also expected to fall in the same proportion. However in the Indian Economy though the share of agriculture in the GDP is falling steadily, there is no corresponding decline in the share of population dependent on agriculture. Because of population growth, the absolute number of people depending on agriculture is increasing, even if there is a decline in percentage terms. This is why NCF has recommended a major integrated Rural Non-farm Livelihood Initiative, so that both on-farm and non-farm livelihoods become skilled and profitable. The population supporting capacity of major ecosystems is under severe stress. Through greater attention to non-farm employment and to the concurrent growth of industry and agriculture, we should aim to bring down the population dependant on agriculture to about 30%.

The Technology strategy for an evergreen revolution should be based on methods of defending the gains already made in the heartland of the green revolution.

Defending the Gains

Punjab farmers provide 60% of wheat and 40% of rice to the PDS and national buffer stocks. The net productivity has increased in the Punjab in rice and wheat from 1.2 t/ha and 1.1 t/ha to 4.3 t/ha and 3.9 t/ha respectively from 1960-61 to 2004-05. However, in recent years there is stagnation in productivity improvement due to a variety of causes of which the following are important:

- Declining farm size and income
- Depleting natural resources base, as for example a steep fall in ground water table and impaired water quality.
- Increasing input costs, particularly diesel and adverse economics of farming
- Deficiency of soil organic matter and micro-nutrients in the soil and deteriorating soil health
- Inadequate post harvest technology
- Uncertain market prospects except for wheat and rice though MSP is announced for the following 25 agricultural commodities:
    - Cereals – paddy, rice, wheat, jowar, bajra, maize, ragi, barley
    - Pulses – gram, tur, moong, urad, masur
    - Oilseeds – groundnut, soyabean (yellow/black), rapeseed and mustard, toria, sunflower seed, safflower, nigerseed
    - Sugarcane, cotton, jute, tobacco
- High indebtedness of farmers – the total debt of Punjab farmers for instance is estimated to be about Rs. 24,000 crores.

Similar conditions prevail in Haryana and Western UP. **Thus, the heartland of the green revolution is in grave trouble**. These areas need conservation farming which will help farm families to conserve and improve soil health, water quantity and quality and biodiversity. Some of the eco-technologies developed by the Punjab Agricultural University are: bed sowing of wheat saving 20 to 25% water, leaf colour chart saving 15% N application in rice, tensiometer based irrigation scheduling, zero tillage technology for wheat, and IPM in cotton saving 40% pesticides. Thus, there is vast scope both to promote Green Agriculture and to reduce the cost of production through enhanced factor productivity. **A course on Sustainability Science should be introduced in all Agricultural Universities. India will not be able to maintain a stable food security system, if the "fertile crescent" (i.e. Punjab, Haryana and Western UP) region is not saved through adequate support for conservation farming. Defending the gains already made in this region is an urgent task.**

An example of the need for support for Conservation Farming is provided by the situation in rice cultivation in the Punjab. At present, nearly 26 lakhs hectares are under rice in the Punjab. Much of the irrigation water used is ground water. The water table in the Central districts of the State producing rice and having 70% of the tubewells is receding at an alarming rate of 2 to 2.5 feet annually. At present about 30% of the tubewells have become submersible and it is estimated that during the next 10 years practically all the centrifugal pumps will become nonfunctional and will have to be converted into submersible pumps. It would therefore be advisable to restrict rice cultivation to 2 million hectares in the Punjab with a yield target of 5 tonnes per hectare. The remaining area can come under maize, pulses and oil seeds which are all at the moment in short supply. **Conservation Farming in the Punjab as well as in other intensive agriculture areas will involve a scientific programme of restructuring farming systems**. Such a restructuring is an urgent need in the interests of the long term livelihood security of Punjab farmers and food security of the nation. The same is true in parts of Haryana, Western Uttar Pradesh as well as in many of the early IADP districts in the country.

Eternal vigilance is the price of stable agriculture. NCF had recommended in its Third Report the establishment of a science-based **National Biosecurity System. The sooner the NCF recommendation is implemented, the safer will be the country from the point of freedom from invasive alien species, which could cause potential harm to crop and animal husbandry, fisheries and forestry**. Our wheat crop for example, now faces threats from new strains of rusts like the 78S84 (similar to Yr 27 virulence) race of yellow rust and Ug 99 strain of stem rust. The steps taken to defend the gains already made should therefore include pest surveillance and management and gene deployment for checkmating the spread of pathogens. This is equally important in the case of poultry and animal enterprises.

In every State, the agricultural "bright spots" and "hot spots" will have to be mapped. The State should develop a strategy for enlarging the extrapolation domain of bright spots. Similarly, every State should develop a **Good Weather Code** to maximize the benefits of adequate moisture availability, a **Drought Code** to minimize the adverse impact of drought, and a **Flood Code** both to prevent excessive distress and damage, and to promote a post-flood production plan. In the desert areas of Rajasthan, the Good Weather Code should include provision for raising nurseries of appropriate plants, so that in years of excessive rainfall, an extensive tree planting and sand dune stabilization drive can be launched. This will help to strengthen the ecological infrastructure of the desert, and gradually convert the desert into an oasis. The Drought Code should include the adoption of crop life saving technologies and contingency plans to change the cropping pattern according to moisture availability. "**Be prepared**" – both to take advantage of a good monsoon and to reduce the impact of adverse seasons – should be our national motto in agriculture.

Our three pronged agricultural strategy should include defending **the gains** already made, **extend in the gains** to dry farming areas and **making new gains** through agro-processing and value addition to primary products.

Overcoming the Technology Fatigue

The Eleventh Five Year Plan document refers to the need for overcoming the prevailing technology fatigue. Technologies, which can help to enhance land, water and labour productivity, are urgently needed. They should lead to an evergreen revolution in small farms, i.e. increase in productivity in perpetuity without associated ecological harm. The smaller the farm, the greater is the need for marketable surplus in order to generate cash income. The small farm can lend itself to higher productivity and profitability, provided the small farmer is enabled to overcome his/her handicaps arising from lack of capital and credit and access to appropriate technologies and inputs and remunerative markets. There is need for a **small farm management revolution**, which can result in conferring the power and economy of scale on small producers both in the production and post-harvest phases of farming; if this does not happen, mounting debts arising from adverse economics will continue to affect them. The strategy for a Small Farm Management Revolution will have to be developed by Panchayati Raj Institutions (PRIs) with technical help from Agricultural, Rural and Women's Universities as well as IITs and IIMs, since much of the action will be location-specific. Cooperative farming, service cooperatives, stakeholder companies, formation of compact production and processing states by Self-Help Groups and farmer centric contract farming can all help to improve the economics of small holdings and thereby foster improved management. To promote group cooperation in a watershed or command area of an irrigation project, it will be useful to introduce attractive group insurance policies. Group-endeavour can be promoted only if cooperation results in a win-win situation for every member of the group.

At the production end, there is need for integrating frontier technologies like biotechnology, information and communication technologies, space and nuclear technologies and renewable energy technologies such as solar, wind, biogas and biomass based energy systems with traditional ecological prudence. Bio-energy based on pyrolysis and gasification of biomass can be a decentralized source of energy. Bio-fuels also offer scope wherever ecological and economic conditions are favourable. Biomass is an under-utilized resource. **Biomass-Parks** can be promoted in every block to convert the available biomass into a wide range of economic products, including energy and manure.

Conservation farming is the pathway to an evergreen revolution. The greatest problem with applying conservation agriculture concepts in dry land areas is the lack of adequate quantities of crop residues. The removal of crop residues for alternative uses accelerates the already fast decline of soil organic matter content in dry land areas. Long term sustainability of dry land soils may be significantly enhanced by reduced tillage that leaves more crop residues on the soil surface.

Besides enhancing soil fertility and soil organic matter, the need for the economic and efficient use of irrigation water cannot be over emphasised. The average yield of cereals can be increased by 30 to 60% annually in dry farming areas by increasing crop water use by 25 to 35 mm. This can be readily achieved by conservation agriculture. High input costs, uncertain rainfall and poor income lead to widespread indebtedness. **The younger generation will be reluctant to take up farming as long as income prospects are poor**. Declining terms of trade between farm and non-farm sections is a matter of concern.

It is in this background that we have to examine the opportunities opened up by new technologies. New agriculture technologies like genomics and information technology together with improved agronomic management should form the cornerstone of increasing agriculture productivity and profitability of small farms both in irrigated and rainfed areas as well as in problem soils and coastal areas. Recombinant DNA technology has already resulted in the breeding of crop varieties possessing tolerance to salinity and drought as well as to serious biotic stresses caused by the triple alliance of pests, pathogens and weeds. It is however essential to have a professionally and socially credible National Biotechnology Regulatory Authority, on the lines recommended by the Swaminathan Committee in 2004. **The bottom line for any biotechnology regulatory policy should be the safety of the environment, the well being of farming families, the ecological and economic sustainability of farming systems, the health and nutrition security of consumers, safeguarding of home and external trade, and the biosecurity of the nation.** The Government of India is currently developing a Biotechnology Regulatory Act and it will be useful if such an Act, which will lead to the establishment of an autonomous and statutory Authority and which can promote the safe and responsible use of biotechnology, is enacted soon.

The Village Knowledge Centre (VKC) or *Gyan Chaupal* will help to bridge the growing gap between scientific knowledge and its field application. **It will also facilitate the removal of many intermediaries from the marketing chain.** The wholesale fruit and vegetable markets are likely to lose their importance under the growing influence of contract farming and direct supply relationships between producers and major market chains. Changes in intermediary relationship will occur as internet based marketing tools are adopted by both producers and suppliers. **Government of India's Bharat Nirman (New Deal for Rural India) has rightly given priority to knowledge connectivity, in addition to physical connectivity through roads**. As a single step, the *Gyan Chaupal* will bring about a transformation in the economic conditions and social relations in our villages. **Bridging the digital divide is a powerful method of bridging the gender divide, since rural women master the ICT technologies with ease**.

India is poised for a major ICT revolution in rural India. The broad strategy proposed by NCF is as follows:

- Establishment of a Village Resource Centre (VRC) in every block with the help of the Indian Space Research Organisation. These VRCs will be linked to satellites and will have telecommunication facilities.
- Every Panchayat headquarter will have a Gyan Chaupal or Village Knowledge Centre (VKC). This will have internet connectivity. Alternatively, the Gyan Chaupal can be established in the village school or any other public space where there will be social inclusion in access to the technology.
- The last mile and last person connectivity will be through FM / Community radio and / or mobile phones. **The internet – radio – mobile phone synergy is a very powerful tool for social inclusion in access to all the needed information, including warning of impending natural disasters**. Villagers give priority to health and marketing information. In addition, an Entitlements Database can empower them with information on all the Government schemes designed for their well being. Gender-specific information is equally important. Every farmer in the village should be issued with an Entitlements Pass Book. Artesenal fishermen can now be assisted with a cell phone, which can provide GPS data on wave heights and location of fish shoals.

We are thus on the threshold of both a biotechnology and information technology revolution. Biotechnology does not imply only GMOs. Non-GMO applications are many, such as tissue culture for multiplying elite germplasm, bio-fertilizers, bio-pesticides and bio-remediation of ground water as well as marker-assisted breeding. In the case of GMOs, safe and responsible use should be ensured. Organic farming procedures permit the use of varieties developed by marker-assisted breeding.

The third technological revolution relevant to agriculture is the ecotechnology movement. This involves the appropriate integration of frontier sciences with the ecological prudence of farming communities. The ecotechnology revolution underpinning the ever-green revolution movement has many pathways as indicated below.

Green Revolution : Commodity – centred increase in productivity	**Ever-green Revolution**: increasing productivity in perpetuity without associated ecological harm Farming system based agronomy
Change in plant architecture, and harvest index	Organic Agriculture : cultivation without any use of chemical inputs like mineral fertilizers and chemical pesticides
Change in the physiological rhythm – insensitive to photoperiodism	Green Agriculture : cultivation with the help of integrated pest management, integrated nutrient supply, integrated natural resource management systems and the cultivation of the most efficient variety whether bred by Mendelian or Molecular breeding

For most small farmers, green agriculture will be the most feasible form of eco-agriculture. Crop-Livestock integrated systems of production will be ideal for organic farming. More research is needed on nitrogen fixing tree species and shrubs, as well as green manure plants. Our soils are hungry and thirsty and they need both nutrients and water.

Training

How can such a technological, ecological and managerial upgrading of small farm agriculture be brought about? This is where training, re-training, re-tooling and redeployment of both farmers and farm graduates become important. Fortunately, we have over 50 Agricultural and Animal Husbandry (including Fisheries) Universities. Nearly 20,000 farm graduates including about 7000 postgraduates are becoming available each year. There is a vast chain of National Research Institutes and Centres, National Bureaus and All India Coordinated Projects under ICAR. There are also a growing number of R & D institutions in the private sector and a number of civil society organisations working on agricultural issues. The Indian National Agricultural Research System (NARS) is thus a formidable one.

NCF has recommended the following steps to bridge the knowledge gap:

- Promote farmer-to-farmer learning by establishing **Farm Schools** in the fields of outstanding farmer-achievers.
- Revitalize and upgrade **Krishi Vigyan Kendras or Farm Science Centres** by adding a post-harvest technology wing.
- Organise Lab to Land demonstrations in the areas of post-harvest processing, marketing and value addition to primary produce all over the country.
- Establish **Gyan Chaupals** in every village based on the integrated use of the internet, cable TV, community radio, cell phone and local language community newspapers. If Mission 2007: Every Village a Knowledge Centre is accomplished, the knowledge deficit currently prevailing in villages can be removed and the "know-how" – "do how" gap can be bridged.
- Establish Capacity Building Centres for those operating Gyan Chaupals.
- Train one woman and one male member in every Panchayat as **Farm Science Managers**
- Establish at the District level a **SHG Training and Mentoring Centre**, in order to build a local level cadre of SHG catalysts, capable of organizing **Sustainable Livelihood Banks** based on micro-credit.
- Establish in coastal areas **Fish for All Training Centres** to provide training in all aspects of fisheries ranging from capture/culture to consumption.

Training of farm and home science graduates also needs revamping. NCF has proposed that the major mission of our Agricultural, Veterinary, Fisheries, Rural and Women's Universities should be to help every scholar to become an entrepreneur. They can then organize Service Co-operatives, Stakeholder companies, Agri-clinics, Agri-business centres, Bio-Parks, Food-Parks and other enterprises, which can help to improve the efficiency and economics of farming. Home

Science Colleges could be restructured as College of Human Sciences, where both men and women learn the science and art of nutrition, agro-processing and home economics.

A reorientation in the mindset of farm graduates can be brought about only by innovative changes in curricula and courses. In all applied areas, business and financial management should be added to the disciplinary training. For example, a course in Seed Technology can be restructured and designated as "Seed Technology and Business". Similarly, nutrition courses could be reorganized as **Food Safety and Nutrition Security** programmes. Courses in Agronomy could be developed into Agronomy and Agri-business Programmes. **If the business, financial and trade aspects are integrated with disciplinary training, such courses will give the farm / home science graduates the self-confidence essential for embarking upon a career of self-employment**. A large number of graduates are now being trained in the field of biotechnology. However, many of them are not able to utilize their training after taking degrees due to lack of appropriate employment opportunities. Agricultural Biotechnology is an area where there are considerable opportunities for remunerative self-employment. It would therefore be appropriate that support is extended to the creation of a **National Association of Genome Entrepreneurs** who could be supported with Venture Capital Fund in order to enable them to convert the rich knowledge available in Government institutions in the field of functional genomics into commercially viable products. They could also undertake work for other countries in the area of preparation of genome maps of the crops of interest to those countries. The National Centre for Plant Genome Research setup by the Department of Biotechnology at New Delhi could organize short-term course on Functional Genomics and Business Development. **Mainstreaming entrepreneurship and business skills in all applied courses, rather than keeping Business Management Course as a separate entity is essential, if small farm agriculture is to become economically sustainable and educated youth are to be attracted to take to a career in agriculture**.

Another urgent need is the establishment of a chain of Regional Institutes for Food Safety and Security. They can be established in appropriate Agricultural, Veterinary or Fisheries Universities. Home Science Graduates can be employed in such Regional Institutes to launch a movement for food safety including awareness of codex alimentarius standards. They should also spread quality literacy among farmers through Gyan Chaupals.

Techno-infrastructure

Bharat Nirman should help to improve the infrastructure essential for a technological upgrading of farm operations. Post-harvest infrastructure, particularly for perishable commodities, is extremely weak. This is why the Swaminathan Committee recommended in 1981 the establishment of a National Horticulture Board, solely devoted to the cause of improving post-harvest infrastructure, processing and marketing. Unless this is attended to on an urgent basis, farmers will not be able to get adequate return for their labour. Fortunately, the National

Horticulture Mission is likely to fill this felt need. Similarly facilities for food safety, water quality, sanitary and phytosanitary measures and biosecurity need to be improved.

Facilities for soil testing, particularly estimation of micro-nutrient status also need considerable strengthening. **Unless more investment is made in strengthening the support services needed by farmers for the scientific upgradation of farming, the average productivity will continue to remain low and youth will not be attracted to farming**. Simple but safe storage bins need to be popularized on a large scale, along with low cost refrigeration facilities for perishable commodities. As recommended by NCF, a Livestock Feed and Fodder Corporation, a Land Use Advisory Service, an Indian Trade Organisation, Living Heritage Gene Banks to conserve unique local breeds of farm animals, internationally recognized certification agencies for organic farm products, an Agricultural Price Stabilisation Fund, integrated insurance products like Parivar Bima Policy and other essential support services are needed to help increase the productivity and profitability of small farm agriculture. The National Fisheries Development Board and the National Rainfed Area Authority recommended by NCF have fortunately come into existence. **It is essential that such bodies should serve as professional powerhouses in their respective fields and help to upgrade the technological, ecological and management aspects of culture and capture fisheries on the one hand, and rainwater harvesting, conservation, sustainable use, aquifer recharge, more crop and income per drop of water and other issues relating to achieving the goal of Jal Swaraj in rainfed agriculture, on the other.**

Trade

Producer oriented market holds the key to remunerative and sustainable farming. Quality and trade literacy should receive high priority in Gyan Chaupals. Facilities for Farmers' Markets need to be expanded rapidly. In commodities essential for maintaining the Public Distribution System, the procurement price should be the market price at the time of purchase. Those providing essential commodities for the PDS should be recognized through the provision of Smart Cards, which will entitle them to certain benefits while purchasing essential farm inputs, including agricultural implements and machinery.

The Commission on Agricultural Costs and Prices (CACP) should be an autonomous statutory organization with its primary mandate being the recommendation of remunerative prices for the principal agricultural commodities of both dry-farming and irrigated areas. The Minimum Support Price (MSP) should be atleast 50% more than the weighted average cost of production. The "net take home income" of farmers should be comparable to those of civil servants. CACP should become an important policy instrument for safeguarding the survival of farmers and farming. Suggestions for crop diversification should be preceded by assured market linkages. The membership of CACP should include a few practicing farm men and women. The terms of reference and status of CACP need review and appropriate revision.

The Pricing Policy for farm commodities should have the following three components:

- Minimum Support Price (MSP) announced before the sowing season; MSP should be C:2 plus 50% (i.e. 50% more than the total cost of production)
- Procurement price at the time of purchase which should be the best available market price
- **Post-procurement adjustment through Smart Cards** issued to those who sell their produce for public good like PDS, ICDS, School Noon-Meal etc., In Case Government has to import food grains at a price much higher than given to our farmers, the Smart Card should enable them to obtain essential inputs at concessional price.

Thus, a farmer friendly integrated MSP, procurement price and post-procurement adjustment system will help our farmers to tap the existing unutilised yield reservoir and thereby improve the productivity and profitability of small farms.

To sum up, imports or exports of food grains may be necessary from time to time, but the **bottom line of our import-export policies must be the livelihood security of both the farm and non-farm populations of rural India who constitute 70% of our population.** Time has come for Government to setup a Pan-political **National Food Security and Sovereignty Board** chaired by the Prime Minister with its membership including the Minister for Agriculture and Food and other concerned Ministers of GoI as well as a few Chief Ministers of food surplus and deficit States, leaders of all national political parties, a few experts including specialists in the gender dimension of agriculture and food security, and mass media representatives. We are confronted with the need to safeguard the food security requirements of both resource poor farmers and resource poor consumers. The bulk of such resource poor farmers are small or marginal farmers and landless agricultural labour in unirrigated areas. It is these linkages, which need to be understood and attended to. The proposed National Food Security and Sovereignty Board can attend to these complex linkages in a holistic manner and develop and implement a transparent national food security policy in the interests of all regions of the country and all sections of our population.

India will remain during most of the 21st century a predominantly agricultural country, particularly with reference to livelihood opportunities. Therefore there is need for both vision and appropriate action in the area of shaping our agricultural destiny. Our major agricultural strengths are our large population of hard working farm women and men, our high scientific capability, our varied climatic and soil resources, abundant sunshine throughout the year, reasonable rainfall and water resources, a long coast line and rich agro-biodiversity. Converting them into jobs and income is the major challenge for scientists and technologists.

We should look upon agriculture not just as a food-producing machine for the urban population, but as the major source of skilled and remunerative employment and the backbone of the rural livelihood system. Unless there is such a change in mindset, the food security situation will remain precarious and rural poverty will persist. Food and fuel are going to be the commodities in greatest demand in the future. We should therefore take full advantage of our vast untapped potential in crop and animal husbandry, fisheries, forestry and agro-processing.

2008 marks the 40th anniversary of the beginning of the Green Revolution in India. We have moved away from the "ship to mouth" existence of the early decades after independence and have entered an era where the fusion of professional skill, political will and farmers' enthusiasm is to taking our Nation nearer to the goal, "Food for All and For Ever".

References

1. Swaminathan, M.S. *Farmers don't know that Crop Yields can be Trebled*, Yojana Article, February 14, 1965, Page 12-13, 26.
2. Swaminathan, M.S. (1965). *The Impact of Dwarfing Genes on Wheat Production* J. I.A.R.I., P.G. School, 3: 57-62.
3. Swaminathan, M.S. (1966). *National Demonstrations in Rice. Indian Farming* 16(6): 67-70.
4. Swaminathan, M S, (1968). *The Evolution and Significance of Jounti Seed Village.* Indian Farming, Jan 1968 : 1-4.
5. Swaminathan, M.S. (Ed.) 1993. *Wheat Revolution: A dialogue.* Macmillan India Ltd., Madras 164 pp.
6. Swaminathan, M.S. (1968), *The Age of Algeny, Genetic Destruction of Yield Barriers and Agricultural Transformation*, Section of Agricultural Sciences, Proc. 55th Ind. Sci. Cong. Part II: Presidential Address Varanasi Indian Science Congress. pp. 235-248.
7. Swaminathan, M.S. 1996. *Sustainable Agriculture: Towards an Evergreen Revolution.* Konark Publishers Pvt Ltd, Delhi. pp 232
8. Wilson E O, 2002. *The Future of Life*, Alfred A Knopf, pp. 229.
9. 59th Round of NSSO (Report No. 497).

SECTION I

FOOD SECURITY AND ECONOMIC DEVELOPMENT — HOW SCIENCE IS APPLIED TO SOLVE PROBLEMS OF POVERTY, DROUGHT AND FAMINE

Section I

Food Security and Economic Development

How science is applied to solve problems of poverty, drought and famine

Commentary by M.S. Swaminathan

Dr Paul Crutzen, one of the world's leading climate experts has recently stated "In order to be safe we would have to reduce our carbon emissions by 70% by 2015. Instead of achieving any reduction, we are currently adding 3% more each year". The consequences of such a situation will be multi-dimensional. First, most of the Himalayan glaciers will melt with adverse repercussions for many of the major rivers in the region. Second, both Bangladesh and South India will suffer from a rise in sea level. It has been predicted that Western Antartica's ice sheets will not survive beyond this century, leading to a sea level rise of atleast 1 to 2 meters. Third, the Global Mean temperature is likely to go up by 4 degree centigrade before the end of the century. Some experts believe that such a disaster may occur even sooner, say by 2050. The major economic powers of the world have agreed to get the rise in average mean temperature pegged at 2°C. Even such a rise will reduce wheat and rice production and thereby post a serious threat to sustainable food security.

Over 20 years ago, my colleague the late Prof S K Sinha and I studied the impact of a one degree rise in mean temperature on wheat yield in North India. We found that such a rise in temperature will lead to a reduction in the duration of the wheat crop and thereby result in a drop in yield. Therefore, global warming and its consequent adverse impact on temperature, precipitation and sea level will cause unprecedented hardship and loss of livelihoods. Food security will be threatened and people will try to migrate towards Northern latitudes. Also such migration may not be politically feasible.

We should therefore initiate anticipatory action to meet the challenges arising from climate change. In 1973, in my Sardar Patel Memorial Lectures over the All India Radio, I suggested that we should develop Drought, Flood and Good weather codes inorder to help local communities to minimize the adverse impact of aberrant weather and to maximize the benefits of a good monsoon. Unfortunately we are yet to develop well planned anticipatory steps to meet such challenges. Only recently, a National Plan of Action has been developed for climate management. I hope that this plan will become operational soon.

The papers in this section deal with different aspects of management of food security including the adverse impact of global warming such as adverse changes in precipitation and temperature and sea level rise.

Third World Lecture 1983
Agricultural Progress—Key to Third World Prosperity

Address delivered by Dr M S Swaminathan, Director-General of the International Rice Research Institute, when receiving the Third World Prize for 1982 on behalf of the Institute. The Prize was presented to Dr Swaminathan by the Prime Minister of the People's Republic of China, H E Zhao Ziyang, at the innaugural session of the South-South Conference on Strategies of Development, Negotiations and Cooperation, in Beijing on 4 April 1983

On behalf of the Board of Trustees and staff of the International Rice Research Institute may I express our sincere gratitude to the Third World Foundation for Social and Economic Studies. You have recognised the role of agricultural science as a catalyst for agrarian prosperity through the selection of IRRI for the 1982 Third World Prize. We accept the Third World Award on behalf of the global family of farm scientists.

Agriculture began in what is now called the Third World. The ancestral homes of the world's most important crops are in Third World countries: rice, wheat, maize, potato, cassava, sweet potato, sugarcane, soybean, pulses, numerous vegetables, most fruit trees, cotton and other fibres, many forage grasses and legumes, and numerous forest trees including most hardwood species.

This is not surprising because agriculture revolves around the harvest of solar energy, and abundant, year-round sunshine is the Third World's greatest asset.

Culture was cradled by the prehistoric shift from food-gathering to food-growing in the Third World. Domestication of crop plants and farm animals and the development of water conservation and irrigation systems led to the civilisations of the Nile Valley, Mesopotamia, the Indus Valley, the Andean highlands and Central America, and, of course, China. No wonder the Chinese peasants are known as 'farmers of 50 centuries.' Their development of efficient methods of restoring soil fertility has enabled them to continuously and intensively harvest crops from limited land for more than 7,000 years.

Thus, the world's most experienced farmers are found in the Third World.

Asia is today the home of more than half of the world's people. But some 500 million Asians live in absolute poverty and go to bed hungry each night. About 10 million or more children die every year from causes associated with hunger and malnutrition.

We thus live in an unequal and paradoxical world. It is an irony that 'God

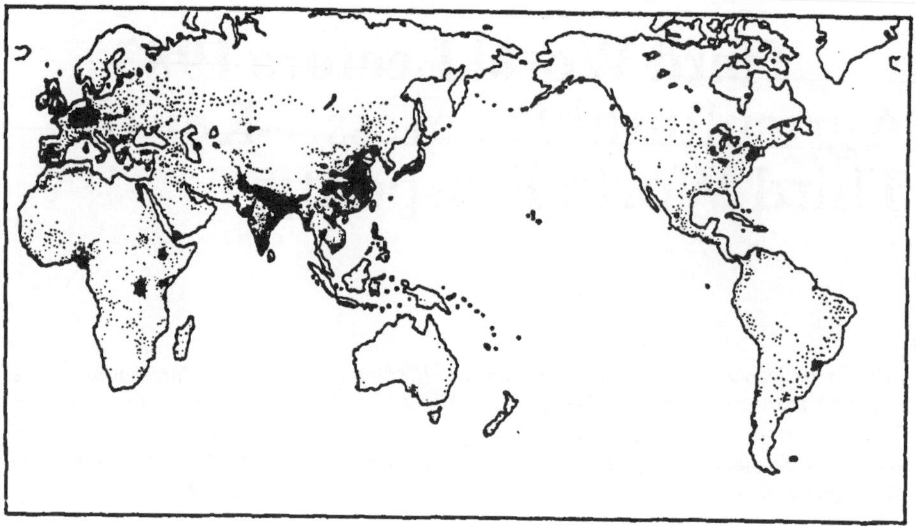

Rice means life itself to the world's most densely populated countries. About 55 per cent of the world's people live in Asia.

continues to be bread' (to quote Mahatma Gandhi) to children, women, and men in many parts of the 'South', while the prayer 'God give us today our daily bread...' has lost historical relevance in many countries of the 'North'. On one hand, mountains of grain are rising in highly developed countries. Some governments are giving farmers incentives to leave land fallow. On the other hand, many Third World countries lead a 'ship to mouth' existence because they are increasingly dependent on grain imports. It is a paradox that many Third World countries—who must balance their food budgets by accepting free grain or purchasing it on concessional or commercial terms—are the countries with the largest reservoirs of untapped agricultural resources.

Why should an abundance of natural resources coexist with low farm productivity and consequent hunger? Can't we take steps to convert the natural blessings of the Third World—its basic life support systems of land, water, sunshine, flora and fauna—into wealth that is vital to its people? Why are many developing countries still giving low priority to agriculture?

In response to this paradox, the Rockefeller and Ford Foundations, in cooperation with the Republic of the Philippines, established the International Rice Research Institute in 1960 on the beautiful campus of the University of the Philippines at Los Baños. IRRI's objective is to increase the production of rice and of food from rice-based farming systems through applied agricultural science.

The choice of rice as the focus of an international research initiative and the choice of the Philippines as its location were appropriate. More than 95 per cent of the world's rice area is in the Third World. For more than 1.5 billion low-income people in Asia, and hundreds of millions in Africa and Latin America, rice is the

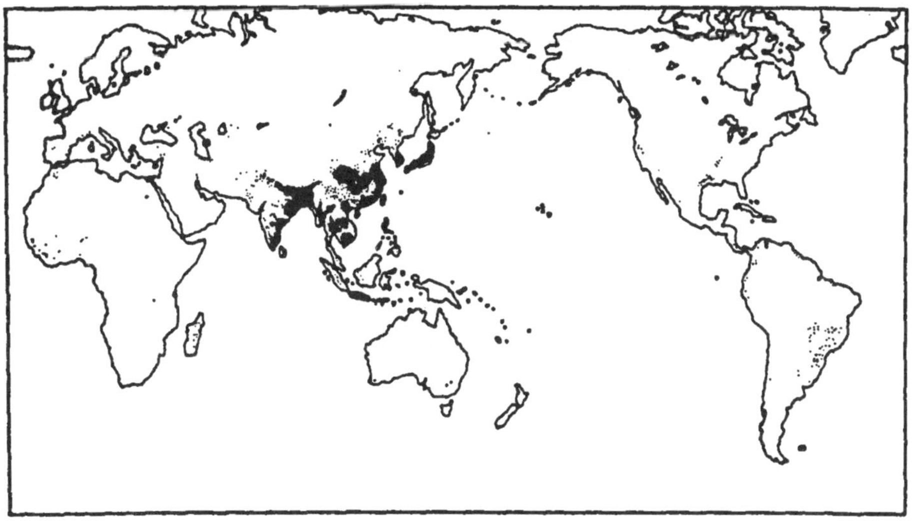

Ninety-two per cent of the world's rice is grown and consumed in Asia.

major source of calories and protein. Rice comprises a third of the area planted to cereals in developing countries. In 36 countries where more than 100,000 hectares are planted to rice, the annual per capita income is less than US$300. They are the lowest income group in the world.

The history of IRRI—appropriately titled *An adventure in applied science*—was recently written by its first director, Dr R F Chandler Jr. Therefore, I shall not give you detailed background on the Institute and its accomplishments. I am content to say that IRRI was blessed with the able guidance of scientists of eminence and vision such as late Dr Sterling Wortman of the Rockefeller Foundation; the earlier directors, Dr Chandler and Dr Nyle C Brady; and of Boards of Trustees formerly headed by the late Dr George Harrar and by Dr Forrest F Hill, and now, Dr Clarence C Gray III. Above all, the staff of the Institute, drawn from more than 20 countries, has worked with single-minded devotion to achieve a scientific elimination of barriers to high and stable yields.

IRRI's success is due to the cooperation of national governments in many developing nations. Not only did the Philippine Government provide the land for IRRI's research centre and experiment farms, it also allowed freedom for IRRI scientists to travel to observe rice problems first hand, and freedom for rice workers everywhere to come to the Philippines to participate in IRRI training programmes.

The neighbouring University of the Philippines at Los Baños awarded affiliate graduate faculty status to IRRI scientists, which gave scholars an opportunity to study for advanced degrees at UPLB and conduct their thesis research at IRRI.

Without such international cooperation, IRRI could have become another

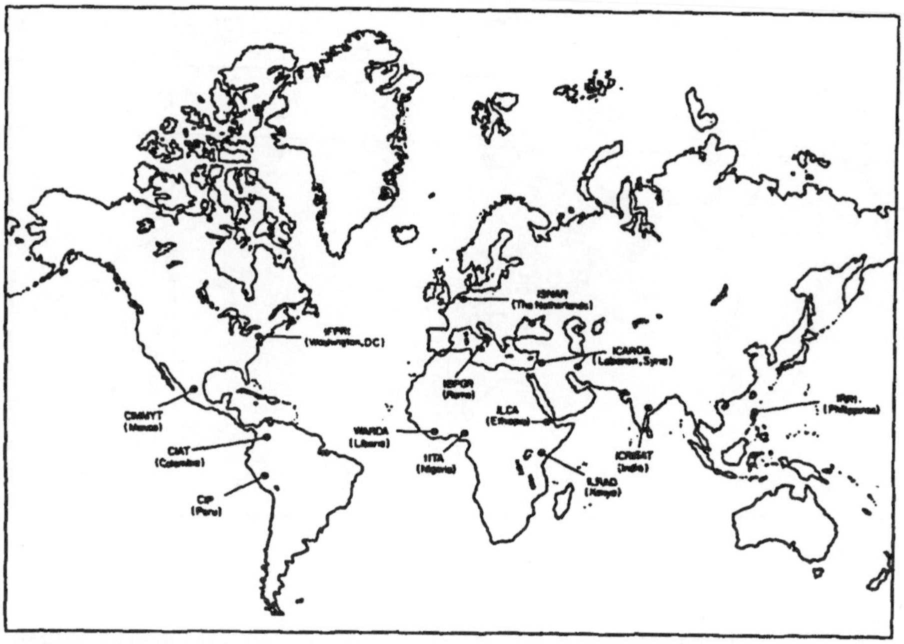

IRRI was the first of the International Agricultural Research Centres. Today a network of 13 Centres is supported through the Consultative Group on International Agricultural Research (CGIAR), an international consortium of 36 government and private agencies dedicated to the support and improvement of agricultural research in the Third World. Employed within this far-flung network are about 7,000 persons, which include more than 600 senior scientists from 40 nations. CIAT = Centro Internacional de Agricultura Tropical, Apartado Aereo 6713, Cali, Colombia; CIMMYT = Centro Internacional de Mejoramiento de Maiz y Trigo, Londres 40, Mexico 6, D. F., Mexico; CIP = Centro Internacional de la Papa, Apartado 5969, Lima, Peru; IBPGR = International Board for Plant Genetic Resources, Food and Agriculture Organisation of the United Nations, Via delle Terme de Caracalla, 00100 Rome, Italy; ICARDA = International Centre for Agricultural Research in the Dry Areas, P.O. Box 114/5055, Beirut, Lebanon; ICRISAT = International Crops Research Institute for the Semi-Arid Tropics, Patancheru P.O. Andhra Pradesh 502324, India; IFPRI = International Food Policy Research Institute, 1776 Massachusetts Avenue, N.W., Washington, D.C. 20036, USA; IITA = International Institute of Tropial Agriculture, P.O. Box 5320, Ibadan, Nigeria; ILCA = International Livestock Centre for Africa, P.O. Box 5689, Addis Ababa, Ethiopia; ILRAD = International Laboratory for Research on Animal Diseases, P.O. Box 30709, Nairobi, Kenya; IRRI = International Rice Research Institute, P.O. Box 933, Manila, Philippines; ISNAR = International Service for National Agricultural Research, P.O. Box 93375, 2509 AJ The Hague, The Netherlands; WARDA = West Africa Rice Development Association, E.J. Roye Memorial Building, P.O. Box 1019, Monrovia, Liberia.

'ivory tower'. Instead IRRI became the hub of an international endeavour to assure food security in countries where rice is the staple. The research investment grew rapidly in both national and international research systems in Asia. The

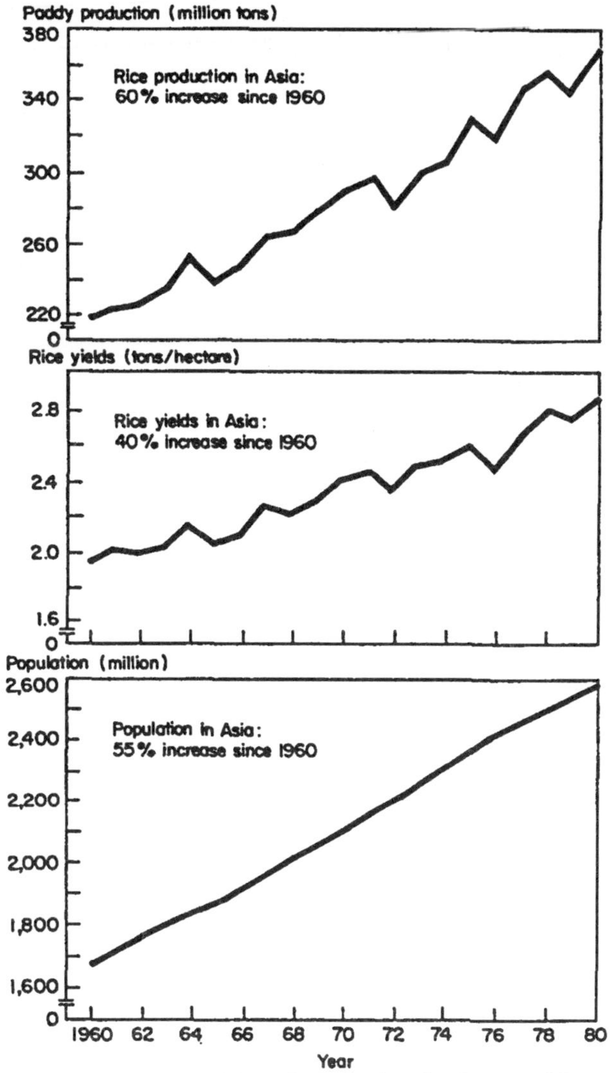

The dramatic increases in rice yields and production in the past 20 years have averted widespread hunger. However, the increased production has barely kept ahead of Asia's population growth. Thus, a continuous flow of improved rice varieties and technology is essential to humankind until the world's population is stabilised.

return on international research in Asia was estimated at from 76 to 102 per cent per year during 1966 to 1975. An eminent group of scientists that examined the work of IRRI in 1981 reported that the rice variety IR36 alone is helping to add US$1.5 billion worth of additional income to farmers each year.

In 1971 IRRI, the first of the International Agricultural Research Centres, became part of a global family of IARCs funded through the newly established Consultative Group on International Agricultural Research (CGIAR), jointly sponsored by the Food and Agriculture Organisation of the United Nations, the United Nations Development Programme, and the International Bank for Reconstruction and Development (World Bank). In 1983, 36 countries and donor organisations will contribute an estimated US$164 million to support the core research programmes of 13 International Centres through the CGIAR. Donor members now include Third World countries such as Brazil, India, Mexico, Nigeria, the Philippines, and Saudi Arabia. The crops and livestock on which the International Centres focus provide 75 per cent of the food for the Third World. The 600 senior scientists who work at these Centres are drawn from 40 nations.

In a world beleaguered by negation, despair and conflict, the CGIAR stands out as a meaningful, affirming flame. Mr Warren C Baum, Chairman, CGIAR and Members of this Group deserve our gratitude for nursing this flame which strives to bring cheer and light to homes engulfed by the degrading darkness of hunger.

IRRI today is more than a research institute. It is also a coordinating body for international networks such as the International Rice Testing Programme, through which global teams of capable, highly dedicated scientists and scholars work to improve and stabilise rice yields under diverse, and often harsh, growing conditions. More than 800 rice scientists in 75 countries on five continents work together in experiments and testing programmes catalysed and coordinated by IRRI. The strengths of individual scientists and countries may vary, but the collective strength of this worldwide rice research community is mighty.

We are honoured that the award ceremony for the Third World Prize is being held in the People's Republic of China, the world's oldest and largest producer and consumer of rice. China grows about 35 per cent of the world's rice and almost half of China's total grain production is rice.

Some of the most important scientific ingredients of modern rice technology originated in China. Chinese scientists were among the original IRRI staff. The semidwarfing gene that was essential to the development of high-yielding rice varieties—and most semidwarfs bred by national rice improvement programmes—came from Dee-geo-woo-gen, a short-statured rice variety of Chinese origin.

IRRI is privileged to have strong cooperative links with China through the Chinese Academy of Agricultural Sciences, the Chinese Academy of Agricultural Mechanisation Sciences, Academia Sinica, and other organisations. This award ceremony gives us an opportunity to salute both farmers and scientific colleagues in China for their contributions to raising the yield ceiling in rice.

Farmers are believed to have domesticated semiaquatic rice varieties shortly after the Neo-Thermal Age (12,000 to 14,000 years ago). Some say rice farming began in CHina, others say India or Thailand. Recent archeological findings show that farmers cultivated rice in China and India at least 7,000 years ago. The civilisations that developed around major Asian river systems—the Ganges, Yangtse, Brahmaputra, Irrawaddy, Mekong—thrived on rice culture.

Yet rice yields were almost stagnant for centuries, levelling at a yield plateau of less than 2 tons per hectare—the level that basic soil fertility could sustain because rice needs about 1 kilogram of nitrogen to yield 15 kilogram of grain.

From 1960 to 1980, average rice yields in Asia rose by 40 per cent and production more than 60 per cent, although we must also consider that Asia's population rose 55 per cent in the same 20 years.

How did the farmers of Bangladesh, Burma, China, India, Indonesia, Japan, the Philippines, and other Asian countries raise rice production more in the past 20 years than in the preceding 7,000 years? Five major advances led to the doubling and tripling of rice production:

1. *The modern semidwarf plant type that increases the efficiency of nutrient, water, and sunshine use.* The traditional varieties 'lodged', or fell over when soil fertility exceeded the capacity to produce more than 2 or 3 tons of rice per hectare. Grain rotted in the water or was eaten by rats. Photosynthesis ceased.

   Modern semidwarf rice varieties yield 5 tons or more because their strong stems—inherited from Chinese parents—enable them to stand erect and convert nutrients, water, and sunshine into grain more efficiently. Many of the newer varieties carry built-in genetic resistance to a host of insects and diseases.
2. *Short growth duration and insensitivity to daylength.* Teams of researchers selectively bred these traits into the modern rice varieties to make it possible for farmers to grow two or three crops where they previously grew only one crop, and to grow superior varieties at different latitudes.
3. *Cultivation of F_1 hybrid rice.* China is the world's pioneer in hybrid rice production. China now grows more than 6 million hectares of hybrid rice; yields are about 30 per cent higher than those of ordinary semidwarfs.
4. *Improved irrigation and water management.* Better availability of water gives farmers yet another opportunity to grow two or more crops per year in the same field.

   An old Chinese saying vividly portrays the *green power* of tropical rice land: 'a field can look gold, black, and green on the same day.' The *gold* represents the mature rice crop, which the farmer and his family harvest early in the morning. The same rice field looks *black* around noon because the family has already plowed the field in preparation for the next crop. Late that afternoon the field looks *green* because the farm family has transplanted a new rice crop!

   In the IRRI Rice Garden, a crop is harvested every Monday and another planted every Friday. Good water management and early maturing varieties allow the planting and harvesting to be staggered throughout the year. The system spreads labour, cash inputs, and risk—and enhances both income and food production for the family. And the intensive production more fully exploits human inputs, water, sunshine, and soil.
5. *Introduction of new farm management techniques.* Yields are higher and farm incomes more secure because farmers now have improved implements, mineral fertilisers and better methods to apply them, integrated procedures for pest control, and improved postharvest technology.

Most Asian countries have assisted farmers to derive benefits from these technological advances by introducing yield-oriented production programmes such as Masagana 99 in the Philippines, BIMAS (Mass Guidance Toward Self-Sufficiency in Food Production) in Indonesia, the Whole Township Programme in Burma, and the High-Yielding Varieties and Small Farmers Development Programmes in India.

These factors now enable farmers to exploit the yield potential of management-responsive crop varieties, multiple cropping, and improved preharvest and postharvest operations. Each development can be traced to the application of science to agriculture.

Mr. Lu Shen Rong, an old rice farmer in a production brigade in Lu Jha Chun, Hangzhou Province, recently described the impact that these advances have made in China. Mr. Lu has grown rice for almost 50 years; hence his deep insight was born from vast experience.

I asked Mr Lu, 'What are the most important farm problems today, and in what areas do you feel we should focus the research in future?' The farmer's immediate reply was *'The care of the soil and the health of the plant.'*

I have worked in agricultural research for almost 35 years, but I could not have better summarised the most urgent challenges for the future.

Asia has little unused land, and agricultural land is a shrinking resource as population multiples and urbanisation forces its way into the countryside. Asia's ratio of land to people—0.21 hectares per person—is the lowest in the world. Future increases in agricultural production must come mainly from higher productivity and cropping intensity per unit of land, time, water, energy, and labour. The Food and Agriculture Organisation estimates that rice production must increase by more than 3 per cent per year during the remainder of the 20th century. This will require a continuous improvement in productivity—and it must be done without harming the long-term production potential of the soil. This challenge requires the intensification of research on all aspects of plant and soil care.

The improvement of soil and plant health motivated IRRI and the International Fertiliser Development Centre to organise a network of cooperative experiments under the International Network on Soil Fertility and Fertiliser Evaluation for Rice (INSFFER).

Proper attention to plant health not only prevents crop damage from the unholy triple alliance of insects, diseases, and weeds but also reduces risk and uncertainty in agriculture. Institutes such as IRRI offer unusual opportunities to develop varieties with multiple pest resistance.

In the IRRI Genetic Resource Centre, more than 65,000 varieties of rice are preserved for the use of today's rice scientists and farmers, and for generations yet unborn.

But the IRRI collection probably includes little more than half of the world's genetic variability in rice. With the cooperation of the Chinese Academy of Agricultural Sciences and other national research systems, we propose to develop a global strategy to collect the remaining genetic strains as rapidly as possible. Many irreplaceable rice strains—the fruits of thousands of years of natural and human selection—grow in endangered habitats and face the threat of extinction. Agricultural organisations everywhere must join forces to collect and preserve this invaluable germplasm for posterity.

But preserving the world's rice germplasm is not enough. More important, we must know the diversity of genetic traits these strains possess so that we can put those traits to work for the benefit of rice farmers everywhere.

Through the multidisciplinary Genetic Evaluation and Utilisation (GEU) programme, IRRI and worldwide cooperators carefully test this vast seed collection for the expression of more than 50 inherited characteristics, such as resistance to the diseases and insects that ravage farmers' crops and tie them to costly chemical protection, and tolerance of adverse soils.

Armed with such information, scientists from many countries withdraw parents from the Genetic Resources Laboratory to crossbreed with rices that have still other traits needed by the small-scale farmer. Such a strategy has led to a worldwide genetic pipeline from which flows a stream of high-yielding rice varieties with the built-in ability to tolerate not only pests but also salinity, alkalinity, or other adverse soil conditions; drought; deep water and floods; or cold temperature.

The best rices from all nations are selected under adverse environments through the International Rice Testing Programme—a worldwide cooperative network. The power that such global cooperation gives all scientists is illustrated by the success of IR36, a rice variety now grown on almost 11 million hectares in Asia. IR36 is probably the most widely grown variety—of any food crop—the world has ever known. To develop IR36, IRRI scientists crossed 13 parents, including a wild species, from six countries. Scientists in the Philippines, Indonesia, and India cooperated in the selection of IR36.

When tungro virus disease reached an epidemic level in South Sulawesi, IRRI rushed a set of genetic material to Indonesia. Indonesian scientists planted the rices in 'hot spots' where the disease was rampant. The survivors included the line that became IR36. Another set of materials was sent to Orissa, where Indian scientists selected plants for resistance to gall midge, a serious pest in much of Asia that was not present in the Philippines.

The insect resistance alone of IR36 saves rice farmers in the Third World untold millions per year—money that would otherwise have been spent on petroleum-based insecticides.

One moral stands clear in a world divided by discord and distrust. International cooperation leads to universal prosperity.

We have no time to relax. Eternal vigilance is the price of a stable, prosperous, and productive agriculture. A new biotype of brown planthopper recently attacked IR36 in North Sumatra. Fortunately, Indonesian scientists were prepared and immediately arranged a new 'genetic barrier'. In February 1982 IRRI airlifted more than 20 tons of seeds of a newer variety, IR56—which is resistant to the new biotype—to Indonesia for multiplication and distribution to farmers.

Most experts agree that fighting the famine of jobs is as important as fighting the food famine during the remainder of the 20th century. The nutrition problem in many parts of the Third World can be improved even more by an increase in person-years of employment than by an increase in grain production. Rapid agricultural development is important to generate the employment necessary to purchase food and increase its availability. In fact, rural development in Asia implies largely the development of land- and water-based occupations such as crop and animal husbandry, horticulture, fisheries, and forestry.

Asia is a continent of small-scale farmers; more than 75 per cent of its farmers cultivate less than 2 hectares of land. Earlier I mentioned that recent scientific advances have enhanced the production potential of small farms in the Third World. But we should recognise that although new technologies have in many cases been used economically by all farmers, regardless of their farm size, the technologies themselves are not 'resource neutral'. Farmers need cash inputs to enhance agricultural output. Modernisation of agriculture involves the increased use of purchased inputs and a greater dependence on marketing opportunities. Public policies that simultaneously help small farmers raise production and help poor consumers increase consumption are vital to advancing agricultural well-being in the Third World.

Because of risks, farmers largely base their decisions to adopt new technology on the likely net returns per hectare as well as on security of income. In many South-East Asian countries, an increase of 1 million tons in rice production requires the active participation of about 2 million farming families. Hence, attention to their problems is vital for success in increasing production.

Inadequate attention to small farm management has increased production costs and risk. Agricultural scientists must find more ways to enhance the resource neutrality of technology by substituting nonmonetary or inexpensive inputs for high-cost inputs and minimising yield fluctuations caused by weather aberrations and pest epidemics. New technologies must ensure that labour demand increases faster than its supply and that food supply increases faster than its demand. These objectives require that social, biological and physical scientists work as teams.

New systems of small farm management must marry individual initiative and group endeavour. Japan has pioneered yield increases in agriculture through integrated systems of research, extension, input supply (including credit) and remunerative pricing. In China, a household responsibility system of crop production has been superimposed over social land ownership. This provides a mechanism for blending collective management of farm operations such as irrigation, plant protection, and postharvest operations with individual enterprise and incentive.

In countries where land is individually owned, can social or group management of key farm operations be superimposed on individual initiative and ownership rights? Obviously, farmers have often done this themselves. Agricultural research organisations can help by demonstrating how enlightened self-interest demands that farmers living in a watershed or a village participate in the management of farm operations that elevate and stabilise yield per unit of cash input. In particular, there is immediate need for increasing returns from irrigated agriculture through equitable systems of water distribution and better on-farm management of water. The major irrigation systems of the world are in Asia, but there is considerable scope for improving productivity per unit of water.

Farmers live in a world of action, not of words, so personal experience is the only meaningful source of conviction. Therefore, they must be shown how blending of individual ownership and group action can better serve the interests of the individual and the community in the same manner that China is demonstrating how social ownership and individual initiative can be combined in a meaningful way.

Third World countries should lose no further time in developing strong *national food security systems* with the following major components:

1. Ecological security to protect basic life support systems such as land, water flora, fauna, and the atmosphere;
2. Technological security at both the production and postharvest phases, which can help to promote accelerated growth coupled with production stability;
3. Social security to provide the needed purchasing power to the rural and urban poor through greater opportunities for gainful employment;
4. Water security, both for drinking and irrigation;
5. Nutrition education; and
6. Population stabilisation.

The most precious resources of the Third World are its intelligent and hardworking rural populations and its abundant sunlight. These assets can be purposefully married through a photosynthetic pathway of development that gives greater attention to green plants. Poverty persists when human resources are undervalued and land and other physical and material resources are overvalued. Human resource development holds the key to national prosperity as well as to peace and progress.

IRRI emphasises manpower development through its education and training programme in which about 3,000 rice scientists, extension specialists, and educators from more than 50 countries have participated; 95 per cent have been from the Third World. Sixty per cent remain active in rice research.

During the next few decades rapidly rising income in many developing countries is likely to fuel a tremendous growth in the demand for food, which can only be met by increased food imports.

From 1970 to 1977, 20 developing countries with some 700 million people had an average growth rate of 4 per cent or more in per capita income. Eight of these countries, all major oil-exporters, had an average per capita income growth rate of 5.6 per cent: Algeria, Indonesia, Iran, Iraq, Mexico, Nigeria, Saudi Arabia, and Venezuela. Such income increases sparked an extraordinary rise in the demand for food in those countries, a demand that their still-immature agricultural sectors were unable to meet. Thus, food imports to those countries grew at an unprecedented rate of 19 per cent per year in real terms.

Eleven other rapid-growth countries—Brazil, Hong Kong, the Democratic People's Republic of Korea, the Republic of Korea, Malaysia, the Philippines, Singapore, Syria, Thailand, Tunisia, and Turkey—had an average growth rate of 5.6 per cent in per-capita income during the same period. In those countries, the demand for food rose by more than 5 per cent per year.

Some of the countries with rapid growth of income may well be able to achieve the high rates of growth in domestic food production needed to satiate demand. But research carried out by the International Food Policy Research Institute indicates that developing countries with the fastest growth rates in food staple production from 1961 to 1976 collectively more than doubled their net food imports during the same period. From 1961 to 1976 the average growth rate for basic food staple production in those countries was 3.9 per cent—but their increases in food imports meant that their self-sufficiency ratio actually declined by 2 per cent. These data show that while it is possible for developing countries with rapid income growth to achieve impressive increases in basic food production, it is still quite difficult for such production to keep pace with the rate of growth in demand for food caused by increases in population and income.

Stagnation in social and economic evolution has resulted in inadequate progress in achieving self-reliance in food production in many Third World countries. Such stagnation was caused by colonialism before World War II. Population explosion facilitated by advances in preventive and curative medicine after World War II, as well as continued low priority to agricultural and rural development after independence aggravated the problems. Neglect of agriculture also led to the continued nonutilisation or underutilisation of the national blessing of these countries for generating more jobs and income. The economic problems of Third World countries are well articulated, but the potentials for overcoming major problems through a careful blend of brain (technology), brawn (hard and dedicated work), and bank (cash inputs) continue to remain neglected or underestimated. It is in this context that the the role of accelerated agricultural growth as a trigger of Third World prosperity must be appreciated.

Priority to the farm sector does not imply neglect of industry. In fact, there are several major linkages between agriculture and industry. Agriculture supplies the raw materials for employment-intensive industries. Agriculture stimulates and sustains industrial output through rural household demands for consumer goods and services. Agriculture influences industry through government savings and public investment. Agricultural growth requires fertiliser, machinery, tools and energy.

The International Food Policy Research Institute has attempted to quantify the impact of such linkages. Using a macroeconomic model, it has been estimated that from 1961 to 1972, a 1 per cent increase in agriculture in India generated a 0.5

per cent increase in industrial growth. Agriculture increased national income by 0.7 per cent.

Poor people spend most of any additional income on food. In India, studies show that the poor spend 60 per cent of increments to income on grain and almost 80 per cent on total food products. Food and employment are intimately related through the multiplier effects of agricultural growth and development.

Technological change in agriculture can stimulate growth in income and employment in many sectors of the economy. It raises the incomes of farmers, who devote a large share of their additional income to locally produced goods and services. IFPRI research in Malaysia, Bangladesh, and India has shown that from 40 to 80 per cent of increments to farmers' income is spent on such local 'tertiaries' as housing, entertainment, and services. Such tertiaries create more employment.

In a recent study in Malaysia, it was found that each additional dollar of value an irrigation project created in the paddy rice sector indirectly generated an additional $0.83 of value in the non-rice economy. Of this $0.83, about two-thirds was attributable to final household demand linkages and only one-third to increased demands for paddy intermediaries. Thus there are symbiotic linkages in the growth of all the three major factors of an economy—agriculture, industry, and services. Agricultural growth necessitates better rural communication and better communication helps to minimise the gap between the city and the village in technological evolution.

Science and technology are important components of the wall dividing poverty and prosperity. Today there are unusual opportunities for developing countries to improve the quality of life of their people and to reduce drudgery in rural areas, particularly in women-specific occupations, through an integration of traditional and emerging technologies. Recent advances in microelectronics and microprocessors, biotechnology, satellite communication and imagery, and solar and renewable energy technology can be harnessed to upgrade traditional skills and occupations. In the past, industrialisation was synonymous with urbanisation, centralisation, automation, and pollution. Today it is possible to promote in rural areas sophisticated agroindustrial complexes based on decentralised infrastructure and production techniques. That will help marry the techniques of science with the culture and skills of the people. With financial support from the Asian Development Bank, IRRI and the University of the Philippines at Los Baños this year intend to initiate a pioneer project to show how productivity can be maximised and income enhanced through the scientific utilisation of every part of the rice biomass.

History shows us that man causes civilisations to blossom... and to decay. Some historic centres of agriculture are useless deserts today. In contrast, some areas that were considered hopeless wastelands in the early 20th century are now fertile farming areas. A proper combination of political will, professionalism, and people's action is essential to harness the power that science gives us for increasing human happiness and welfare.

During the 1980s, the scientists and scholars of IRRI, and of the world's rice research family, will continue their relentless struggle for higher and more stable income for farmers through a multipronged research strategy that consists of:

1. Sustaining and expanding production gains in irrigated areas;
2. Extending frontiers of high-yield technology to areas of moisture stress and/or excess;
3. Increasing productivity in problem-soil areas;
4. Improving the income and employment potential of rice farming systems;
5. Adding a dimension of resource neutrality to scale neutrality in technology development;
6. Improving methods of manpower development, information dissemination, technology transfer;
7. Identifying and helping to remove the constraints responsible for the gap between potential and actual yields at the farm level; and
8. Monitoring the consequences of new technology.

We live in this world as guests of green plants and of the farmers and fishermen who cultivate them and harvest their products. Today an eminent group dedicated to Third World development has recognised the pivotal role of science in agricultural progress by choosing an agricultural research centre for a distinguished award. I assure you, on behalf of my colleagues in national agricultural improvement programmes and at IRRI and other CGIAR institutions, that we shall do everything possible to help farmers—who will in turn help humankind make hunger a problem of the past.

Changing Nature of the Food Security Challenge: Implications for Agricultural Research and Policy

M.S. Swaminathan

Sir John Crawford Memorial Lecture
November 1, 1990
Washington, D.C.

Published by the Consultative Group on International Agricultural Research, CGIAR Secretariat, 1818 H St., N.W., Washington, D.C., 20433, United States. November 1990.

Sir John Crawford Memorial Lecturers

1985 Robert S. McNamara, United States
1986 Bukar Shaib, Nigeria
1987 Amartya Sen, India
1988 Helen Hughes, Australia
1989 Jacques Diouf, Senegal
1990 M. S. Swaminathan, India

The Sir John Crawford Memorial Lecture has been sponsored by the Australian government since 1985 in honor of the distinguished Australian civil servant, educator and agriculturalist who was one of the founders of the Consultative Group on International Agricultural Research (CGIAR). Crawford (1910-84) was also the first chairman of the CGIAR's Technical Advisory Committee.

Changing Nature of the Food Security Challenge: Implications for Agricultural Research and Policy

M.S. Swaminathan

One day early in 1965 I had a call from an officer of the Planning Commission of the Government of India asking whether I could show a distinguished visitor from the World Bank around the wheat plots at the Indian Agricultural Research Institute (IARI), New Delhi. I asked what exactly the visitor would like to see and how much time he could spare. In reply, Sir John Crawford came on the line himself and said he would like to see the semi-dwarf wheat varieties in order to understand their yield-enhancing potential. I asked whether he would like to see them in the experimental station or on farmers' fields. He immediately replied, "In farmers' fields." I took him to the Jounti Seed Village of IARI where he spent a whole day squatting in farmers' homes drinking the sugarcane juice and eating the pearl millet bread they gave him and enjoying himself thoroughly. At the end of the day when I was taking him to his hotel, he said, "I now see light where there was only darkness."

Thus began a strong personal friendship which lasted until his death. For over 15 years, he was a friend, philosopher and guide to me and my family. I am hence grateful for this opportunity to pay homage to the memory of a truly remarkable person, whose genius lay in his being simultaneously a vertical and a horizontal man. This combination made his summing up of complex facts a combination of clarity, precision, and a delightful mixture of wit and wisdom.

M.S. Swaminathan, laureate of the World Food Prize and former Director General of the International Rice Research Institute, heads the M.S. Swaminathan Research Foundation in Madras, Tamil Nadu, India.

The recommendations he made on behalf of the Technical Advisory Committee (TAC) to members of the CGIAR were always listened to with great respect and evoked a positive response. He had no peer in foresight about the emerging developments in global agriculture.

I had the good fortune of working with Sir John during my term as Vice-Chairman of TAC during 1971-77. One of the things I learned from him was the continuous quest for excellence and relevance in research and a concern for maintaining sustainable dynamism in older institutions. I am glad the traditions he set up have been not only maintained but have been strengthened further over the years.

Sir John's major obsession both in his work in India and in TAC was the elimination of hunger and poverty through appropriate packages of technology, services and public policies. He felt that a high priority should be accorded to the development of technologies which would be meaningful for ecologically and socially handicapped farmers. The World Development Report 1990 on Poverty by the World Bank underlines the continued importance of this emphasis.

Thanks to advances in agricultural production, consumption per capita in the developing world went up by almost 70 percent between 1968 and 1985. However, according to the World Development Report 1990, in spite of impressive achievements in improving both income and consumption, more than one billion people in the developing world are living in poverty. This is the number of people who are struggling to survive on less than $370 a year. The Report rightly stresses that "progress in raising average incomes, however welcome, must not distract attention from this massive and continuing burden of poverty."

Between 50 to 80 percent of the population in developing countries depend on agriculture for their livelihood. Although the contribution of agriculture to GDP is going down in many developing countries as a result of the diversification of the economy, the burden of providing jobs still lies largely with the farm sector. The 1990 report of UNDP on
2 Human Development clearly shows that countries with a low

per capita GNP can achieve a low under-5 mortality rate and high life expectancy through placing a high priority on resource allocation for human development, a point which an earlier lecturer in this series, Professor Amartya Sen, has been making for several years. This illustrates the critical role of public policy in rescuing millions of children, women and men from sub-human living conditions. Professor Amartya Sen has often stressed that we cannot reap what we do not sow and that the persistence of chronic hunger is due to the inadequate political attention paid to eliminating it.

The World Bank's 1990 Report on Poverty as well as studies conducted by the International Labour Organization (ILO) indicate that the number of people living in extreme poverty is likely to rise in Africa in the coming years (Table 1). It has hence been appropriate that several of the earlier lectures devoted to the memory of Sir John Crawford have specifically addressed issues in Africa. In this lecture, I wish to consider some of the emerging challenges in agricultural research and public policy as a result of the changing nature of the food security challenge.

Table 1

Number of people living in extreme poverty
(Millions)

	1980	**1985**	**1995 (projection)**
Africa	210	278	405
Asia	562	538	450
Western Hemisphere	47	65	58
Total	819	881	913

Source: International Labour Organization (ILO), "High-Level Meeting on Employment and Structural Adjustment," Geneva, 23-25 November 1987, p. 17, cited in Jean Mouly, "Reviving the world's economic growth: chances and risks." Mimeo, February 1989, p. 4.

Food Security Challenges Today

FAO defines *Food Security* as physical and economic access to food to all people at all times. I have been pleading for replacing this concept with that of *Nutrition Security* which involves physical and economic access to balanced diet and safe drinking water to all people at all times. The emphasis on clean drinking water is particularly important in my view. Nutrition Security is essential for giving children an opportunity for the full expression of their innate genetic potential for physical and mental development.

It is now widely accepted that chronic hunger today is due more to the lack of purchasing power than to the non-availability of food in the market. In other words, to win the battle against hunger, we have to fight the "famine" of jobs. An integrated programme of work, wage (ie: the enforcement of minimum wage) and welfare measures in the fields of public health, sanitation, drinking water and education are essential for this purpose.

Compounding the problem of gross economic inequity is the growing damage being done to the ecological foundations of sustainable agriculture. Diminution of the biological potential of soils, loss of biological wealth, deforestation, depletion of ground water resources, pollution, and the growing imbalance between carbon emissions and carbon absorption are leading to a situation where safeguarding the ecological base of agriculture may become the most important food security challenge. It is clear that economic entitlements and ecological obligations must go together for achieving sustainable food and nutrition security.

The Food Security Act of 1985 of the United States is a fine example of an integrated approach to production, conservation and consumption. It would be advisable for every country to adopt similar legislation which can provide the legal framework essential both for sustainable advances in biological productivity and for eliminating chronic hunger. Since the World Bank and CGIAR are deeply concerned with the public policy decisions necessary to stimulate higher production by small farm families based on ecologically sound technologies and to provide increased access to better quality diets for the rural and urban poor, it would be useful to organize a seminar for political leaders and policy makers on the role of legal instruments for the achievement of national goals in food and ecological security. The implications and impact of the US Food Security Act of 1985 could provide the basis for such a discussion.

Technology and Food and Ecological Security

The Green Revolution of the sixties and seventies in wheat and rice resulted in higher production through an improvement in yield per hectare and to that extent helped to save land. India, for example, produced 12 million tonnes of wheat in 1964 from 14 million ha. This was before the advent of high-yielding wheat varieties.

In 1990, Indian farmers harvested about 55 million tonnes of wheat from about 23 million ha. To do this at 1964 yield levels, India would have needed over 40 million ha of additional land. Globally, UNEP estimates that 90 percent of the annual deforestation of about 15 million ha is due to the spread of agriculture. FAO places the estimate even higher — at 17 million ha per year. Thus, the ecological necessity for improving production through higher productivity rather than through area expansion is obvious.

In their article on "Feeding the World in the Nineties" (published in the State of the World — 1990 — Report of the World Watch Institute) Lester Brown and John Young estimate that the world could be losing 14 million additional tonnes of grain output due to land degradation and crop damage caused by factors such as flooding and pollution. According to them, environmental damage could lead to the net gain in global food output remaining at about 1 percent per year, while population growth will be close to 2 percent. They hence conclude that the food supply position will be tight and prices will rise in the coming years.

The position with regard to animal nutrition is equally serious. If under-nutrition or calories deprivation is the major cause of malnutrition in the human population of many **5** developing countries, under-nutrition is even more serious in animal populations. Since grazing land is scarce, the only pathway open for improving the yield of animal products is stall-feeding. The technology of stall-feeding should be based on enriched agricultural raw material and not on food gains. In other words, what we need in population-rich but land-hungry countries is land-saving agriculture and grain-saving animal husbandry.

The needs of land-saving agriculture and grain-saving animal husbandry can be met only by further improvements in technology. At the same time, persistent environmental degradation which is threatening the natural resource base underpinning sustainable advances in biological productivity will have to be arrested. This will call for intensification of efforts in the blending of traditional and frontier technologies in such a manner that the ecological and economic strengths of both are combined. The frontier technologies of particular interest are biotechnology, space technology such as weather satellites and remote sensing, information technology including computer-aided instruction and extension and management technology which helps to introduce a systems approach to all aspects of production, processing, storage and marketing.

A major factor having a bearing on the development and dissemination of new technologies is the nature of the demographic profile of countries in terms of age composition as well as dependence on the primary farm sector for livelihood security. In most developing countries, the population is predominantly young. Over 50 percent of India's current population of about 850 million is below the age of 21. Also, nearly 70 percent of the population derive their livelihood from agriculture and other rural occupations. The future of agriculture in such countries will hence depend on their ability to attract and retain youth in farming and other rural occupations. Otherwise urban slums will multiply and social tensions will grow.

For educated youth to be attracted to work in rural areas, farming and allied rural work must become both economi-
6 cally rewarding and intellectually satisfying. In other words,

agriculture should become an occupation which requires brain as much as brawn. Land ownership and tenancy rights and input and output pricing policies should be such that farmers feel encouraged to save and invest surplus funds in strengthening the ecological infrastructure essential for sustainable advances in biological productivity. Since capital is scarce, knowledge must become to the extent possible a substitute for capital. Fortunately, most ecologically sound technologies like integrated pest management and integrated nutrient supply are also knowledge-intensive.

Technology and trade have been prime movers of economic prosperity in the countryside in industrialized countries. Non-farm and off-farm employment opportunities grow when post-harvest technologies and trade opportunities improve. The primary, secondary and tertiary sectors of the economy then get symbiotically inter-linked.

To achieve this goal, we would need a faster rate of agricultural growth. In a recent study of India's development experience conducted for WIDER (World Institute for Development Economics Research, Helsinki) Yoginder Alagh, who was until recently a member of India's Planning Commission, observed, "A high rate of agricultural growth was a pre-condition of faster employment growth. In any employment strategy, anywhere between two-thirds to three-quarters of the desired employment growth would come from achievement of higher agricultural output levels. Thus, widespread agricultural growth was a great equalizer in terms of employment generation and poverty reduction in India."

Nearly half the world's population live in coastal areas. Both the ecological security of coastal regions and the livelihood security of coastal communities need greater attention. There is need for Coastal Systems Research Programmes designed to promote the linked development of capture and culture fisheries, and coastal forestry and agroforestry including the protection of mangroves, coral reefs, sea grasses, and associated flora and fauna. We also need methodologies for an ocean productivity classification on
the model of land capability studies. As much carbon is fixed **7**

in the ocean as on land and we must derive better benefit from such aquatic carbon fixation.

Inadequate purchasing power in developing countries often leads to uncomfortable food gluts even when there is an increase of only 5 to 10 percent in food grain production. Ironically, in countries characterized by a high incidence of poverty, marketing problems become serious when efforts in increasing agricultural output succeed. Violent undulations between acute scarcity and uncomfortable glut are not uncommon under such conditions. For example, Mary Anne Fitzgerald in an article on hunger in Ethiopia (Financial Times, London, May 5-6, 1990) states, "There are 50,000 tons of grain available in Western and Southern Tigray where the harvest has been good. The Relief Society of Tigray has asked for $18 million to purchase it but so far only $4 million has been pledged." Quite often in Africa, the World Food Programme (WFP) has extended assistance in purchasing locally grown food grains. But for such interventions by WFP, farmers would be cursing themselves for having produced good crops. Imaginative efforts in improving food consumption and in eliminating chronic hunger are essential for accomplishing economically sustainable advances in production.

International cooperation in trade and trade-related matters, in the sharing of environmentally-friendly technologies and in long-term commitments of development assistance to meet the infrastructure and institutional investments required for a more sustainable and equitable agriculture, needs considerable strengthening. All this will call for a revolution, both in patterns of international cooperation and in information systems. Such information systems should be capable of integrating and analyzing the interactions between bio-physical, environmental, economic, social and demographic trends of rural livelihood systems. Unless the substrate requirements for new technologies to strike roots and confer economic and social benefits are clearly understood and attended to, the availability of useful technologies and stagnant production conditions will continue to co-exist.

Response to the Challenges Ahead

By 2020, we may need an annual production of at least 3000 million tonnes of food grains and 200 million tonnes of aquatic products besides larger quantities of vegetables and fruits to provide balanced diets for over 8 billion human beings, many of whom will have enhanced purchasing power. In addition, considerable additional quantities of fuelwood, fodder, fibre and other agricultural commodities will be needed. Even now, fuelwood occupies the second position in relation to the gross value of production in developing countries, as shown below:

Commodity	*Gross Value* (in US $ million)
Rice	70,465
Fuelwood (non-coniferous)	45,769
Sawlogs	44,430
Milk	38,970
Wheat	27,542

(Source: FAO Data 1983-85)

According to FAO statistics, the agricultural self-sufficiency rates of developing countries are likely to drop to 100 percent in the year 2000 from 101 percent in 1983-85. Most of the additional population will be in developing countries and according to a World Bank projection, the proportion of global population living in developing countries in the year 2025 will be 83 percent.

In contrast to the expanding need for food, fuel and other commodities, many developing countries are already confronted with a growing loss of farm land, diminishing resources of water and biological diversity and expanding biotic and abiotic stresses. In addition, all nations may face changes in temperature, precipitation, sea levels and a higher incidence of ultraviolet radiation.

What should be our instruments in handling such a situation? First, we have to tap the untapped technologies on the 9 shelf through appropriate social engineering, extension, and training programmes and public policies in land ownership, input and output pricing and rural infrastructure development. An idea of the extent of opportunities available for taking advantage of existing know-how in rain-fed crops is provided by the data of the International Crops Research Institute for the Semi-Arid Tropics (ICRISAT) cited in Table 2. Second, we must intensify agricultural research, making effective use of emerging scientific opportunities provided by biotechnology as well as space, information and management technologies. We have to duplicate what has been achieved on the production front during the past 12,000 years in the next 40 years. This cannot be done without the help of new technologies and management tools. Third, we must ensure that ecological sustainability becomes the foundation for all efforts in the development and dissemination of technologies.

Table 2

Grain yields under improved and traditional technologies on deep Vertisols at ICRISAT Center*; Hyderabad, India.

		Grain yield (t ha) Improved systems; double cropping			Traditional system single crop	
Year	**Cropping period rainfall (mm)**	**Sorghum/ Maize**	**Sequential chickpea/ intercropped pigeonpea**	**Total**	**Sorghum**	**or Chickpea**
1976/77	708	3.2	0.72	3.92	0.44	0.54
1981/82	1073	3.19	1.05	4.24	0.64	1.05
1986/87	585	4.45	0.38	4.83	0.37	1.27
1988/89	907	4.64	1.23	5.87	0.61	1.18

*Average rainfall for Hyderabad (29 kms away from ICRISAT Center) based on 1901-84 data is 784 mm with a CV 28%.

CGIAR Centers and TAC have been giving serious attention to sustainability issues. I would plead for greater efforts in the following areas:

a) Standardization of measurement tools for (i) disaggregating diverse components of sustainability such as the ecological, the economic and the socio-cultural and (ii) estimating the impact of production techniques on environmental capital stocks such as land, water, genetic diversity and greenhouse gas emission.

b) Selection of participatory research sites characterized by a large extrapolation domain; this is essential for developing location-specific technologies.

c) Organization of demonstration and training centers to explain the implications of adding the prefix "Sustainable" before "farming system". The President of Guyana has offered 400,000 ha of tropical rain forest in his country for developing techniques for the sustainable management of rain forests and a team set up by the Commonwealth Secretariat has prepared proposals for taking advantage of this offer. Such work will need a long-term commitment of financial and technical resources.

d) Diversification of cropping systems and food habits, with particular emphasis on vegetables and fruits; and

e) Promotion of a Sustainable Agriculture Matrix approach on the lines indicated in Table 3. This will call for the integration of considerations of ecology, economics and equity by all concerned with the development of packages of technology, services and public policies.

I realize that existing uni-dimensional structures for policy formulation and task implementation are not conducive to the growth and spread of such a symphonic approach. Often decisions have to be taken on the basis of trade-offs between immediate economic gains and long-term ecological considerations. It will, however, be difficult to make the concept of sustainable agriculture a reality without such coordinated thinking and action.

Table 3

Sustainable Agriculture Matrix
Illustrative components

Basic Principle	**Action Points**		
	Technology	Services	Public Policy
I. Ecology	Conservation and enrichment of environmental capital stocks	Organization of group activities on a watershed basis	Legal, promotional and educational instruments for the sustainable management of natural resources and for population stabilization
II. Economics	Promotion of biological and farm grown inputs — higher yields at lower cost-diversification of employment and income sources	Organization of input supply and marketing services through farmers' organizations	Incentives for group endeavor and for the adoption of knowledge-intensive farm practices
III. Equity			
i) Intragenerational	Technology options for resource poor farmers — attention to gender-based occupations	Empowerment of the rural poor through organizational structures designed to take new technologies to the unreached	Steps to protect the livelihood security of the poor through measures in the areas of work, wage and welfare
ii) Intergenerational	Sustainable use of basic agricultural assets	Education and early warning systems	Political commitment to sustainable development

Biotechnology: Opportunities

There has been considerable debate in recent years on the potential impact of new biotechnologies on agriculture. According to Ellen Messer and Peter Heywood ("Trying Technology: neither sure nor soon", Food Policy 15, 336-345, 1990), the impact of biotechnology in overcoming hunger may have to await the next millennium. Since I do not share this view, I wish to discuss a few issues relating to research and extension in the field of biotechnology at some length.

The term biotechnology is currently being used to connote a wide variety of biological manipulations such as cell and tissue culture, embryo transplantation, transfer of DNA material across sexual barriers, microbiological enrichment of cellulosic material, fermentation and various forms of biomass utilization. There are immediate opportunities for the multiplication of superior clones of fruit and forest tree species, as well as plantation crops like cardamom and oil palm through tissue culture methods.

The hard core of biotechnology is recombinant DNA technology resulting in transgenic micro-organisms, plants and animals. The first transgenic plants expressing engineered foreign genes were produced in tobacco in 1983 by the use of *Agrobacterium tumefaciens* vectors. Since then, transgenic material has been produced in a wide range of plants, animals and micro-organisms. The Animal and Plant Health Inspection Service of the U.S. Department of Agriculture issued nearly 100 permits for testing genetically engineered material in the field between November 1987 and September 1990. The plants with new characters now under testing include maize, cotton, soybeans, potato, tomato, tobacco, alfalfa, cucumber, cantaloupe, squash, rice, walnut and poplar. We can expect even more rapid progress in the nineties as a result of the work of research networks like that supported in rice by the Rockefeller Foundation.

The basic research underpinning the techniques of biotechnology has largely been carried out in universities and public-funded laboratories. However, the work on the 13 conversion of scientific information into economically viable technologics has largely been undertaken in the private sector. This had led to the question whether the fruits of such research will be available only to those who can afford to pay adequately for them. For example, in agriculture, some experts have stated that while the "green revolution" technologies arising from research funded by philanthropic foundations like Rockefeller and Ford Foundations and by governments of developing and industrialized countries were available to all farmers who could derive benefit from them, the "gene revolution" technologies associated with biotechnological research may not likewise be available, since they owe their origin by and large to investments made by private companies and may be protected by patent rights. Where should the line be drawn between private profit and public good, particularly in a world characterized by glaring economic inequities?

The results of genetic engineering research in medicine, such as the production of insulin, interferon and different kinds of vaccines, are being disseminated by the pharmaceutical industry. Likewise, applications in animal health care and production can be expected to be spread by private industry. Similar may be the case of enzymes like rennin used for making cheese. In the case of crop improvement, there is an on-going debate about methods of integrating the principles of equity with those of economic profitability. The basic dilemma arises from the fact that while developing nations often represent centres of biological diversity and have rich endowments of biological wealth, the capacity to convert biological diversity into biological productivity through science and technology resides predominantly in industrialized countries, where such conversion work is increasingly in the hands of private industry.

The significance of biotechnology for a better biofuture of the Third World can be illustrated by taking the example of Asian agriculture. Asia has over 50 percent of the global population, over 70 percent of the world's farming families, but only 25 percent of the world's arable land. At the begin-
14 ning of the 21st century, the per capita land availability will

be 0.1 ha. in China and 0.14 ha. in India. The average Asian population growth rate is 1.86 percent.

The only pathway open to countries like China and India for feeding their growing human populations is continuous improvement in yield. This involves research which can further raise the yield ceiling. China has gone into the large scale exploitation of hybrids in rice for this purpose. The tools of biotechnology can help in raising the productivity of major crops through an increase in total dry matter production which can then be partitioned in a way favorable to economic value.

Water availability will be another serious constraint in many developing countries. In several of them the availability of even adequate drinking water is a serious problem. There has been an almost threefold expansion in global irrigated area since 1950 and much of this expansion has been in developing countries. Scope for further expansion is getting gradually reduced. In particular, ground water is already being utilized in an unsustainable manner in several parts of the world. Improved irrigation water management and the incorporation of genes for drought-tolerance in major crop plants are urgent necessities.

In animal husbandry also, the needs of developing nations, particularly in South and Southeast Asia differ from those of the industrialized countries. Most of the productive animals in India, for example, are stall-fed. This enables government to provide farm animals to landless labor families for increasing their household income. Such resource-poor animal-rearing families have to be assisted in running the enterprise as efficiently and economically as possible, by providing services in the areas of genetic improvement, health care, nutrition and marketing. Nutrition has to come from high-yielding fodder legumes and grasses grown in crop rotations and from enriched cellulosic material.

A market research report entitled "Biological Products for Aquaculture-A Worldwide Market Study on Vaccines, Therapeutics, Diagnostics, Hormones and Genetic Minipulations" published recently by the Technology Management 15

Group, New Haven, U.S.A., suggests that as aquaculture farms increase their production per unit space, effective disease and stress control will assume greater importance. The markets for vaccines, diagnostics, hormones and new feeds will increase. It is anticipated that by the year 2000, 25 percent of world-wide seafood consumption is likely to be produced by aquaculture. Vaccines are seen as a growth area, since vaccines are still needed for many major diseases. In Scandinavia and parts of the U.S.A., nearly all trout and salmon, produced by aquaculture, were vaccinated in 1989 as compared to 5 percent in 1984.

Many companies are developing aquaculture therapeutics to meet the growing demand. It is estimated that over 50 percent of the total global production of fish, shellfish and molluscs is lost to disease. Breeding programmes and genetic engineering have led to the production of new "boneless" breeds of trout that have a better feed conversion rate, and salmon which possess an antifreeze gene to enable them to survive in colder waters. Further research in fish breeding is expected to focus on growth acceleration, sex identification and determination, flesh quality, disease resistance, sea water adaptation, and the ability to utilize specific dietary components. Thus, biotechnological research is opening new windows of opportunity both in terrestrial and aquatic farming systems.

Biotechnology can make a useful contribution for integrating brain and brawn in rural professions. For example, Kerala State in India is planning to develop the district of Ernakulam as a **Biotechnology District,** for taking advantage of its rich educated human resources, particularly educated women, who often tend to be inappropriately employed. The programme will include extensive tissue culture propagation of forest tree species, banana, cardamom and ornamental and medical plants, genetic improvement of cattle and poultry and the establishment of biomass refineries. The cause of educated unemployment is often not the lack of employment opportunities per se, but the paucity of employable skills in educated youth. The prevailing mismatch between
16 the skills needed for the sustainable conversion of natural

endowments into economic wealth should be ended through a carefully planned learning revolution.

Biotechnology and Third World Concerns

A reference needs to be made to a few of the major concerns of Third World scientists and political leaders relating to current global trends in the objectives and organization of biotechnological research. First, the farm sector is a major export-earning enterprise for Third World countries. Therefore, there is genuine concern about the potential adverse impact of genetic engineering research directed at finding substitutes for natural products. Some examples are: high-fructose corn sweetener as a substitute for sugarcane sugar and substitutes for vanilla, cocoa and diosgenin extracted from *Dioscorea* species.

A second major concern relates to the safety aspects of genetic engineering research. Will tests be done in the Third World which are not permitted in the industrialized countries? Will "super weeds" arise from research aimed at the development of pesticide and herbicide resistant crop varieties? Will the ecological groundrules underpinning the field testing of transgenic material be the same everywhere?

Third, the nutritive quality and food safety issues relating to genetically engineered strains and growth promoting agents like bovine growth hormones need careful study, using criteria more relevant to conditions where under-nutrition and malnutrition are widespread. Will crop varieties with multiple resistance to pests contain toxins which will ultimately affect the health of the human beings or animals which consume their economic parts? What kinds of safety evaluation procedures are needed for food ingredients produced by microorganisms, single chemicals and simple chemical procedures and whole foods and other complex mixtures?

Fourthly, will the biotechnology revolution help resource-poor farmers increase productivity largely with the help of farm-grown inputs? How can we design mutually reinforcing packages of technology, services and public policies which can ensure that all rural people — rich or poor, land owners or landless labour families — can derive economic and social benefit from new biotechnologies?

Fifthly, what will be the impact of the extension of intellectual property rights to individual genes and genotypes on the availability of such improved material to developing countries and resource-poor farm families? Also, will intellectual property rights be exclusively reserved for rewarding formal innovations and neglect the pivotal role the informal innovation system has played and is playing in the identification and conservation of plant and animal genetic resources? What are the rights of the farm families who have conserved and selected genetic diversity in contrast to the rights of the breeders who have used them to produce novel genetic combinations? How can the concept of genetic diversity as a common human heritage be promoted, if only a few can derive economic benefit from such a diversity?

Sixthly, will priorities in biotechnology research be solely market-driven or will they also take into consideration the larger interests and the long-term well-being of humankind, whether rich or poor? In other words, will **orphans remain orphans** in the choice of research priorities and investment decisions? For example, rice is the staple of nearly half the human population most of whom live in Asia. Yet, the application of biotechnological know-how to solve some of the important problems in rice production would not have received the financial and scientific support they needed but for the decision of the Rockefeller Foundation to make a major long-term investment in this area.

Finally, there are environmentalists who believe that each technological fix to an ecological problem will ultimately generate new levels of ecological catastrophe. They therefore caution against venturing into unknown territories.

The above concerns can be met only by a proactive analysis of the potential beneficial and adverse impacts of biotechnological research, not only from the economic angle but also from the ecological and equity perspectives. Social scientists and ecologists should be involved in project design teams right from the beginning and should not just come at the end to make a post-mortem analysis. For biotechnology to lead to a better future for human-kind, we need a systems approach, keeping in mind Albert Einstein's exhortation that human well-being should be the ultimate objective of all scientific endeavor.

Several CGIAR centres have integrated the new research tools in their scientific work. UNIDO has sponsored an International Centre for Genetic Engineering and Biotechnology for spreading the benefits of research in this field to Third World countries. Other initiatives of this kind are now in various stages of development. I would suggest that CGIAR should consider organizing, in collaboration with the governments of developing countries, UN organizations, private industries and universities, a network of **Genetic Enhancement Centres** at IARCs and appropriate national research institutes in developing and developed countries for the purpose of evolving breeding material containing new gene combinations. Such material should be readily available for use by practical breeders in the development of location-specific varieties. Location-specific varieties and technologies are essential for achieving ecologically sustainable advances in biological productivity. A global network of Genetic Resource Conservation Centres linked to a similar network of Genetic Enhancement Centres will ensure that genetic diversity is not only conserved but is utilized for the benefit of all. Only such steps can help to bridge the growing gap in economic well-being between nations that have achieved mastery in biological technology and those rich in biological wealth but relatively weak in converting such blessings into economic wealth.

Biodiversity and Biofuture

Our biological future depends on our ability to conserve and utilize the rich genetic diversity occurring in living organisms on our planet. The extent of ignorance on the number of species existing on earth came out clearly at a Conference on the Ecological Foundations of Sustainable Agriculture organized by the Commonwealth Agricultural Bureaux International at London in July 1990. Some experts felt that more than 50 billion species may be occurring, while less than 2 million have been described so far. This underlines the importance of training more biosystematists.

Unfortunately, there is much controversy on methods of saving and sharing the global biological wealth. Discussions on this topic are in progress in various international fora such as FAO and UNEP. The Keystone International Dialogue Series on Plant Genetic Resources has tried to throw light on methods of resolving opposing viewpoints. Terms such as "Farmers' Rights" and "Breeders' Rights" are freely used to indicate the importance of according recognition to the informal innovation system in conjunction with the rights already accorded to plant breeders in the 20 developed nations which have so far adhered to the rules of the International Union for the Protection of New Varieties of Plants (UPOV). The ongoing discussions at the General Agreement on Tariffs and Trade (GATT) on Trade-related Intellectual Property Rights (TRIPs) are also important in the context of North-South relationships in germplasm conservation and exchange. Fourteen developing nations have proposed to the Negotiating group on TRIPs at the Uruguay Round of Multilateral Trade negotiations that plant or animal varieties or essentially biological processes for the production of plants or animals should not be subjected to patent protection.

Farmers and breeders are allies in the common task of advancing biological productivity. Therefore, I feel that their rights should be presented not as mutually antagonistic rights but as mutually reinforcing ones. I would like to suggest that UPOV should evolve into an International Union for the Protection of Breeders' and Farmers' Rights, with its membership including all countries — industrialized and developing.

The UPOV convention is now undergoing revision. A draft revision introduces the concept of "dependence" which would ensure that a variety "essentially derived from another variety protected by PBR cannot be used commercially without the permission of the breeder of the protected variety." It should not be difficult to develop a methodology under the dependency clause which enables recognition and reward for informal innovation. The financial reward in this case will have to go to a special fund which can help to finance conservation and plant breeding activities in the country from which the key genetic material came.

CGIAR institutes have rendered valuable service in the *ex situ* conservation and utilization of plant genetic resources. They can assist in promoting the evolution of a system of recognition which is universally regarded as equitable. A better biofuture for all will depend upon shared goals and action in conserving and using biological diversity.

Food and Jobs

The World Employment Review (1988) published by ILO stated:

"There can be little disagreement that employment is a critical issue around the globe — for young people seeking to establish themselves in a career; for those employed who fear redundancy or redeployment; for employers trying to decide on acquiring new technologies or hiring more workers; for trade unions attempting to protect the security of employment and maintain real wages; for those engaged in precarious work in the informal sector; for older women seeking to re-enter the labor force; for the long term unemployed; for training institutions preparing workers and managers for an uncertain future; for governments trying to manage substantial wage bills and to improve efficiency, and for the self-employed in urban and rural areas who have little in the way of support systems."

Most developing countries facing serious debt burdens and other economic difficulties caused by high prices of petroleum products are unable to provide the support systems essential for encouraging educated youth to take to self-employment. Much, however, can be done to enhance returns from scarce resources through synergy among diverse programmes connected with rural development. A consultation convened by the International Fund for Agricultural Development (IFAD) in 1988 suggested that the ongoing Farming Systems Research (FSR) Programme could be developed, where appropriate, into a Rural Systems Research Programme (RSR), specifically for the purpose of achieving a diversification of employment opportunities in rural areas. The essential features of a RSR programme will be the following:

a) The programme should foster the following three groups of linkages:

   i) Backward linkages with modern science and technology, particularly biotechnology, space technology and computer sciences

   ii) Forward linkages with markets, agro-industries and the services sector, and

   iii) Lateral linkages among crops, farm animals, fisheries, forestry and rural industries;

b) While fostering such linkages, emphasis should be, on the one hand, on ecologically sound technologies, and on the other, on strengthening growth linkages among the primary (farm), secondary (industries) and tertiary (services) sectors of the rural economy. Traditional technologies should be preserved, whenever they are still the best from the ecological and employment-generation points of view. They should be enriched and their economic viability and consumer acceptance enhanced by suitably blending them with frontier technologies;

c) The RSR programme should be designed and implemented jointly by scientists, the village community, non-governmental organizations and input supply and marketing agencies. Research on delivery systems should receive adequate attention;

d) Social scientists should be involved right from the planning stage, so that they are able to play a proactive
22 rather than a reactive role;

e) A new approach should be developed to component technology development by starting with the priority problems confronting small farmers and working towards scientific approaches and solutions. Priority should go to activities which can lead to the sustainable enhancement of the quality of life of the rural poor; and

f) The RSR Programme should include specific attention to the training of women in new skills. The training methodology should be "learning by doing." A well planned RSR programme will help to give concurrent attention to the production of more food as well as the generation of more jobs and income.

Research for Tomorrow

Sir John Crawford advocated anticipatory research where new problems or new opportunities may arise. To him, today's economic progress should not be at the expense of tomorrow's prospects for similar advances. The Second World Climate Conference is currently in progress at Geneva, Switzerland. At this conference strategies for avoidance and adaptation in relation to potential changes in global climate are being discussed. Until now, we were more concerned with a study of the impact of climate on human activity. A matter of equal concern now is the impact of human activity on climate. The international research networks operated by IRRI, CIMMYT and other international centers provide insights into genotype x environment interactions across the globe. CGIAR can take the lead in organizing, in collaboration with WMO, UNEP, FAO, UNDP, IBRD and other interested organizations an *International Research Network on Climate Change and Food Security* with the following aims.

a) Estimation and monitoring of the emission of greenhouse gases as a result of farm operations under different agro-ecological and technological conditions.

b) Standardization of techniques both for minimizing agriculture's contributions to the accumulation of green- 23 house gases and for withstanding additional biotic and abiotic stresses on crops and farm animals arising from climate change.

c) Promotion of research and training in the field of restoration ecology, in the afforestation of degraded forests and in the sustainable management of tropical rain forests.

d) Anticipatory research and development measures in coastal areas to avoid or minimize the adverse impact of potential changes in sea levels.

e) Standardization of methods for deriving benefit from higher CO_2 concentration in the stratosphere.

f) Standardization of post-harvest technologies for perishable agricultural commodities based on non-CFC dependent refrigeration methods, and

g) Stimulating policy research designed to strengthen the public policy back-up both for avoiding adverse changes in climate and for adapting to new growing conditions.

This will be another example of the pathway of sustainable dynamism in agricultural research which Sir John Crawford wanted CGIAR to nurture.

I thank you again for this opportunity to pay homage to the memory of one of the greatest scholars and humanists of our time.

 *Printed on recycled paper.

SECURITY

M.S.SWAMINATHAN

Illustration by VISHWAJYOTI GHOSH

Bridging the Nutritional Divide

BUILDING COMMUNITY CENTRED NUTRITION SECURITY SYSTEMS

UNDP's *Human Development Report 2001* has introduced a Technology Achievement Index (TAI), an aggregation of four groups of indicators relating to the creation of technology, diffusion of recent innovations, diffusion of old innovations and human skills. Creation of technology has been measured by the number of patents per capita and receipts per capita of royalty and licence fees from abroad. The emphasis is thus on the intellectual property rights (IPRs) of nations, evidenced by the power of proprietary science. The other indicators relate to digital, extension and educational divides. This report, titled 'Making New Technologies Work for Human Development', has, however, not drawn attention to the fact that bridging the expanding nutritional divide is fundamental to bridging the other divides, particularly that relating to IPRs.

The Commission on the Nutrition Challenges of the 21st Century, in its report titled 'Ending Malnutrition by 2020: An Agenda for Change in the Millennium', has pointed out that some 30 million infants are born every year in developing countries with intra-uterine growth retardation, representing about 24 per cent of all new births in these countries (*Philip et al*, 2000). Low birth weight (LBW) children are characterised by mental impairment. Worldwide, there are more than 150 million underweight pre-school children and more than 200 million stunted children. At current rates of progress in fighting these maladies, about one billion children will be growing up by 2020 with impaired mental development. What will be the impact on the intellectual property of a nation of such a denial to the child of opportunities for the full expression on its innate genetic potential for mental and physical development? Denying the child an opportunity for mental and physical development even at the foetal stage is the cruellest form of inequity. In contrast, excess weight is the major health problem among children in most industrialised countries and some developing ones *(Table 1)*. Thus, bridging the nutritional divide is the first requisite for a more equitable and humane world.

Growing disparities

The nutritional divide is increasing between the rich and the poor within and among nations. The situation is particularly alarming in developing countries. The nutritional paradox of South Asia lies in the coexistence of grain mountains and hungry millions. This is largely due to inadequate purchasing power, arising from the lack of sustainable livelihood opportunities. Famine of income is becoming the most important cause of a famine of food at the household level. Pregnant and nursing mothers and children belonging to the families living below the poverty line (the World Bank poverty line is an income of one US dollar per capita per day or below) suffer the worst. For example, severe anaemia during pregnancy is associated with very high relative risk of maternal death. Maternal mortality rates are as low as 3 to 4 per 100,000 births in industrialised countries, while in many developing countries they are at least 100 to 200-fold higher. Protein-energy malnutrition (PEM) affects nearly 30 per cent of children under five years of age in countries in Sub-Saharan Africa. A comparison of the nutritional status of populations in three Asian countries — China, India and Sri Lanka — provides some interesting insights into the impact of public policy on the nutritional well-being of the population.

Nutrition profile among a few nations in Asia: Role of non-nutritional factors

Four parameters — underweight, stunting, wasting and low birth weight — reflect the nutritional status of children below five years of age. The comparative profile of Sri Lanka, China and India is given in Table 2. The data show the importance of non-nutritional factors like education and healthcare in the nutritional well-being of an individual.

A. Body Mass Index. This indicates the nutritional status of adults. Adults with a BMI under 18.5 are considered to be chronically energy deficient. Body Mass Index over 25 indicates excess weight. Obese persons have a BMI over 30. The situation in Sri Lanka is given in Table 3.

B. Iron deficiency anaemia. In Sri Lanka, 58 per cent of children in the 5-10 age group suffer from iron deficiency anaemia, which affects their cognitive capacity and academic performance. In the case of adults, 45 per cent suffer from iron deficiency anaemia. The proportion of pregnant mothers affected is less — 39 per cent.

C. Mortality rates. Thanks to advances in preventive and curative medicine, mortality has declined in the periods 1970-75 and 1999-2000 in China, India and Sri Lanka *(Table 4)*. IMR and MMR are still high, although there is considerable variability among states within the country. The Indian state of Kerala, for example, has figures similar to those of Sri Lanka.

D. Female literacy and child health. A rapid increase in the rate of female literacy has been achieved in Sri Lanka as a result of the introduction of free education in 1945. It enabled girls to have as much access to education as boys. The situation is similar to that observed in the Indian state of Kerala.

Both men and women have achieved high literacy rates with 83 per cent for women and 90 per cent for men. They also have very low dropout rates: 4 per cent for girls and 6 per cent for boys. There is a significant impact of mothers' education on the nutritional status of children *(Table 5)*.

E. People power revolution in nutrition. Ultimately, the success of various nutrition-related programmes depends on the efficacy of delivery systems. Hence, Sri Lanka is attempting a community-based nutrition intervention programme. Called the participatory nutrition improvement project (PNP), this programme was started in 1993 with the help of UNICEF. The guiding principle was to mobilise the energies of the community and people's commitment to their own and their families' nutritional well-being. PNP is a people-focused project, enhancing the ability of mothers and fathers, through group formation and strengthening, to identify or explore their nutritional problems, identify their nutritional needs and maximise their potential in meeting those needs. Countries like Cuba, China and India also have rich and varied experience in the development of effective delivery systems. In Sub-Saharan Africa, Ghana has made rapid progress in overcoming PEM through community-based nutrition *(Gardner and Halweil, 2000)*. **Mobilising people power in the cause of nutritional security is the most effective and sustainable strategy**. The example of Thailand illustrates this fact.

F. Thailand's Nutrition Security Compact. Over the past 10 years, Thailand has achieved remarkable progress in reducing maternal mortality as well as the incidence of LBW children. The strategy consisted of the following components:

- Eliminate severe, moderate and mild protein-energy malnutrition.
- Monitor growth among all pre-school children and provide food supplements where needed.
- Mainstream nutrition in health, education and agricultural policies.
- Retrain and retool existing staff and mobilise community volunteers. Choose one community volunteer for every 10 households and build their capacity.
- Encourage breast-feeding and organise school lunch programmes.
- Promote home gardening, consumption of fruits and vegetables, aquaculture and food safety standards.
- Introduce an integrated food safety net with an emphasis on household food and nutrition security.

The positive impact of the above

Table 1: Share of underweight children and overweight adults
Selected countries, mid-1990s

Country	Share Underweight (per cent)	Country	Share Overweight (per cent)
Bangladesh	56	United States	55
India	53	Russian Federation	54
Ethiopia	48	United Kingdom	51
Vietnam	40	Germany	50
Nigeria	39	Colombia	41
Indonesia	34	Brazil	36

Source: Gardner and Halweil, 2000

Table 2: Comparative Nutrition Profiles of Children
In percentages

Parameter	Sri Lanka	China	India
Undernourished	25	11	21
Underweight	38	10	53
Stunting	24	17	52
Low birth weight	25	6	30

Source: HDI Report 2001 and Planning Department of Sri Lanka

Table 3 - Body Mass Index (BMI) in Sri Lanka

BMI	Men	Women
< 18.5	36.2	33.4
> 25	9.0	12.8
> 30	0.8	1.5

Table 4: Mortality Rates

Category	Period	Sri Lanka	China	India
Life expectancy in years	1970-75	65.1	63.2	50.3
	1995-2000	71.9	69.8	62.3
Infant mortality	1970-75	65.0	85.0	127
	1996-2000	17.0	33.0	70.0
Under-5 mortality	1970-75	100	120.0	202.0
	1995-2000	19.0	41.0	98.0
Maternal mortality*	1995-2000	60.0	55.0	408.0

Source: Human Development Report 2001 — ** Per 100,000 live births*

Table 5: Mothers' education and child malnutrition in Sri Lanka

Category	Stunting	Wasting	Underweight
All categories	23.8	15.5	37.7
No education	46.0	16.7	57.9
Primary	33.6	18.7	37.8
Secondary	22.6	16.8	39.1
More than secondary	13.0	11.3	24.6

Source: Planning and Development Department, Sri Lanka

Nutrition Security Compact is evident from the decline of maternal mortality from 230 per 100,000 live births in 1992 to 17 in 1996 (*Philip et al 2000*). Thailand's initiative in organising a Community Volunteer Corps for Household Nutrition Security is worthy of emulation by other nations.

CHALLENGES AHEAD

Among the nutritional challenges facing the countries in transition, the following need priority attention.

Low birth weight: For the reasons already mentioned, governments and civil society organisations in developing countries should accord high priority to overcoming maternal and foetal under-nutrition and malnutrition. Future intellectual attainments of nations will depend very much on success in this area.

Under-nutrition and stunting among children: Because of its linkages to mental impairment, stunting should be addressed through an integrated package of healthcare, education and nutritional measures. Early under-nutrition accentuates adult chronic diseases, including diabetes, heart disease, hypertension and cancer.

Undernourished adults: Judged by a body mass index of less than 17 kg/m2, over 240 million adults in developing countries are severely undernourished. The nutritional safety net for this category could include programmes like food for eco-development i.e. food for work and food for nutrition.

Vitamin A and iodine deficiencies: Subclinical Vitamin A deficiency still affects nearly 200 million pre-school children in developing countries. Sustained efforts are also needed to eliminate iodine deficiency disorders.

Pandemic anaemia: Maternal anaemia is pandemic and is associated with high MMR; anaemia during infancy, compounded by maternal under-nutrition, leads to poor brain development.

Lack of access to clean drinking water: This is a serious nutritional problem since contaminated water is a major cause of intestinal infections and diarrhoea in children. Access to clean drinking water is becoming a luxury in many developing countries.

Access to sustainable livelihoods: Ultimately, the lack of purchasing power is responsible for poor access to a balanced diet. In India, the poverty line is defined in nutritional terms. The estimation of poverty is based on the consumption expenditure level below which a household of 5.5 persons, on an average, cannot meet the recommended intake of 2,400 kcal for adults in rural areas and 2,100 kcal in urban areas. In poor households, over 70 per cent of the daily income goes towards buying food. Even by this austere yardstick, over 250 million people in India live below the poverty line. In the area of income poverty, South Asia is the hot spot *(Table 6)*.

TABLE 6: REGIONAL COMPARISON OF INCOME POVERTY IN DEVELOPING COUNTRIES

	1987	1990	1993	1996	1998
East Asia and Pacific	417.5	452.4	431.9	265.1	278.3
South Asia	474.4	495.1	505.1	531.7	522.0
Asia and Pacific	891.9	947.5	937.0	796.8	800.3
Europe and Central Asia	1.1	7.1	18.3	23.8	24.0
Latin America and Caribbean	63.7	73.8	70.8	76.0	78.2
Middle East and North Africa	9.3	5.7	5.0	5.0	5.5
Sub-Saharan Africa	217.2	242.3	273.3	289.0	290.9
Asia & Pacific as % of world total	75.4	74.2	71.8	66.9	66.8
Total	1183.2	1276.4	1304.3	1190.6	1198.9

Source: IFAD (2001)

TABLE 7: VITAMIN CONTENT OF SOME LEAFY VEGETABLES
(PER 100G GREEN EDIBLE PART)

LEAFY VEGETABLE	CAROTENE (MG)	VITAMIN C (MG)	IRON (MG)	CALCIUM (MG)
Spinach	5580	28	1.14	73
Amaranth	5520	99	3.49	397
Bathos	1740	35	4.20	150
Drumstick	6780	220	85	40
Fenugreek	2340	52	1.93	395
Radish	5295	81	0.09	265

Source: R.S. Paroda (2001)

MEETING THE CHALLENGES

A food-based approach to nutrition security would involve the following steps.

Food availability: This is a function of both home production and imports. In many developing nations, the gap between potential and present yields is high in most farming systems, even with the technologies available off the shelf. High priority should hence be accorded to bridging the productivity gap through a mutually reinforcing blend of technologies, services and public policies. Also, mainstreaming the nutritional dimension in the design of cropping and farming systems is essential. There is no time to relax on the food production front. The present global surplus of food grains is the result of inadequate consumption on the part of the poor, and should not be mistaken as a sign of over-production. Developing nations should aim to achieve revolutions in five areas to sustain and expand the gains already achieved. These are:

1. PRODUCTIVITY REVOLUTION: The scope is great since average yields are still low in most cropping and

farming systems. However, the production techniques should be environmentally sustainable, so that high yields can be obtained in perpetuity.

2. QUALITY REVOLUTION: This can be achieved through greater attention to post-harvest technologies and bio-processing, as well as to sanitary and phytosanitary measures and *Codex Alimentarius* standards.

3. INCOME AND EMPLOYMENT REVOLUTION: This will call for integrated attention being directed at on-farm and non-farm livelihoods and at farming systems intensification, diversification and value addition. Post-harvest processing offers scope for generating additional livelihoods through micro-enterprises supported by micro-credit.

4. SMALL FARM MANAGEMENT REVOLUTION: Institutional structures which will confer upon farm families with small holdings the advantages of scale at both the production and post-harvest phases of agriculture are urgently needed. For example, thanks to the cooperative method of organisation of milk processing and marketing, India now occupies the first position in the world in milk production. Strategic partnerships with the private sector will help farmers' organisations to have access to assured and remunerative marketing opportunities.

In relation to factors of production, water is likely to become the key constraint during this century. Hence, every effort should be made to enhance productivity and income for every drop of water.

5. ENLARGING THE FOOD BASKET: During the last century, there has been a rapid decline in the number of crops contributing to global food security. In the past, local communities depended upon a wide range of crops for their food and health security. It is important that we revive the old dietary traditions. Particular attention needs to be paid to leafy vegetables, which are rich in micronutrients *(Table 7)*.

Food access: Inadequate livelihood opportunities in rural areas result in household nutrition insecurity. Today, India has over 65 million tonnes of wheat and rice in government godowns; yet poverty-induced hunger affects over 200 million persons. Macro-economic policies, at the national and global level, should be conducive to fostering job-led economic growth based on micro-enterprises supported by micro-credit. Where poverty is pervasive, suitable measures to provide the needed entitlement to food should be introduced.

Food absorption: Lack of access to clean drinking water, as well as poor environmental hygiene and health infrastructure, lead to a poor assimilation of the food consumed. Nutrition security cannot be achieved without environmental hygiene, primary healthcare and clean drinking water security. Culinary habits also need careful evaluation. Some methods of cooking may lead to the loss of vital nutrients.

Transient hunger: Ferro-Luzzi *et al* (1994) have carried out a detailed study of seasonal cycling in body weights related to changes in weather. Any strategy for nutrition security should provide for steps to meet such transient hunger. Almost 25 years ago, The Indian state of Maharashtra introduced an Employment Guarantee Scheme to help the poor earn their

daily bread during seasons when opportunities for wage employment are low. Similarly, there is a need for mainstreaming considerations of gender, age and occupation in the national nutrition strategy.

Fortification and synergy among dietary component: Our understanding of low-cost and high-synergy nutritional systems is growing. Fortification of flour with folic acid and genetic enrichment of staple grains with beta-carotene and iron are now receiving attention.

Knowledge relating to the metabolic interrelationships among micronutrients is also growing, as for example among Vitamin A, protein, zinc, iron, folic acid and Vitamin C. However, in the absence of dietary interventions, iron-folate supplementation often fails to bring about a complete correction of anaemia. Hence, the attack on ***under-nutrition induced hunger*** and ***micronutrients deficiency-caused hidden hunger*** should be an integrated one. Such an integrated strategy should accord concurrent attention to food availability, access and absorption. In addition, there should be provisions in the strategy for overcoming seasonal or transient under-nutrition caused by loss of opportunities for livelihood during seasons of drought, floods or other natural calamities.

Genetic enrichment of nutritional quality: While the problems relating to the food and environmental safety aspects of genetically modified foods are yet to be fully resolved, there is little doubt that an integrated approach involving Mendelian and molecular breeding is likely to make a food-based approach to nutrition even more effective in the future. The quantity and quality of proteins, carbohydrates, fats, vitamins and minerals can all be improved now. The scope for the genetic enhancement of nutritional quality will be evident from the following examples:

■ QUALITY PROTEIN MAIZE (QPM). Scientists have long had an interest in improving maize protein quality. Quality protein maize (QPM) refers to enhanced levels of the two 'essential' amino acids, lysine and tryptophan, in the endosperm protein. Using Mendelian breeding methodologies supported by rapid chemical analysis of a large number of samples, scientists led by

S. Vasal and Evangelina Villegas of the International Maize and Wheat Research Centre (CIMMYT) in Mexico were able to slowly accumulate modifier genes to convert the original soft opaque-2-endosperm into the vitreous hard endosperm type *(Vasal et al, 1984)*. This conversion took nearly three decades. These remarkable new varieties look and taste like normal maize but the nutritive value of their protein is nearly equivalent to cow's milk. They also produce yields as much as 10 per cent higher than the best local hybrid maize varieties and are more tolerant of biotic and abiotic stresses. QPM, which is a product of Mendelian breeding, promises improved nutritional value and cost savings for a wide array of products ranging from infant food to corn chips and feed for non-ruminant livestock. The impact of this breakthrough is likely to be felt throughout the food industry and has great promise in the developing world, both for human and animal nutrition.

■ BETA-CAROTENE RICH RICE. A promising development in the field of genetic engineering is the success in breeding a nutritionally enriched rice variety now popularly referred to as 'golden rice'. This genetically modified rice contains genes that produce high levels of beta-carotene and related compounds, which are converted in the human body into crucially needed Vitamin A. Vitamin A deficiency (VAD) causes more than a million childhood deaths each year and is the single most important cause of blindness among children in developing countries. Rice plants do produce cartenoid compounds (that our body converts into Vitamin A) but only in the green parts of the plant, not in the part of the grain normally eaten. Dr Ingo Potrykus and Dr Peter Beyer of the Swiss Federal Institute of Technology inserted genes from a daffodil (*Narcissus pseudonarcissus*) and a bacterium (*Erwinia uvedovora*) into rice plants to produce the modified grain, which has sufficient Beta-carotene to meet total Vitamin A requirements in a typical Asian diet *(Ye et al, 2000)*. If golden rice, which is still in the laboratory stage, becomes a success in the field, it will help to strengthen the food-based approach to nutrition security.

■ IRON ENRICHMENT. Iron-deficiency anaemia is the most widespread nutrient deficiency in the world, affecting an estimated 2 billion people worldwide. Between 40 and 50 per cent of children under the age of five in the developing countries are iron-deficient and iron deficiency accounts for up to 20 per cent of all maternal deaths. It also impairs immunity and reduces the physical and mental capacities of people of all ages. In short, iron deficiency is a major public health problem worldwide with enormous social and economic costs. Rice fortified with iron was created through the introduction of proteins from kidney beans (*Phaseolus vulgaris*) by the same researchers at the Swiss Federal Institute of Technology *(Lucca et al, 2000)*. It is reported that the iron content increased two-fold in the modified crop, which is now in the testing stage. Japanese scientists have also succeeded in enriching the rice grain with iron. The International Rice Research Institute (IRRI) has developed rice-breeding lines high in iron and zinc using traditional plant breeding techniques. This rice is currently being tested by novitiates at a convent in the Philippines to see how well the nutrients are absorbed in the human body.

■ DESIGNER POTATOES. Advances in plant tissue culture techniques and gene transfer technology have opened up possibilities for modifying the amino acid contents of plants. The potato, which is the most important non-cereal food crop, ranks fourth in terms of total global food production. Besides, it is used as animal feed and as raw material for the manufacture of starch, alcohol and other food products. This crop was genetically modified using the seed albumin gene *Ama*1 from *Amaranthus hypochondriacus* by researchers of Jawaharlal Nehru University (JNU), New Delhi, India *(Chakraborty et al, 2000)*. The *Ama*1 protein is non-allergenic and rich in all essential amino acids. Its composition corresponds well with WHO standards for optimal human nutrition *(Raina and Datta, 1992)*. The JNU team was able to use a seed albumin gene with a well-balanced amino acid composition as a donor protein to developing a transgenic potato. The genetic enrichment of protein quantity and quality in potatoes can make a significant contribution to child and adult nutrition, since mashed potatoes can be fed to young children.

The above are a few examples of the work in progress in improving, through conventional and molecular breeding techniques, the quantity and quality of protein in important food crops. Consumer confidence based on an appreciation of the scientific evidence and the regulatory checks and balances will ultimately decide whether or not genetically modified foods (GMOs) will make a significant contribution to feeding the 8 billion people who are likely to inhabit our planet by 2020. Marker-aided selection and transgenic approaches are two powerful tools to accelerate plant breeding to produce crop varieties with improved nutritional traits and qualities. An intelligent integration of Mendelian and molecular breeding techniques will help to enhance the nutritive value of staples. By integrating pre-breeding in laboratories with participatory breeding in farmer's fields, it will be possible to breed location-specific varieties and maintain genetic diversity in crop fields.

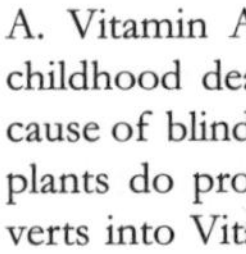

Consumer confidence based on an appreciation of the scientific evidence and the regulatory checks and balances will ultimately decide whether or not genetically modified foods will make a significant contribution to feeding the eight billion people who are likely to inhabit our planet by 2020

BUILDING A SUSTAINABLE COMMUNITY NUTRITION SECURITY SYSTEM

Conferring the right to food — and thereby an opportunity for a productive and healthy life — on those who go to bed undernourished now, is the fundamental duty of the State and the more fortunate sections of the population. Thanks to technological advances and the spread of democratic systems of governance at the grassroots level, we now have a rare opportunity to foster a community-centred and controlled nutrition security system. Such decentralised community management will help improve delivery of entitlements, reduce transaction and transport costs, eliminate corruption and cater to the twin needs of introducing a life-cycle approach to nutrition security and meeting the challenge of seasonal fluctuations in nutritional status. The basic guidelines for such a system are:

Adopt a whole life-cycle approach to nutrition security

■ PREGNANT MOTHERS. Overcoming maternal and foetal under-nutrition and malnutrition is an urgent task, since nearly 30 per cent of children born in South Asia are characterised by low birth weight (LBW), with the consequent risk of impaired brain development. Ramalingaswamy *et al* (1997) have pointed out that half the world's malnourished children are in India, Pakistan and Bangladesh. LBW is a proxy indicator of the low status of women in society, particularly in terms of their health and nutrition status during their entire life-cycle *(Rama Narayanan, 2001)*.

■ NURSING MOTHERS. Appropriate schemes will be necessary to provide support to enable mothers to breast-feed their babies for at least six months, as recommended by WHO. Policies at workplaces, including the provision of appropriate support services, should be conducive to achieving this goal.

■ INFANTS (0-2 YEARS). Special efforts will have to be made to reach this age group through their mothers, since they are the least served at present. Eighty per cent of brain development is completed before the age of two. The first four months in a child's life are particularly critical, since the child is totally dependent on its mother for food and survival.

■ PRE-SCHOOL CHILDREN (2-6 YEARS). A well-designed, integrated child development service will help cater to the nutritional and healthcare needs of this age group *(Measham and Chatterjee, 1999)*.

■ YOUTH (6 TO 20 YEARS). A nutrition-based noon meal programme in all schools (public and private, rural and urban) will help to improve the nutritional status of this group. However, a significant percentage of children belonging to this age group are not able to go to school due to economic reasons. Such school 'push-outs' or child labourers need special attention.

■ ADULTS (20 TO 60 YEARS). The nutrition safety net catering to this category should consist of both an entitlements programme like food stamps and the Public Distribution System (PDS), as well as a food for eco-development programme (also called 'food for work' programme). The food for eco-development programme can promote the use of food grains as wages for the purpose of establishing water harvesting structures (Water Banks) and for the rehabilitation of degraded lands and ecosystems. Thus, many downstream benefits and livelihood opportunities will be created. In designing a nutrition compact for this age group, persons working in the organised and unorganised sectors will have to be dealt with separately. Also, the intervention programmes will have to be different for men and women, taking into account the multiple burdens on a woman's daily life.

■ OLD AND INFIRM PERSONS. This group will have to be provided with appropriate nutritional support, as part of the ethical obligations of society towards the handicapped.

The above whole life-cycle approach to nutrition security will help ensure that the nutritional needs of everyone in the community and of every stage in an individual's life are satisfied.

Adopt a Holistic Action Plan to achieve sustainable nutrition security at the level of each individual

The major components of such an integrated action plan are the following:

▶▶ IDENTIFICATION: Identify those who are nutritionally insecure through local communities. Trained Community Volunteers of the kind mobilised in Thailand will be useful for this purpose.

▶▶ EDUCATION AND INFORMATION EMPOWERMENT: Empower those who are not aware of their entitlements about the nutritional safety nets available to them and also undertake nutrition education. An entitlements database can be developed for each area and household entitlement cards can be issued, indicating how to access nutritional, healthcare and educational programmes. The educational programmes should also stress the conservation of essential nutrients in cooked food.

▶▶ OVERCOME PROTEIN-CALORIE UNDER-NUTRITION: The various steps indicated under the whole life-cycle approach will have to be adopted. The problems of child labour and of persons working in the unorganised sector will need specific attention.

▶▶ ELIMINATE HIDDEN HUNGER CAUSED BY THE DEFICIENCY OF MICRONUTRIENTS IN THE DIET: Introduce an integrated approach, including the consumption of vegetables and fruits, millets, grain legumes and leafy vegetables and the provision of fortified foods like iron and iodine-fortified salt and oral doses of Vitamin A. The basic approach should be a food-based one, with an emphasis on home and community nutrition gardens, wherever this is socially and economically feasible *(Gopalan, 2001)*.

▶▶ DRINKING WATER, HYGIENE AND PRIMARY HEALTHCARE: Attend to the provision of safe drinking water and to the improvement of environmental hygiene. Also, improve the primary healthcare system.

▶▶ SUSTAINABLE LIVELIHOODS: Improve economic access to food through market-linked micro-enterprises supported by micro-credit. Also, create an economic stake in the conservation of natural and common property resources. Ensure that agreements under the World Trade Organisation (WTO) provide a level playing field for products coming from decentralised small-scale production (production by masses or farmers' farming) as compared to those emerging from mass production technologies or factory farming. Promote job-led economic growth and not jobless growth.

▶▶ PAY SPECIAL ATTENTION TO PREGNANT AND NURSING MOTHERS AND PRE-SCHOOL CHILDREN: Measure progress through monitoring MMR, IMR, incidence of LBW children and male-female sex ratio. Iron-folate supplements during prenatal care should be accompanied by steps to overcome protein-energy deprivation. Mina Swaminathan (1998) has proposed a maternity and childcare code, which if adopted, will help to bring down speedily MMR, IMR, LBW and stunting. The sex ratio is a good index of the mindset of a society in relation to the girl child.

Community Food Bank as an instrument of sustainable food and nutrition security

Community Food Banks (CFB) can be started at the village level, with initial food supplies coming as a grant from governments and donor agencies like the World Food Programme. Later, such CFBs can be sustained through local purchases and from continued government and international support for food for eco-development and food for nutrition programmes. CFBs can be the entry point for not only bridging the nutritional divide, but also for fostering social and gender equity, ecology and employment. They can also be equipped to cater to emergencies like cyclones, floods, drought and earthquakes.

The CFBs can be organised with the following four major streams of responsibilities.

1. **ENTITLEMENTS:** The benefits of all government and bilateral and multilateral projects intended for overcoming under-nutrition and malnutrition can be delivered in a coordinated and interactive manner (as for example those intended to overcome deficiencies of macro and micro-nutrients.)
2. **ECOLOGY:** Food for eco-development with particular reference to the establishment of Water Banks, land care, control of desertification and afforestation. Thus, grain can be used to strengthen local-level water security.
3. **ETHICS:** This group of activities will relate to nutritional support to old and infirm persons, pregnant and nursing mothers and infants and pre-school children.
4. **EMERGENCIES:** This activity will relate to the immediate relief operations following major natural catastrophes like drought, floods, cyclones and earthquakes, as well as to meet the challenge of seasonal slides in livelihood opportunities.

Each of the above four streams of activities can be managed by four separate self-help groups of local women and men. This will help to generate a self-help revolution in combating hunger. The overall guidance and oversight may be provided by a multistake-holder Community Food Bank Council.

A famine of income, which arises from a famine of jobs or sustainable livelihood opportunities, is currently our major food security challenge. This is where the huge grain stocks afford rare opportunities for eliminating endemic hunger, strengthening ecological security, reducing the number of school 'push-outs' and minimising the incidence of diseases like TB and leprosy

Resource centres for CFBs

For the CFB movement to succeed there is need for training managers of such food banks and for building the capacity of the Community Oversight Council to plan and monitor the different programmes. Training modules will have to be prepared for this purpose. Accounting and monitoring software will have to be developed and the members of the self-help groups will have to be trained in the use of the software and in managing computer-aided knowledge centres linked to CFBs. Four training modules relating to entitlements, eco-development, ethics and emergencies will have to be developed, so that each SHG is headed by a professionally trained woman or man. A network of institutions which will provide the necessary managerial, technical and training support to managers of self-help groups and CFBs will have to be organised in every country where there is a strong political commitment to ending the nutritional divide as soon as possible.

AGENDA 2007: A HUNGER-FREE INDIA

The Prime Minister and the Government of India are to be commended for three important recent initiatives for dealing with the mounting grain stocks in a socially and environmentally meaningful manner. First, the Prime Minister announced on August 15 the initiation of the Sampoorna Gramin Rozgar Yojana, with an initial allocation of 5 million tonnes of food grains for organising food for work programmes. Second, a million tonnes of food grains has been offered to Afghanistan through the World Food Programme, the largest assistance of this kind given by any nation to a people in deep distress. Third, the government has decided to launch a Grain Bank Scheme (GBS) in tribal areas, with an initial allocation of one million tonnes of food grains and Rs 66 crore in cash for meeting transportation costs.

India is home to the largest number of poor in the world, judged by the World Bank's poverty line of a per capita income of US $1 or less per day (about Rs 48 per capita per day). Food stocks are growing, while chronic protein-energy under-nutrition caused by poverty, hidden hunger resulting from micronutrient deficiencies, and transient hunger triggered by natural calamities like drought, still prevail to an unacceptable extent. A famine of income, which in its turn arises from a famine of jobs or sustainable livelihood opportunities, is currently our major food security challenge. This is where the huge grain stocks afford rare opportunities for eliminating endemic hunger, strengthening ecological security, reducing the number of school 'push-outs' who are victims of unfavourable economic and social circumstances, and minimising the incidence of diseases like tuberculosis and leprosy, where complete cure needs prolonged treatment.

While the manner of use of the grains offered to Afghanistan is not in our hands, we can shape the impact of the Sampoorna Gramin Rozgar Yojana and the tribal food and health security programmes in a manner that both economic development and ecological regeneration are accelerated. This will, however, call for a community-centred and controlled Food Bank movement. The term 'Food Bank' is preferable to 'Grain

Gouache on paper by GANESH HALOI

Bank', since in many tribal and rural areas, tuber crops serve as life-saving foods. Local grains like *ragi, samai, bajra* and several other millets are more nutritious than wheat or rice and they, along with other life-saving crops like tubers, can also be purchased and distributed once the CGBs are established with the initial grant of grains offered by the government. The CGBs should not be conceived as an emergency operation, but should be structured in a manner that they become the hubs of a sustainable and replicable community nutrition and ecological security system.

As mentioned earlier, the Community Grain Banks can perform multiple functions depending upon local needs and opportunities. In the area of health, the control of tuberculosis and leprosy can be speeded up if food grains can be given to those whose economic circumstances do not permit taking drugs regularly over many months. The Tuberculosis Research Centre at Chennai is initiating an imaginative programme in collaboration with the UN World Food Programme, for using food grains to encourage the regular intake of the needed drugs. Similarly, in the area of education, economic and social conditions prevent many children, belonging to the category of child labour, as well as adolescent girls, from continuing their school education. By introducing carefully designed food for education programmes, substantial progress can be made in reducing the prevalence of such school push-outs.

Another area of nutrition security which can be strengthened through CGBs is the introduction of a whole life-cycle approach to overcoming malnutrition during various stages in one's life, ranging from pregnant mothers and infants to old and infirm persons. This will help to bring down speedily the incidence of low birth weight children and infant and maternal mortality rates. This will call for steps which can foster the integrated implementation of numerous ongoing nutrition intervention programmes.

Strengthening ecological security will be another lasting benefit of the use of grains for the conservation and enhancement of natural resources. The new tribal area grain bank programme should be structured in such a manner that food grains are used to establish field-level gene banks to conserve local agro-biodiversity, seed banks and water banks. CGBs can then promote concurrent progress in achieving desirable goals in conservation, education and nutrition. The storage bins can be designed according to local climatic conditions and separate bins can be fabricated locally for different grains. The Save Grain programme of the government can be used for training and capacity-building in storage methods. The Rural Godown Scheme can be integrated with CGBs.

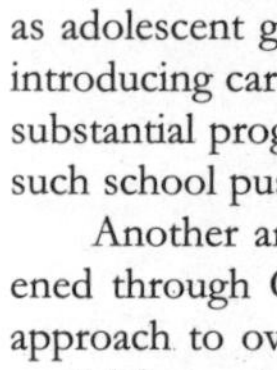

Another area of nutrition security which can be strengthened through community grain banks is the introduction of a whole life-cycle approach to overcoming malnutrition during various stages in one's life, ranging from pregnant mothers and infants to old and infirm persons

Grain Banks already exist in different parts of the country, set up and operated by both government and non-governmental agencies. The Madhya Pradesh government has institutionalised the grain bank (Anna Kosh) programme. The government of Rajasthan is planning to integrate the CGB initiative with the ongoing Gandhi Gram Yojana, which has several social and ecological objectives, including local-level water security. The available experience stresses the need for community control and involvement to ensure sustainability and replicability. Otherwise, the grain banks will vanish when government support ends.

CGBs should become central to community-managed nutrition and ecological security programmes. Only then will transaction costs become affordable and the programme replicable. CGBs provide opportunities for achieving convergence and synergy among the numerous *yojanas* initiated in recent years, including the Annapoorna and Antyodaya Anna Yojana. The time is opportune for unifying all of them under the supervision of Gram Sabhas into a community nutrition security system based on a whole life-cycle approach.

The US operates the following three programmes involving the distribution of grains.

FOOD FOR PEACE (PL-480) — started in 1949
FOOD FOR PROGRESS — started in 1985
FOOD FOR EDUCATION — started in 2000

Other than the United States, India is the only country in a position to launch such programmes. We have reached a stage in our agricultural evolution when our production will increase only if we can improve consumption. In this context, the Sampoorna Gramin Yojana, the Tribal Area Grain Bank programme and Annapoorna are extremely important steps in using surplus food grains for peace, progress and development.

This unique initiative may be continued and consolidated with the launching of three programmes soon:

- FOOD FOR COASTAL ECOLOGICAL SECURITY: One million tonnes of grains may be made available to coastal states as well as to the Andaman and Nicobar and Lakshadweep administrations for food for work programmes designed to restore mangrove wetlands, coral reefs and sea grass beds, and to control sea erosion and organise agro-aqua and agro-forestry programmes.
- FOOD FOR THE SECURITY OF MOUNTAIN ECOSYSTEMS: About 2 million tonnes of food grains may be made available to states in the Himalayan (including Northeastern states) and Western and Eastern Ghats regions for the eco-restoration of hydrologic and biodiversity "hot spots", for preventing genetic and soil erosion and for establishing field gene banks through *in situ* on-farm conservation.
- NAGARPALIKA ROZGAR YOJANA: Two million tonnes of food grains may be made available to urban local bodies for undertaking scientifically designed treatment and recycling of all solid and liquid wastes, including conversion of wastes into organic manure, urban water harvesting and for the bio-environmental

management of mosquitoes. With a slump in construction activities, urban unemployment leading to urban crime is increasing and it will be useful to use the opportunity provided by grain availability for improving sanitation and environmental hygiene in towns and cities, in addition to converting a public health problem into public wealth.

Thanks to the impressive grain stocks, which may go up to 100 million tonnes next year, if the consumption capacity of the poor does not improve, we have for the first time in the history of independent India an opportunity to leapfrog in achieving freedom from poverty-induced hunger, illiteracy and ill-health. I hope we will not miss this opportunity, since it may not occur again. tlm

This article is based on a plenary lecture delivered at the 17th International Congress on Nutrition, Vienna, August 2001

M.S. Swaminathan, a founding father of the Green Revolution in India, worked to develop the nation's food security as Secretary for Agriculture. He has also been Director General of the Indian Council of Agricultural Research and the International Rice Research Institute, Independent Chairman of the FAO Council, and President of the International Union for the Conservation of Nature and Natural Resources. He is a member of the Royal Society, the US National Academy of Sciences, the Russian Academy and the Italian and Chinese Academies. His honours include the World Food Prize, UNEP's Sasakawa Award and the Tyler and Honda Prizes. He lives in Chennai, where he heads the Centre for Sustainable Agricultural and Rural Development.

REFERENCES

1. Chakraborty S., Chakraborty N., Datta A. (2000): 'Increased Nutritive Value of Transgenic Potato by Expressing a Non-allergenic Seed Albumin from *Amaranthus hypochondriacus.*' *Proc. Natl. Acad. Sci.* USA, 97: 3724-3729.
2. Ferro-Luzzi, A., Branca F., and Pastore G. (1994): 'Body Mass Index Defines the Risk of Seasonal Energy Stress in the Third World.' *European Jour. Clinical Nutrition* 48 (Suppl. 3): 165-178.
3. Gardner, Gary and Halwell, Brain (2000): 'Overfed and Underfed: the Global Epidemic of Malnutrition.' World Watch Paper 150, World Watch Institute, Washington, DC 68 pp.
4. Gopalan, C. (2001): 'Combating Vitamin A Deficiency and Micronutrient Malnutrition through Dietary Improvement', MSSRF, Chennai (Mimeographed).
5. IFAD (2001): 'Assessment of Rural Poverty in Asia and the Pacific Region', August 2001.
6. Lucca P., Hurrell R., Potrykus I. (2000): 'Development of Iron-rich Rice and Improvement of its Absorption in Humans by Genetic Engineering.' *Journal of Plant Nutrition* 23: 11-12
7. Measham, A.R. and Meera Chatterjee (1997): *Wasting Away: The Crisis of Malnutrition in India.* The World Bank, Washington, DC. Directions on Development series, 78 pp.
8. Swaminathan, Mina (1998). *Maternity and Child Care Code*: Forum for Creche and Child Care Services (FORCES), New Delhi.
9. Paroda, R.S. (2001): *Plant Genetic Resources for Food and Nutritional Security.* Indian Society of Plant Genetic Resources, New Delhi.
10. Philip, James *et al,* (2000): 'Ending Malnutrition by 2020: An Agenda for Change in the Millennium.' *Food and Nutrition Bulletin,* United Nations University, Tokyo. Vol 21 (13), supplement, 88 pp.
11. Quality Protein Maize 1964-1994: *Proceedings of the International Symposium on Quality Protein Maize.* Edited by Brian Larkins and Edwin T Mertz. Sete Lagoas, MG, Brazil, Dec 1-3, 1994.
12. Raina A., Datta A. (1992): 'Molecular Cloning of a Gene Encoding a Seed-specific Protein with Nutritionally Balanced Amino Acid Composition from *Amaranthus.*' *Proc. Natl. Acad. Sci.* USA, 89: 11774-11778.
13. Ramalingaswami, V., Johnson, U. and Rohde, J. (1997). 'Malnutrition: A South Asian Enigma.' In *Malnutrition in South Asia,* Ed. Stuart Gillespie, Rosa Publication No.5, UNICEF, Nepal.
14. Rama Narayanan, 2001. 'How Secure are our Mothers and Children?' Paper presented at a workshop on Human Health Security, Chennai, India.
15. UNDP *Human Development Report (2001)*: Making New Technologies Work for Human Development, Oxford University Press, 264 pp.
16. Vasal S.K., Villegas E., Tang C.Y., Werder J., Read M. (1984): 'Combined Use of Two Genetic Systems in the Development and Improvement of Quality Protein Maize.' *Kulturpflanze* 32: 171-185.
17. Ye X., Al-Babili S., Kloti A., Zhang J., Lucca P., Beyer P., Potrykus I. (2000): 'Engineering Provitamin A (Beta carotene) Biosynthetic Pathway into (Carotenoid-free) Rice Endosperm.' *Science,* 287: 303-305

FEATURE

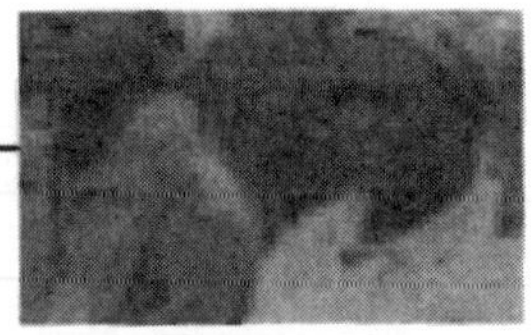

AFRICA'S RAINBOW REVOLUTION

WILL AFRICA BE ABLE TO FEED ITSELF? THE ANSWER IS 'YES' – BUT ONLY IF IT ADOPTS A MULTIFACTED STRATEGY THAT RECOGNIZES THE CONTINENT'S UNIQUE ECOLOGY AND CULTURE. THAT'S THE CONCLUSION OF A RECENT REPORT BY THE INTERACADEMY COUNCIL (IAC).

Africa is rich in both natural and human resources. Yet nearly 200 million of its people are undernourished because of insufficient food supplies. Strategies must be adopted across the continent to harness the power of science and technology in ways that boost agricultural productivity, profitability, and sustainability – ultimately ensuring that all Africans have access to enough safe and nutritious food to meet their dietary needs. TWAS Founding Fellow M.S. Swaminathan, one of the co-chairs of the InterAcademy Council's (IAC) panel that prepared the recent report, Realizing the Promise and Potential of African Agriculture, *highlights the main elements of the action plan recommended to increase the productivity of African agriculture.*

There have been numerous studies, symposia and books on African agriculture in recent decades, each proposing strategies to ensure sustainable food security in Africa.

Several of these studies have shown that a 10 percent increase in crop yield would lead to a 6 to 10 percent increase in the annual survival rate of African women and men earning US$1 or less per day. Because more than 80 percent of Africa's population depends on crop and animal husbandry, fisheries, agro-forestry and agro-processing for its livelihood, enhancing farm productivity provides the best safety net against poverty and hunger.

Agriculture in Africa serves as the backbone of the food, livelihood and ecological security systems. What Africa needs is science and technology that helps to produce more food, more jobs and more income.

Building on the vast volume of data and information available and placing total confidence in the wisdom and innovativeness of African farm families and farm scientists, the InterAcademy Council's (IAC) panel report, *Realizing the Promise and Potential of African Agriculture*, calls for the launch of an ever-green revolution in African agriculture driven by the enhanced productivity, profitability, stability and sustainability of the major farming systems of this diverse and resource-rich continent.

The IAC report, however, refers to the productivity-based progress of African agriculture not as an ever-green

[CONTINUED PAGE 72]

TWAS Newsletter, Vol. 16 No. 3/4, 2004

revolution but as a 'rainbow revolution' because unlike Asia, where wheat and rice are the dominant food crops, Africa does not have a dominant farming system on which food security largely depends.

In recent decades, Africa's agricultural progress has been hampered by the spread of the HIV/AIDS pandemic; uncertain market access and fluctuating market prices associated with the trade and subsidy policies of industrialized countries; political instability; and ethnic conflicts. Nevertheless there have been several success stories on the continent that can serve as signposts for the road that lies ahead. These include:

- Soil health enhancement through nitrogen-fixing shrubs and rock-phosphate applications.
- Biological control of the cassava mealy bug.
- Cultivation of maize with improved protein quality.
- Development of tissue culture for the disease-free multiplication of banana.
- Development of New Rice for Africa (Nerica).

The steps that led to the conversion of unpromising 'hot spots' into appealing 'bright spots' were:

- Indigenous research by African scientists.
- Strategic partnerships with farm women and men, as well as with advanced research institutions, including institutions belonging to the network of the Consultative Group on International Agricultural Research (CGIAR).
- Farmer participatory research and knowledge management systems leading to the demystification of technologies.
- Location-specific technologies supported by such services as credit and seed inputs.
- End-to-end strategies involving all steps in the technology development, verification, adaptation and adoption process.
- Political support at the highest level that ensures access to remunerative markets and accords greater social prestige to agricultural scientists.
- Sustained funding for agricultural research, education and development.

Unlike Asia, Africa does not have a dominant farming system on which food security largely depends.

The formation of the Forum for Agricultural Research in Africa (FARA) and the New Partnership for Africa's Development (NEPAD) also constitute organizational 'bright spots' that can help accelerate the pace of agricultural progress.

ACTION

Let me highlight the principal components of the action plan devised by the IAC panel for leading Africa's agricultural future in the right direction.

Most African soils are both hungry and thirsty – that is, short of nutrients and water. Therefore an indispensable priority must be the nurturing of soil fertility and the development of measures for water harvesting, conservation and efficient use.

Rain-fed agriculture is dominant in Africa. That makes community water harvesting and watershed management as essential as the cultivation of high-value, low-water crops. Our panel has therefore placed considerable stress on the adoption of an ecological production approach in land and water use planning. Of the 17 distinct farming systems identified in different parts of Africa, we concluded that the following four systems offer immediate promise for increasing African food security:

- Maize-mixed system based primarily on maize, cotton, cattle, goats and poultry. The system recognizes the importance of off-farm work.
- Cereal-root crop mixed system based primarily on maize, sorghum, millet, cassava, yams, legumes and cattle.
- Irrigated farming system based primarily on rice, cotton, vegetables, cattle and poultry.
- Tree-crop-based system involving cocoa, coffee, oil palm, rubber, yams and maize. This system also recognizes the importance of off-farm work.

In addition, coastal areas hold promise for agroforestry systems involving aquaculture and such halophytic plant species as mangrove and saltbush. Ethiopia's highlands also hold promise for significant gains in productivity if the proper agricultural systems are put in place.

The IAC panel suggests methods for strengthening national agricultural research systems and establishing African Centres of Agricultural Research Excellence (ACARE) by building on the strengths of existing institutions. Because ecological agriculture is more knowledge-intensive than capital- or chemical-intensive, we

STUDY PANEL

Responding to a request from the Secretary-General of the United Nations, in 2002, the InterAcademy Council (IAC) appointed a study panel on agricultural productivity in Africa. The panel was composed of co-chairs Speciosa Wandira Kazibwe, past vice-president of Uganda and former minister of agriculture, animal industry and fisheries; Rudy Rabbinge, dean, Wageningen Graduate School, and professor of sustainable development and systems innovation, Wageningen, Netherlands; and M.S. Swaminathan, chairman of the M.S. Swaminathan Research Foundation in Madras and father of India's green revolution in the 1960s. The panel also included 15 other distinguished members mostly from Africa. Jim Ryan of Australia, former director-general of the International Crops Research Institute for the Semi-Arid Tropics (ICRISAT) in Hyderabad, India, served as the study director. The panel's mandate was to produce a consensus report for the UN that (1) addressed how science and technology could help improve agricultural production in Africa and (2) identified the larger economic, social and political conditions that are essential for effective use of science and technology in both the public and private sectors. The report was presented to the UN Secretary General in New York on 25 June 2004. The full text may be browsed at www.interacademycouncil.net.

TWAS Newsletter, Vol. 16 No. 3/4, 2004

[CONTINUED PAGE 74]

> **SUCCESS WITH RICE**
>
> *Such cultivated rice species as* Oryza sativa, *which are indigenous to Asia, have been subjected to many cycles of selection and improvement by both farmers and scientists and, as a result, now yield up to 12 tonnes per hectare. In contrast, Africa's only native rice,* O. glaberrima, *despite having been domesticated centuries ago, yields only one tonne per hectare. It does, however, possess strong resistance to weeds and local pests – the reasons why Asian rice varieties are not widely grown in Africa. Until now, these two species of rice were genetically incompatible. Using a technique developed in the 1980s, known as 'embryo rescue', scientists have now combined the high-yielding traits of elite Asian rice with the pest-resistant properties of African rice. The product, labelled Nerica (New Rice for Africa), yields up to 2.5 tonnes per hectare and matures quickly, allowing farmers to grow a second crop – beans, for example, that fix nitrogen and help build soil fertility. Farmers throughout the upland rice growing regions of west Africa are currently cultivating Nerica varieties. Indeed Guinea saved some US$13 million in rice imports in 2003. Nerica rice is also being promoted in Uganda, where some 6,000 hectares have been planted. Thanks to Nerica, Africa is expected to be self-sufficient in rice by 2010.*

recommend the extensive application of modern information, education and communication technologies involving the integrated use of the internet, cable television, community radio and local press. And we call for the establishment of an African Virtual Academy for Agricultural Progress and Rural Prosperity to stimulate the spread of a rainbow revolution in rural areas.

Women play a key role in both the production and post-harvest phases of African agriculture. That's why our panel suggests that a 'women in agriculture programme' be initiated to promote both knowledge enhancement and skill empowerment of women farmers and women farm labourers. Programmes for bridging the 'digital divide' should also aim at narrowing the gender divide in technological capacity building. The panel also recommends that, rather than lamenting the brain drain, efforts should be made to support educated professionals who continue to live and work in their native countries. "Care for the brains at home" should be the motto.

Monetary compensation alone, although important, may not be able, on its own, to attract and retain creative scientists. Indeed such non-monetary benefits as well-equipped laboratories, a nurturing environment, academic freedom and efficient research management that encourage initiative and invention are just as important as high salaries in arresting the brain drain.

Finally, our panel recommends an African 'grid' of participatory science and technology pilot projects to introduce farm women and men to the opportunities offered by modern science and technology. The goal is to enhance agricultural productivity and income on an ecologically sustainable basis in a wide range of farming systems.

These pilot programmes should be set up where the production-processing-marketing-consumption chain can be assessed – and subsequently implemented – and should include:

- Indigenous technologies relevant to the improvement of productivity and food security.
- Market potentials and constraints for existing and prospective commodities in farming systems.
- New technologies to enhance productivity and food security, including:
  - – Integrated nutrient and soil fertility enhancement.
  - – Integrated pest management.
  - – Small-scale water harvesting and use of micro-irrigation systems for delivery of water and nutrients.
  - – Applications of improved genetic strains, biofertilizers and biopesticides.
  - – Use of improved farm implements and appropriate mechanization for increasing labour productivity, reducing drudgery and ensuring timely farm operations.
  - – Introduction of appropriate post-harvest processing, storage and marketing techniques.
  - – Promotion of non-farm employment through introduction of technologies adding economic value to primary products and through agri-business.
  - – Development of communication programmes to provide location-specific information related to meteorological, management and marketing factors and to

promote genetic and trade literacy among rural farm families.

– Establishment of farmer field schools for integrated pest, disease and weed management, integrated water and fertility management, and other aspects of production and post-harvest technologies based on the principle of learning-by-doing.

– Promotion of such institutional structures as co-operatives and self-help groups that can confer the power of scale to small landholders at the production and post-harvest phases of farm operations.

- For each pilot programme, the scope for other institutional innovations should also be explored, including:

In African nations, as in other developing countries, if agriculture goes wrong, nothing else has the chance to go right.

– Promotion of a participatory knowledge coalition led by small landholders, who are encouraged to become involved with universities, national agricultural research institutions and extension agencies for the purposes of exploring new modes of partnership.

– Identification of candidates for ACARE centres to serve the interests of small landholders.

– Stimulation of public-private partnerships to address priority constraints that cannot be alleviated by independent activities and that are aimed at building convergence and synergies.

– Identification of the constraints that prevent the realization of the promise and potential of the pilot pro-

SUCCESS WITH CASSAVA

More than 60 percent of the world's cassava is grown in sub-Saharan Africa. In the early 1970s, production of this root crop was threatened by the accidental introduction into west Africa of the cassava mealybug, Phenacoccus manihoti. *In the absence of natural enemies, the cassava mealybug spread at a pace of more than 300 kilometres per year. By the 1980s, it could be found throughout tropical Africa, endangering the food supply of 200 million people who relied on cassava as a subsistence crop. Among the several natural enemies introduced by scientists of the Nigeria-based International Institute for Tropical Agriculture (IITA) and collaborating agencies was a parasitoid wasp,* Apoanagyrus lopezi, *from Paraguay, the native home of the mealybug. Techniques were developed for mass-rearing the wasps, which were then distributed from aeroplanes. This large-scale strategy quickly led to the control of the pest over a wide area. The parasitoid wasp has now established itself in 26 African countries and continues to provide safe and effective control of the cassava mealybug throughout the region.*

[CONTINUED PAGE 76]

SUCCESS WITH BANANA

Banana is a major staple crop and source of income for more than 20 million people in east Africa, many of them resource-poor women farmers. Yet production is less than half the level it could be due to pests and diseases. The problem lies in the way bananas are traditionally propagated – simply by uprooting a sucker and transplanting it. Using this method transfers all the diseases that are infecting the ageing mother tree. Tissue culture propagation, on the other hand, can be used to produce thousands of tiny plantlets free of bacterial, fungal and viral diseases. Apart from being pest- and disease-free, tissue-cultured plantlets have several other advantages over traditional planting material. For example, trees reach maturity earlier and bear fruit bunches up to 50 percent heavier than traditionally raised trees. In addition, each tree produces double the number of suckers than traditionally propagated trees, which provides a rapid means of multiplying and disseminating improved planting material. A project launched in 1996 brought together the Kenya Agricultural Research Institute (KARI), the Institute of Tropical and Sub-Tropical Crops (ITSC) in South Africa and others to encourage the production, testing, dissemination and adoption of tissue-cultured banana plantlets. The private sector in South Africa is already producing more than 4 million banana plantlets each year, mainly for export, and private laboratories have also been set up in Kenya. An estimated 500,000 small-scale farmers in east Africa now grow bananas derived from tissue culture. In Kenya, as a result of farmers planting healthier banana plants, household income has reportedly increased from 700 Kenyan shillings (less than US$10) to 5,000 Kenyan shillings (US$60) per harvest.

grammes to improve agricultural productivity and food security at the local level.

– Introduction of a horizontal dimension to the different vertically structured programmes relating to the realization of the UN Millennium Development Goals.

The IAC panel suggests that interdisciplinary teams from national agricultural research systems, universities, extension services and farmers' organizations be constituted to prepare business plans and research agendas in each of the priority farming systems described above.

Nothing succeeds like success. Therefore the sites for the initial rainbow revolution pilot programmes should be developed where there is a socio-economic, political, scientific and ecological environment conducive to the achievement of the programmatic goals. For each pilot programme, a local farmers' advisory council, involving both men and women, should be constituted to assume ownership and undertake monitoring and evaluation.

In African nations, as in many other developing countries, if agriculture goes wrong, nothing else has the chance to go right. Bilateral and multilateral donors and national governments should lose not a minute more in implementing IAC's recommendations. Just as African societies take pride in rainbow coalitions at the political level, African agriculture can lead the world in demonstrating the power of a science-and-technology-based rainbow revolution in agriculture for eliminating poverty and hunger. ■

M.S. Swaminathan *(TWAS Founding Fellow)*
Chairman
National Commission on Farmers, India

Co-chair, InterAcademy Council's (IAC)
Panel for Harnessing Science and Technology to Increase the Productivity of African Agriculture

For the full text of the IAC report, see *www.interacademycouncil.net*

Millennium Project

Hunger in Africa: the link between unhealthy people and unhealthy soils

Lancet 2005; 365: 442–44

Pedro A Sanchez, M S Swaminathan

Columbia University Earth Institute, Lamont-Doherty Campus, 61 Route 9W, PO Box 1000, Lamont Hall, 2G, Palisades, NY10964, USA (Prof P A Sanchez PhD); and M S Swaminathan Research Foundation, Chennai (Madras), India (Prof M S Swaminathan PhD)

Correspondence to: Prof Pedro A Sanchez sanchez@iri.columbia.edu

Malnutrition is the biggest risk factor for illness worldwide. Various dimensions of malnutrition (eg, underweight, zinc deficiency, iron deficiency, vitamin A deficiency) account for seven of the 13 leading risk factors associated with the global burden of diseases.[1] For both children and adults, malnutrition reduces the body's natural defences against a vast range of diseases. The death rate from diseases such as lower respiratory infection, malaria, and measles, which account for a large proportion of children's deaths, are much higher in children who are underweight or have specific nutrient deficiencies than in those who are not.[2–4] Undernourished people infected with HIV/AIDS develop the full symptoms of the disease more quickly than people who are well fed.[5] Yet one of the earliest side-effects of AIDS is reduced consumption of food in affected households.

Inadequate sanitation, repeated bouts of infectious disease, chronic parasitic infections, poor health facilities, and unsafe water sources contribute significantly to malnutrition by increasing the burden of illness for both children and adults. Households that are dependent on well or surface water for drinking are more likely to have underweight children because the water is more likely to be contaminated.

Panos Pictures

The fact that malnutrition has many causes, including infectious disease itself, has led to the recognition that multiple, synergistic interventions are needed to fight malnutrition, including improved diets, safe drinking water and sanitation, special attention to vulnerable groups (pregnant women, lactating mothers, young children, unemployed people, those with HIV/AIDS, socially excluded groups), and access to health services to prevent and treat infections. Disease prevention and treatment is both an end in itself and an input to improved long-term absorption and use of food.[1,4,6]

Yet because the world as a whole produces enough food for the entire world population, and because countries such as India with massive malnutrition also produce enough food to feed their entire populations, recent approaches to combat malnutrition have de-emphasised food production itself in the interventions needed to combat malnutrition. The emphasis instead has been placed on increasing access to food by increasing the purchasing power of the poor together with better health care, sanitation, education, hygiene, and nutritional practices.

These generalisations, however, do not apply to tropical sub-Saharan Africa (henceforth Africa). The UN Millennium Project Task Force on Hunger has identified Africa as the region facing the greatest challenge in attaining the Millennium Development Goal for hunger—reducing the proportion of people who suffer from hunger by half between 1990 and 2015.[7] The prevalence of hunger in Africa is pervasive and rising. By contrast with most of Asia, Latin America, and the middle east, however, Africa is experiencing a decline in overall food production per capita. Its farmers generate the lowest food output per hectare of any major region in the world, and its farm households account for most of the continent's hungry population. The task force has concluded that for most of Africa, as well as some remote parts of Asia and Latin America, increasing food production has to be included as an essential part of the synergistic interventions to fight malnutrition.[7,8]

Understanding Africa's food productivity crisis

Africa, unlike Asia, did not benefit from the green revolution, one of the key successes of humankind in the latter third of the 20th century. Food production in developing countries tripled during the past 30 years, the number of rural poor decreased by half, the proportion of malnourished people dropped from 30% to 18%, and the real prices of the main cereal crops fell by 76%. Several decades ago, a small group of determined scientists and policymakers identified the key entry point for producing such a green revolution—the need for high-yielding varieties of rice and wheat. When that goal was attained other necessary factors were put in place, such as enabling government policies, fertilisers and irrigation, better marketing, infrastructure, national research institutions, strong agricultural universities, and the international agricultural research system.

This success did not extend to Africa. Although the rate of adoption of improved crop varieties has been uniformly high in all developing regions, their contribution to crop yield increases has been 70–90% in Asia, Latin America, and the middle east, but only 28% in Africa.[9] Overall improvements in food yields in Africa have lagged far behind the rest of the developing world.

In Africa, soil nutrients and water management are the major limiting factors, not improved crop varieties. Fertilisers are almost unused by Africa's smallholder farmers on staple food crops. The main result is a massive and widely unrecognised deterioration of soil health in Africa, by contrast with other developing regions. Moreover, almost all crops in Africa are rain-fed rather than irrigated, because of the absence of major river systems for irrigation, electricity for irrigation pumps, and other factors that limit irrigation. Rainfall is highly variable and often inadequate over much of the continent. Therefore the key entry point for African food production is not improved crop varieties but replenishment of soil fertility, coupled with improved water management at the lowest possible cost. No matter how good genetic improvement is, crops cannot grow well without sufficient nitrogen and phosphorus. This is a biological imperative that transcends socioeconomic and political ones.

The crisis in soil health is a quiet catastrophe. Over the decades, Africa's small-scale farmers have removed large quantities of nutrients from their soils without returning them in sufficient quantities as manure or fertiliser. Additionally, almost all crop residues have been removed. These practices have resulted in a very high average annual depletion rate—22 kg of nitrogen, 2·5 kg of phosphorus, and 15 kg of potassium per hectare of cultivated land over the past 30 years in 37 African countries. This annual loss is the equivalent of US$4 billion in fertiliser.[10] This widespread nutrient mining has decreased soil organic matter, decreased the ability of microorganisms to recycle nutrients, and the soil's water holding capacity.

The major economic reason for lack of fertiliser input has been the poor infrastructure in rural areas of Africa. Road density per person and per area is an order of magnitude lower than in south and east Asia, and railways are non-existent or decrepit. The farm-gate price of fertiliser is often two to six times that of the rest

Still Pictures

of the world because of very high overland transport costs after the fertiliser arrives at an African port. Moreover, rural poverty spirals in a vicious circle. With rural hunger and poverty, rural markets are weak, credit is very limited, and access to information is poor. When a particular region enjoys an increase in food production, for example because of a temporary increase in fertiliser use, local prices of food tend to plummet because of the lack of regional transport and storage facilities. This fall in turn discourages farmers from maintaining a high level of farm inputs in the next season.[7]

Time for a 21st century green revolution for Africa

The time is right to increase dramatically the productivity of African agriculture, with a new and highly focused plan that UN Secretary-General Kofi Annan called "a uniquely African green revolution for the 21st century".[11] New techniques in farming can help make this change possible. The low soil nitrogen can be replenished not only by chemical fertilisers but also by agroforestry, through the planting of leguminous trees that fix atmospheric nitrogen in the soil. New grain-and-tree-crop systems have proved to be highly reliable and economic in many parts of impoverished rural Africa, offering farmers not only far higher food yields but also the added benefit of fuel wood that can be harvested from the leguminous trees. Low-cost water harvesting techniques can be combined with agroforestry to promote a highly reliable package of soil and water interventions. These novel approaches should be complemented with increased use of conventional approaches as well, including chemical fertilisers and irrigation where appropriate.

The Hunger Task Force's plan, together with the other Millennium Development Goals recommendations,[12] brings forth the nexus between agriculture and nutrition, health, water, and sanitation, to fulfill the secretary-general's vision[11] and that of task force members. The task force emphasises the role of increased donor support in restoring soil health, expanding water harvesting and small-scale irrigation, and investing in critical rural infrastructure (roads, power, telecommunications) to integrate rural farm communities with regional, national, and international markets. These increased investments should be facilitated by a new cadre of paraprofessional extension workers at the villages, trained to deal with agriculture, nutrition, markets, health, and energy, in turn supported by professional extension services and greatly enhanced research institutions. The agricultural interventions aim to boost the quantity and quality of food output and raise rural incomes. To fight hunger, these agricultural interventions would be closely integrated with the critical interventions in primary health care, access to safe drinking water and sanitation, and micronutrient supplementation that have been successfully deployed in many parts of the developing world, but not yet in rural Africa.

In Asia, hunger was reduced decisively within a generation. It is now time, and feasible, to do the same in Africa. The survival, health, and prosperity of Africa's children require nothing less.

Contributors

PAS and MSS co-chair the UN Millennium Project Task Force on Hunger.

Acknowledgments

We thank the members of the Hunger Task Force. The task force's work was funded by the Millennium Project Trust Fund, the Canadian International Development Agency (CIDA), UK Department for International Development, The Earth Institute at Columbia University, The Irish Development Council, The Rockefeller Foundation, Swedish International Development Association (SIDA), the World Bank, the United Nations World Food Programme (WFP) and the United Nations Development Program (UNDP). Numerous other organisations also provided similar support.

References

1 de Onis M, Blössner M. WHO global database on child growth and malnutrition. Geneva: WHO, 2003.

2 Food and Agricultural Organization of the United Nations. The State of Food Insecurity in the World: 2001. Rome: FAO, 2001.

3 Pelletier DL, Frongillo EA, Schroeder DG, Habicht JP. The effects of malnutrition on child mortality in developing countries. *Bull World Health Organ* 1995; 73: 443–48.

4 United Nations Standing Committee on Nutrition. 5th report on the world nutrition situation: nutrition for improved development outcomes. Geneva: United Nations, 2004.

5 Kadiyala S, Gillespie S. Rethinking food aid to fight HIV/AIDS. Food Consumption and Nutrition Division Discussion Paper 159. Washington DC: International Food Policy Research Institute, 2003.

6 United Nations Administrative Committee on Coordination and Subcommittee on Nutrition (ACC/SCN). Fourth report on the world nutrition situation: nutrition throughout the life cycle. Geneva: ACC/SCN in collaboration with IFPRI, 2000.

7 UN Millennium Project, Task Force on Hunger. Halving hunger: it can be done. London, Sterling, VA: Earthscan, 2005.

8 Sanchez PA, Swaminathan MS. How to cut world hunger in half. *Science* (in press).

9 Evenson RE, Gollin G, eds. Crop variety improvement and its effect on productivity. Wallingford, UK: CABI Publishing, 2003.

10 Sanchez PA. Soil fertility and hunger in Africa. *Science* 2002; 295: 2019–20.

11 UN. Secretary-General calls for "Uniquely African green revolution in 21st century to end the continent's plague of hunger" http://www.un.org/News/Press/docs/2004/sgsm9405.doc.htm (accessed Dec 21, 2004).

12 UN Millennium Project, Task Force on Hunger. Investing in development: a practical guide to achieve the Millennium Development Goals. London, Sterling, VA: Earthscan, 2005.

POLICY FORUM

PUBLIC HEALTH

Cutting World Hunger in Half

Pedro A. Sanchez and M. S. Swaminathan

The Millennium Project was commissioned by the United Nations Secretary-General to recommend the best strategies for meeting the Millennium Development Goals (MDGs) (*1*). In October 2002, the Hunger Task Force was established to determine how to meet the hunger MDG—to reduce the proportion of hungry people in half from 1990 to 2015. Task Force members came from diverse backgrounds in science, policy, the private sector, civil society, U.N. agencies, and government, with broad representation from developed and developing countries (2). After analysis, stakeholder consultations, and observation, the Task Force has just produced its report (*2*), which is summarized here.

Enhanced online at www.sciencemag.org/cgi/content/full/307/5708/357

Diagnosis

There are 854 million people in the world (about 14% of our population) who are chronically or acutely malnourished. Most are in Asia, but sub-Saharan Africa is the only region where hunger prevalence is over 30%, and the absolute numbers of malnourished people are increasing (3). More than 90% are chronically malnourished (*4*), with a constant or recurrent lack of access to sufficient quality and quantity of food, good health care, and adequate maternal caring practices. Acute hunger (the wasting and starvation resulting from famines, war, and natural disaster) represents 10% of the hungry yet receives most of the media coverage and attention. In addition, hidden hunger from micronutrient deficiencies affects more than 2 billion people worldwide. Chronic and hidden hunger deserve much more global attention and support.

Roughly 50% of the hungry are in smallholder farming households; 20% are the landless rural; 10% are pastoralists, fishers, and forest dwellers; and 20% are the urban hungry. The Task Force has identified hunger hot spots, defined as the subnational units where the prevalence of underweight children (*4*) less than 5 years of age is at least 20%. The 313 hunger hot spots identified (see the figure on page 358) indicate priority regions, as they cover 79% of the hungry.

The importance of different causes of hunger varies among regions. Low agricultural productivity is likely to be the primary reason in tropical Africa and remote parts of Asia and Latin America, whereas poverty and unemployment are the main causes in most of South and East Asia, Latin America, Central Asia, and the Middle East.

Economically, hunger results in annual losses of 6 to 10% in foregone Gross Domestic Product (GDP) due to losses in labor productivity. Economic growth alone is insufficient for eliminating hunger, because so many hungry people live in deep poverty traps, beyond the reach of markets (*5*). People affected by HIV/AIDS become unable to grow food or work for a living. Malnourishment weakens their immunity and strength, making them succumb more quickly to disease (*6*). Similarly, nearly 57% of malaria deaths are attributable to malnutrition (*7*). The challenge of halving hunger is, therefore, closely linked with that of achieving other MDGs.

Recommendations

The Task Force calls for simultaneous action at global (recommendation 1), national (recommendation 2), and local levels (recommendations 3 to 7) (see the figure on page 359).

1. *Move from political commitment to action.* A commitment to halving world hunger was made by all member countries of the United Nations at the World Food Summits of 1996 and 2001, the Millennium Summit of 2000, the 2002 World Summit on Sustainable Development, and the 2002 Monterrey Summit on Development Finance. The message for political leaders is that halving hunger is within our means; what has been lacking is action to implement and scale up known solutions.

The Secretary-General of the United Nations reinforced this message when he called for a "uniquely African green revolution for the 21st century" (*8*).

2. *Reform policy and create an enabling environment for hunger reduction.* Government policies in poor countries can make or break efforts to end hunger. Good governance, including the rule of law, low levels of corruption, and respect for human rights, is essential for achieving food security. Policies conducive to ending hunger and poverty need to be put in place at all levels, from the local to the national.

The Task Force proposes that poor countries integrate hunger reduction action plans into their Poverty Reduction Strategies or equivalent national planning process. Poor countries need to adopt a multisectoral approach to hunger reduction. African governments should invest at least 10% of their national budgets specifically in agriculture and nutrition, in addition to making investments in rural energy, infrastructure, health, education, and other sectors. Building capacity at all levels should be the central goal of national government and donor-funded activities. Linking nutritional and agricultural interventions, which are so often implemented separately, would be a powerful means of creating more effective hunger reduction programs.

Clearly assigned and enforceable rights for women to own, inherit, and trade land must be guaranteed. Women and girls need better access to services such as credit, health care, and education, as well as to technologies that will ease the workload of rural women, such as rooftop water harvesting and growing trees for firewood close to home.

Agricultural research has been a major driver of hunger reduction. The Task Force recommends doubling investments in national research to at least 2% of agricultural GDP by 2010. It also recommends that donors increase funding to the Consultative Group on International Agricultural Research to US$1 billion by 2010.

3. *Increase agricultural productivity of food-insecure farmers.* Small-scale farming families represent about half the hungry worldwide and probably three-quarters of the hungry in Africa. Raising the productivity of their crops, livestock, fish, and trees is a major priority.

Restoring soil health is often the first entry point for increasing agricultural productivity, because soil nutrient depletion is extreme in most areas where farmers have small holdings, as in Africa (*9*). Applying appropriate combinations of mineral and organic fertilizers, using leguminous green manures and agroforestry fertilizer trees, returning crop residues to the soil, and using improved methods of soil conservation can

P. A. Sanchez is at the Earth Institute at Columbia University, New York, NY 10964, USA. E-mail: sanchez @iri.columbia.edu; M. S. Swaminathan chairs the M.S. Swaminathan Research Foundation, Taramani, Chennai 600 113, India. E-mail: msswami@mssrf.res.in. The authors cochair the U.N. Millennium Project Task Force on Hunger.

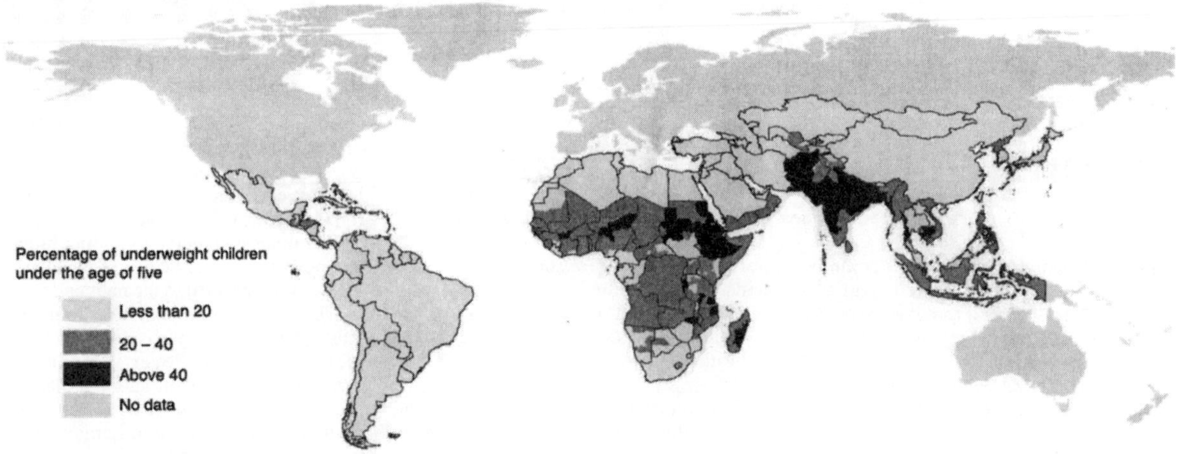

Hot spots of world hunger.

restore soil health and double or triple yields of the cereal staple crop. Making mineral fertilizers available at affordable prices and using them efficiently remain major challenges. As an emergency short-term measure, targeted subsidy programs should be designed to supply mineral and organic fertilizers (as seeds) to farmers. Tamper-proof "smart cards" redeemable at private agrodealers are one promising way of administering targeted subsidies, avoiding many of the pitfalls of past fertilizer subsidy schemes. When combined with similar vouchers for farmers to sell their products to school and community feeding programs, the demand side can be also addressed, avoiding price crashes when production increases.

In subhumid and semiarid areas, improving water management can be at least as important as improving soil fertility. Various water harvesting and small-scale irrigation techniques can be used to transform crop and livestock production in these regions. Investments in small-scale water management can also be financed with targeted subsidies.

The provision of genetically superior crop, pasture, tree, livestock, and fish germ plasm can greatly increase the productivity of small-scale farms. The Task Force supports both conventional breeding and transgenic research with appropriate biosafety measures. The traits that will benefit poor farmers in more marginal areas are tolerance to stresses (drought, salinity, poor soil fertility, pests, and diseases) and improved nutritional value. .

After farmers attain food security, they can begin to diversify their farming systems to produce high-value products. Livestock, farm trees, aquaculture, and vegetables are attractive options for diversifying their diets and sources of income. Increases in milk production, for example, can reduce malnutrition in rural and urban settings. In South Asia and Africa, farming systems integrating crops and livestock are very important in strengthening household nutrition and income. Small-scale farmers could emerge as major timber suppliers of the 21st century in many tropical regions.

Breathing new life into the moribund extension services of many poor countries is vital if the benefits of new knowledge and improved technology are to reach farmers. The Task Force recommends that every village in a hunger hot spot have paraprofessional extension workers trained in agriculture and nutrition, with counterparts in health and energy. They should be supported by professional services and enhanced research institutions.

4. *Improve nutrition for chronically hungry and vulnerable groups.* Adequate nutrition lies at the heart of the fight against hunger. As the primary care providers for children and families, women are particularly important in improving nutrition for vulnerable groups. Particular attention should be focused on children under the age of two and on supplemental feeding for pregnant and lactating mothers. The Task Force recommends that, where possible, locally produced foods be used, rather than imported food aid.

To break the intergenerational cycle of undernutrition, the Task Force recommends supplemental feeding for underweight pregnant women and nursing mothers. Exclusive breastfeeding up to 6 months of age is the best way of ensuring optimum nutrition for babies, although the decision may be complicated by the risk of transmitting HIV through breast milk.

To reduce malnutrition in children under five, the Task Force recommends providing fortified or blended supplementary foods, clean drinking water, and therapeutic care for all seriously malnourished children and women, especially in remote rural areas. Community extension workers should take the lead in raising awareness and implementation.

The Task Force recommends that malnutrition be reduced among school-age children and adolescents by providing free, nutritionally balanced school meals from locally produced foods for all poor children. This will improve learning, attract the 40% of primary school age children who are currently out of school in Africa (mostly girls), empower girls with good nutrition and knowledge before they become mothers, and create a steady demand for local foods. We estimate that if this is practiced in half of the primary schools in Africa, the local demand for maize alone could increase by as much as 25%. Systematic deworming; micronutrient supplementation; education about HIV/AIDS, health, nutrition, and hygiene; and provision of safe drinking water and take-home rations should be part of the program.

Vitamin and mineral deficiencies should be reduced by increasing consumption of micronutrient-rich foods such as fruits and vegetables; improving food fortification; and increasing micronutrient supplementation when necessary. Village extension workers should promote these mutually reinforcing actions.

Parallel health measures are also needed to eliminate the diseases that rob people of nutrients. All children should be fully immunized and receive prompt treatment for common infections such as diarrhea, pneumonia, malaria, and helminthes, as well as appropriate nutritional care provided for people living with or affected by HIV/AIDS.

5. *Reduce vulnerability of the acutely hungry through productive safety nets.*

While investing in agriculture, education, and health remains critical to long-term food security, past gains can be threatened if people's vulnerability to short-term disasters and shocks are not addressed. To address acute hunger, the Task Force recommends strengthening (i) national and local early warning systems to take advantage of advances in climate prediction; (ii) the capacity to respond to emergencies; and (iii) investments in productive safety nets (food for work; cash for work).

The Task Force recommends, whenever possible, the substitution of cash for program food aid, so that governments can invest more flexibly in reducing hunger. The additional resources needed to reduce vulnerability to shocks must not draw funds away from long-term development.

Safety nets should be both an effective protector of last resort during shocks and economically productive in years without crisis. This involves investing in community activities that reduce vulnerability while increasing productive potential. Large injections of cash or food aid can distort the local economy unless they are targeted toward development objectives.

6. *Increase incomes and make markets work for the poor.* Properly functioning markets are critical in ensuring that farmers are able to earn a decent income, obtain the inputs they need to raise crop yields, and sell their produce at fair prices.

The Task Force proposes that major investments be made in developing and maintaining market infrastructure. Markets will not develop without public investment in transport and communications. A major effort is needed to increase road building, including paved roads and all-weather feeder roads, in large parts of Africa where there is high prevalence of malnutrition. Every village should have a vehicle for transporting products to markets and health emergencies. Effective grain storage capacity at the local level would enable farmers to obtain fairer prices for their crop surpluses and would reduce postharvest losses to pests. Investments in small-scale processing should quickly yield benefits in terms of increased employment opportunities.

Networks of trained rural agrodealers are needed to allow essential agricultural inputs to reach remote areas, especially in Africa. Access to credit and other financial services is particularly problematic for food-insecure farmers. Community groups established to take on loans on behalf of their members could mitigate risk and make lending more attractive to financial institutions.

Lack of market information negatively affects the terms of trade for poor farmers. Governments and donors should continue investing in information technology, including combinations of mobile phones, radio, and the Internet to bring information to producers. For example, fishermen in India are now using mobile phones to seek the best price from dealers before deciding where to land their catch.

There are opportunities for increasing on-farm and off-farm income by encouraging farmers to switch part of their farms from staple food crops into higher value livestock, vegetable, and tree products and to add value through processing. Farmers can grow crops for large-scale producers. Supermarkets are becoming dominant buyers in much of the developing world. Governments should encourage them to pursue socially responsible policies and to stimulate local production.

7. *Restore and conserve natural resources essential for food security.* Degradation of natural resources directly threatens the food security and incomes of poor people. Reversing degradation requires both community- and national-level interventions. Local ownership, access, and management rights should be secured for forests, fisheries, and rangelands. Natural resource–based "green enterprises" should be developed. Poor farmers should be paid for environmental services they provide, including biodiversity protection, watershed stability, and carbon sequestration.

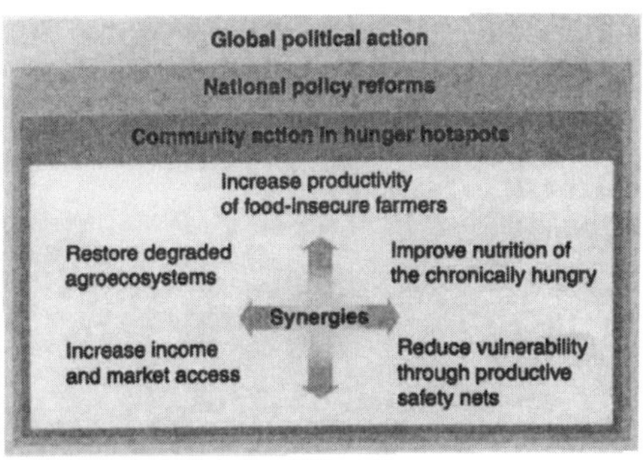

Hunger Task Force recommendations (2).

Entry Points

Community nutrition programs, homegrown school feeding programs, and investments in soils and water are local initiatives that can serve as "entry points" in the battle against hunger. A combination may constitute an attractive new integrated program in rural areas facing the dual challenge of high chronic malnutrition and low agricultural productivity. The increased local production will have a ready market in the homegrown feeding programs, and the joint facilitation by community extension workers will create a virtuous cycle. The resulting synergies will open the way for other interventions.

Resources Needed

The Millennium Project estimates that hunger reduction interventions to increase agricultural productivity and address chronic malnutrition will cost about 6 to 10% of the additional development assistance envisioned for attaining all the MDGs (*1*). That amounts to about $8 billion a year for 2005, between $10 and $11 billion a year for 2010–15, or an average of 60 U.S. cents per month for every person living in a developed country.

It Can Be Done

The Task Force concludes that the hunger MDG can be achieved by 2015 and hunger can eventually be eliminated. This will require focused and unprecedented levels of effort that are well within our financial and technological capability. Currently, more than 5.5 million children are dying of malnutrition-related causes each year. The actions outlined here, taken up by a broad coalition of stakeholders and applied in every poor country, can change that.

References and Notes

1. U.N. Millennium Project, *Investing in Development: A Practical Plan to Achieve the Millennium Development Goals: Overview* (Earthscan, London and Sterling, VA, 2005); available at www.unmillenniumproject.org/html/about.shtm.
2. U.N. Millennium Project, *Task Force on Hunger, Halving Hunger: It Can Be Done* (Earthscan, London and Sterling, VA, 2005); available at www.unmillenniumproject.org/html/tf2docs.shtm.
3. Food and Agricultural Organization of the United Nations (FAO), *The State of Food Insecurity in the World* (FAO, Rome, 2004).
4. A child is underweight if his/her weight is more than two standard deviations below the median of the international reference population used for analysis by the World Health Organization, the U.S. National Center for Health Statistics, and the Centers for Disease Control and Prevention.
5. J. D. Sachs *et al.*, *Brookings Pap. Econ. Act.* **1**, 117 (2004).
6. S. Kadiyala, S. Gillespie, *Rethinking Food Aid to Fight HIV/AIDS* (Food Consumption and Nutrition Division Discussion Paper 159, International Food Policy Research Institute, Washington, DC, 2003).
7. D. L. Pelletier *et al.*, *Bull. WHO* **73**, 443 (1995).
8. Secretary-General calls for "Uniquely African green revolution in 21st century to end the Continent's plague of hunger," U.N. Press Release; available at www.unmillenniumproject.org/html/addis/documents.shtm
9. P. A. Sanchez, *Science* **295**, 2019 (2002).
10. We acknowledge dedicated work of Task Force members and secretariat, particularly P. Dobie, N. Yuksel, L. Dreier, and R. Flor; superb leadership of J. Sachs and J. McArthur of the Millennium Project; and financial support from the Millennium Project Trust Fund, Canadian International Development Agency, U.K. Department for International Development, Earth Institute at Columbia University, Irish Development Council, Rockefeller Foundation, Swedish International Development Association, the World Bank, the World Food Program, and the U.N. Development Programme.

10.1126/science.1109057

Can Science and Technology feed the World in 2025?

M.S. Swaminathan

Swaminathan, M. S. (2007). Can Science and Technology Feed the World in 2025? *Field Crops Research*. 104: 3–9.

A combination of improved genetic strains, appropriate agronomic practices, irrigation facilities and assured and remunerative marketing opportunities led to the birth of the green revolution in many parts of Asia. This resulted in the growth rate in food production exceeding the growth rate in population. However during the last decade there has been a deceleration in food production growth rates largely due to a combination of adverse meteorological, ecological and marketing factors. This has been referred to as the "fatigue of the green revolution". The challenge now is how to reverse the decline and ensure that there is adequate food for the growing population.

Food security has three dimensions, viz., endemic hunger caused by poverty induced under- and malnutrition, hidden hunger caused by the deficiency of micro nutrients like iron, iodine, zinc and vitamin A in the diet, and transient hunger caused by natural calamities or civilian conflicts. A National Food Security System will have to take into account all these kinds of hunger. It is also necessary to shift from the concept of food security based on the availability of the needed food grains in the market to one of nutrition security which involves economic, social and physical access to balanced diet, clean drinking water, environmental hygiene and primary healthcare. The UN Millennium Development Goal No.1 aims to bring down the number of children, women and men going to bed hungry from the current 850 million to about 425 million by 2015. Unfortunately, even this modest goal is not being achieved as revealed by a review conducted in 2005 to assess the progress made during the first five years of this millennium to achieve this goal.

By 2025, the world population is likely to exceed 7 billion. About 2 billion tonnes of food grains will be needed to meet the needs of this population. This is not difficult since there is a large untapped production reservoir available in countries like India even with the technologies currently on the shelf. If these technologies could be transferred to farmers' fields through appropriate packages of services and public policies, the food needs in physical terms can be met. However, even today the food security challenge is not just increasing production but providing jobs or livelihoods which can lead to economic access to food. In general, where there is work, there is money and where there is money there is food. Therefore, it becomes important to look at agriculture not just as a food producing machine, but as an important source of livelihood generation both in the farm and non-farm sectors.

The persistence of widespread under- and malnutrition in many developing countries arises from policies which fail to recognize that the farming population including landless agricultural labour constitute the majority of consumers. Unfortunately, the term, "Consumer" seems to cover only the urban population in the minds of the policy makers. This is one of the reasons why we are off-track in achieving the UN Millennium Development Goal of reducing hunger by half by 2015. **Enhancing small farm productivity and profitability, as a single step, will make a major contribution to reducing hunger and poverty. This in turn will depend on the ability to assure remunerative prices for their produce.**

In industrialized countries, farmers constitute 2 to 4 per cent of the population. The per capita income of farmers is high both because of the size of the farm operated and the extensive support extended by government. They are technology, capital and subsidy rich. Public policies concurrently promote conservation, cultivation, consumption and commerce. Extensive support is given to promote conservation farming. **The collapse of the Doha round of negotiations in agriculture is an indication that farming cannot survive in industrialized countries without substantial support from public funds to ensure its economic viability**.

What then is the future for farmers and farming in developing countries? The following four areas needs urgent and integrated attention: technology, training, techno-infrastructure and trade. Technological, ecological and management upgradation of small farms is the need of the hour.

Technology

Technologies which can help to enhance land, water and labour productivity are urgently needed. They should lead to an evergreen revolution in small farms, i.e. increase in productivity in perpetuity without associated ecological harm. The smaller the farm, the greater is the need for marketable surplus in order to generate cash income. The small farm can lend itself to higher productivity and profitability, provided the small farmer is enabled to overcome his/her handicaps arising from lack of capital and credit and access to appropriate technologies and inputs and remunerative markets. There is need for a **small farm management revolution**, which can result in conferring the power and economy of scale on small producers both in the production and post-harvest phases of farming; if this does not happen, mounting debts arising from adverse economics will continue to affect them. Cooperative farming, service cooperatives, stakeholder companies, formation of compact production and processing Estates by Self-Help Groups and farmer centric contract farming can all help to improve the economics of small holdings and thereby foster improved management.

At the production end, there is need for integrating frontier technologies like biotechnology, information and communication technologies, space and nuclear technologies and renewable energy technologies such as solar, wind, biogas and biomass based energy systems with traditional ecological prudence. Bio-energy based on pyrolysis and gasification of biomass can be a decentralized source of energy. Bio-fuels also offer scope wherever ecological and economic conditions are favourable. Biomass is an under-utilized resource. **Bio-Parks** can be promoted in every block to convert the available biomass into a wide range of economic products, including energy and manure.

Conservation farming and green agriculture are the pathways to an evergreen revolution. The greatest problem with applying conservation agriculture concepts in dry land areas is the lack of adequate quantities of crop residues. The removal of crop residues for alternative uses accelerates the already fast decline of soil organic matter content in dry land areas. Long term sustainability of dry land soils may be significantly enhanced by reduced tillage that leaves more crop residues on the soil surface. Green Agriculture involves the development and adoption of environmentally beginning technologies like integrated nutrient supply and integrated pest management.

Besides enhancing soil fertility and soil organic matter, the need for the economic and efficient use of irrigation water cannot be over emphasised. The average yield of cereals can be increased by 30 to 60% annually in dry farming areas by increasing crop water use by 25 to 35 mm. This can be readily achieved by conservation agriculture. High input costs, uncertain rainfall and poor income lead to widespread indebtedness. **The younger generation will be reluctant to take up farming as long as income prospects are poor**. Declining terms of trade between farm and non-farm sections is a matter of concern.

It is in this background, that we have to examine the opportunities opened up by new technologies. New agriculture technologies like genomics and information technology together with improved agronomic management should form the cornerstone of increasing agriculture productivity and profitability of small farms both in irrigated and rainfed areas as well as in problem soils and coastal areas. Recombinant DNA technology has already resulted in the breeding of crop varieties possessing tolerance to salinity and drought as well as to serious biotic stresses caused by the triple alliance of pests, pathogens and weeds. It is however essential to have a professionally and socially credible National Biotechnology Regulatory Authority. **The bottom line for any biotechnology regulatory policy should be the safety of the environment, the well being of farming families, the ecological and economic sustainability of farming systems, the health and nutrition security of consumers, safeguarding of home and external trade, and the biosecurity of the nation.**

The Village Knowledge Centre (VKC) based on the integrated application of the internal and community radio or mobile phone will help to bridge the growing gap between scientific knowledge and its field application. **It will also facilitate the removal of many intermediaries from the marketing chain.**

The wholesale fruit and vegetable markets are likely to lose their importance under the growing influence of contract farming and direct supply relationships between producers and major market chains. Changes in intermediary relationship will occur as internet based marketing tools are adopted by both producers and suppliers. **Knowledge connectivity is vital in addition to physical connectivity through roads**. As a single step, the *VKC* will bring about a transformation in the economic conditions and social relations in our villages. **Bridging the digital divide is a powerful method of bridging the gender divide, since rural women master the ICT technologies with ease**. The last mile and last person connectivity will be through FM / Community radio and / or mobile phones. **The internet – radio – mobile phone synergy is a**

very powerful tool for social inclusion in access to all the needed information, including warning of impending natural disasters. Villagers give priority to health and marketing information. In addition, an Entitlements Database can empower them with information on all the Government schemes designed for their well being. Gender-specific information is equally important. Every farmer in the village should be issued with an Entitlements Pass Book.

We are thus on the threshold of both a biotechnology and information technology revolution. Biotechnology does not imply only GMOs. Non GMO applications are many, such as tissue culture for multiply elite germplasm, bio-fertilizers, bio-pesticides and bio-remediation of ground water as well as marker-assisted breeding. In the case of GMOs, safe and responsible use should be ensured. Organic farming procedures permit the use of varieties developed by marker-assisted breeding.

The third technological revolution relevant to agriculture is the ecotechnology movement. This involves the appropriate integration of frontier sciences with the ecological prudence of farming communities.

The ecotechnology revolution underpinning the ever-green revolution movement has many pathways as indicated below.

Green Revolution: Commodity-centred increase in productivity **Ever-**	**Ever-green Revolution**: increasing productivity in perpetuity without associated ecological harm
Change in plant architecture, and harvest index	Organic Agriculture: cultivation without any use of chemical inputs like mineral fertilizers and chemical pesticides
Change in the physiological rhythm-insensitive to photoperiodism Green	Agriculture: cultivation with the help of integrated pest management, integrated nutrient supply and integrated natural resource management systems
	Ecoagriculture: based on conservation of soil, water and biodiversity and the application of traditional knowledge and ecological prudence
	EM Agriculture: system of farming using effective microorganisms (EM)
	White Agriculture: system of agriculture based on substantial use of microorganisms, particularly fungi
	One-straw Revolution: system of natural farming without ploughing, chemical fertilizers, weeding and chemical pesticides and herbicides

For most small farmers, green agriculture will be the most feasible form of eco-agriculture. Crop-Livestock integrated systems of production will be ideal for organic farming. More research is needed on nitrogen fixing tree species and shrubs, as well as green manure plants. Our soils are hungry and thirsty and they need both nutrients and water.

In addition to BT, ICT and Ecotechnology, there are opportunities in space application, nuclear techniques and GIS and GPS based precision farming. What is important is the training of Farm Science Managers (atleast 1 woman and 1 man) in every village, so that there is a new dawn in Indian agriculture, which capitalizes on both traditional wisdom and frontier science and technology.

Also, like the Silicon Valley, Biovalleys could be organized in mega-biodiversity areas to enable the local population to convert biodiversity into bio-wealth. **We can then end the irony of the co-existence of prosperity of nature and poverty of people**. Organic farming also requires support from strategic research in the areas of feeding for high yields and pest management. An area of technology of great importance to the survival of small scale agriculture is proactive advice on land use based on anticipated meteorological and marketing conditions. A Land Use Advisory Service, using the latest meteorological and computational tools is badly needed. We cannot abandon farm families with small holdings to their fate in a globalized economy without adequate support services based on the best in modern science. A Market Intelligence Service should be setup which can monitor crop trends in the country and advise farmers what to plant in the coming season, so as to prevent gluts and price crashes. This can be disseminated through the Village Knowledge Centre.

Training

How can such a technological, ecological and managerial upgrading of small farm agriculture be brought about? This is where training, re-training, re-tooling and redeployment of both farmers and farm graduates become important. India has 47 Agricultural and Animal Husbandry (including Fisheries) Universities. Nearly 20,000 farm graduates including about 7000 postgraduates become available each year. There is a vast chain of National Research Institutes and Centres, National Bureaus and All India Coordinated Projects under ICAR. There are also a growing number of R & D institutions in the private sector and a number of civil society organisations working on agricultural issues.

Training of farm and home science graduates also needs revamping. **The major mission of Agricultural, Veterinary, Fisheries, Rural and Women's Universities should be to help every scholar to become an entrepreneur.** They can then organize Service Co-operatives, Stakeholder companies, Agri-clinics, Agri-business centres, Bio-Parks, Food-Parks and other enterprises which can help to improve the efficiency and economics of farming. **Home Science Colleges could be restructured as College of Human Sciences, where both men and women learn the science and art of nutrition, agro-processing and home economics**.

Service cooperatives by farm and home science graduates can help to upgrade speedily the efficiency and economic viability of small farms, since they can facilitate highly productive decentralized production supported by key centralized services. Cooperatives should be organized on a stakeholder rather than on a shareholder principle.

A reorientation in the mindset of farm graduates can be brought about only by innovative changes in curricula and courses. In all applied areas, business and financial management should be added to the disciplinary training. For example, a course in Seed Technology can be restructured and designated as "Seed Technology and Business". Similarly, nutrition courses could be reorganized as **Food Safety and Nutrition Security** programmes. Courses in Agronomy could be developed into Agronomy and Agri-business Programmes. **If the business, financial and trade aspects are integrated with disciplinary training, such courses will give the farm / home science graduates the self-confidence essential for embarking upon a career of self-employment**. Attention should be given to imparting a business orientation to all the applied courses in Agricultural Universities. A large number of graduates are now being trained in the field of biotechnology. However, many of them are not able to utilize their training after taking degrees due to lack of appropriate employment opportunities. Agricultural Biotechnology is an area where there are considerable opportunities for remunerative self-employment. It would therefore be appropriate that support is extended to the creation of a **National Association of Genome Entrepreneurs** who could be supported with Venture Capital Fund in order to enable them to convert the rich knowledge available in Government institutions in the field of functional genomics into commercially viable products. They could also undertake work for other countries in the area of preparation of genome maps of the crops of interest to those countries. **Mainstreaming entrepreneurship and business skills in all applied courses, rather than keeping Business Management Course as a separate entity is essential, if small farm agriculture is to become economically sustainable and educated youth are to be attracted to take to a career in agriculture**.

Another urgent need is the establishment of a chain of Regional Institutes for Food Safety and Security. They can be established in appropriate Agricultural, Veterinary or Fisheries Universities. Home Science Graduates can be employed in such Regional Institutes to launch a movement for food safety including awareness of codex alimentarius standards. They should also spread quality literacy among farmers through Gyan Chaupals.

Training of all engaged in agricultural administration in the basic principles and economics of farming is essential. In the United States, practising farmers often occupy leading positions in Agricultural Departments for specific periods. **It would be useful to begin posting active and accomplished farm/fisher women and men as Directors in State Departments of Agriculture, Horticulture, Animal Husbandry, Fisheries, etc on a 5 year tenure. Unless there is an upgradation in the practical knowledge of those responsible for developing agricultural programmes and policies, there is no hope for developing country agriculture in a globalised economy**.

Techno-Infrastructure

Post-harvest infrastructure, particularly for perishable commodities, is extremely weak in most developing countries. Unless this is attended to on an urgent basis, farmers will not be able to get adequate return for their labour. Similarly facilities for food safety, water quality, sanitary and phytosanitary measures and biosecurity need to be improved.

Facilities for soil testing, particularly estimation of micro-nutrient status also need considerable strengthening. **Unless more investment is made in strengthening the support services needed by farmers for the scientific upgradation of farming, the average productivity will continue to remain low and youth will not be attracted to farming**. Simple but safe storage bins need to be popularized on a large scale, along with low cost refrigeration facilities for perishable commodities. A Livestock Feed and Fodder Corporation, a Land Use Advisory Service, an National Trade Organisation, Living Heritage Gene Banks to conserve unique local breeds of farm animals, internationally recognized certification agencies for organic farm products, an Agricultural Price Stabilisation Fund, integrated insurance products like Parivar Bima Policy and other essential support services are needed to help increase the productivity and profitability of small farm agriculture. **There is need to upgrade the technological, ecological and management aspects of culture and capture fisheries on the one hand, and rainwater harvesting, conservation, sustainable use, aquifer recharge, more crop and income per drop of water and other steps relating to sustainable water security on the other.**

Trade

Producer oriented market holds the key to remunerative and sustainable farming. Quality and trade literacy should receive high priority in VKCs. Facilities for Farmers' Markets need to be expanded rapidly. In commodities essential for maintaining the Public Distribution System, the procurement price should be the market price at the time of purchase. Those providing essential commodities for the PDS should be recognized through the provision of Smart Cards, which will entitle them to certain benefits while purchasing essential farm inputs, including agricultural implements and machinery.

Shaping our Agricultural Future — A three pronged Strategy

Most developing countries will remain during the 21st century predominantly agricultural nations country, particularly with reference to livelihood opportunities. Therefore there is need for both vision and appropriate action in the area of shaping our agricultural destiny. Our major agricultural strengths are our large population of hard working farm women and men, our varied climatic and soil resources, abundant sunshine throughout the year, reasonable rainfall and water resources, a long coast line and rich agro-biodiversity. Converting them into jobs and income is the challenge.

We should look upon agriculture not just as a food producing machine for the urban population, but as the major source of skilled and remunerative employment and global outsourcing hub.

Just as IT industries have specialized in handling outsourcing assignments efficiently, we must enable our farm graduates and farmers to take up outsourcing jobs, in areas where we have a comparative advantage. Some examples are hybrid seed production, tissue culture propagated plants, organic farm products, biological software for sustainable agriculture like biopesticides, biofertilizers, pheremones as well as herbal products, fruits, flowers and vegetables, vaccines and sero-diagnostics and veterinary pharmaceuticals based on medicinal plants. There is also scope for becoming a global outsourcing hub in the areas of plant and animal genomics and ICT for rural poor. Farm, Veterinary, Fisheries and Home Science Graduates should be trained to become genome and digital entrepreneurs. To start with, a few of our Agricultural, Animal Sciences and Fisheries Universities could setup **Bureaus for Outsourcing Business in Agriculture** to facilitate contacts between farmers' organizations as well as agri-business centers operated by farm and home science graduates and external agri-business enterprises. Outsourcing should not only be from other countries to urban India, but also from urban to rural India, so that educated youth continue to live in villages.

We need a new vision for agriculture. That vision should aim to spread happiness among farm and rural families. **Bio-happiness through the conversion of our bio-resources into wealth meaningful to our rural families should be the goal of our national policy for farmers**. The hidden and unrecognized opportunities for creating more skilled jobs and income in the farm and non-farm sectors need to be tapped through appropriate public policies and programmes. Technology Missions should be revamped and revitalized on the lines envisaged by Bharat Ratna Rajiv Gandhi when he first proposed them as organizational and management tools to help the nation leapfrog in the production of essential commodities like oilseeds and pulses.

A structurally progressive economy should reduce the share of people dependent on a sector as the share of that sector falls in the GDP. As the share of agriculture in the GDP falls, the share of people dependent on agriculture is also expected to fall in the same proportion. However in the Indian Economy though the share of agriculture in the GDP is falling steadily, there is no corresponding decline in the share of population dependent on agriculture. Because of population growth, the absolute number of people depending on agriculture is increasing, even if there is a decline in percentage terms. A major integrated Rural Non-farm Livelihood Initiative, should be promoted so that both on-farm and non-farm livelihoods become skilled and profitable.

The Technology strategy for an evergreen revolution should have the following three components.

Defending the Gains

Punjab farmers provide 60% of wheat and 40% of rice to the PDS and national buffer stocks. The net productivity has increased in the Punjab in rice and wheat from 1.2 t/ha and 1.1 t/ha to 4.3 t/ha and 3.9 t/ha respectively from 1960-61 to 2004-05. However, in recent years there is stagnation in productivity improvement due to a variety of causes of which the following are important:

- Declining farm size and income
- Depleting natural resources base, as for example a steep fall in ground water table and impaired water quality.

- Increasing input costs, particularly diesel and adverse economics of farming
- Deficiency of micro-nutrients in the soil and deteriorating soil health
- Inadequate post harvest technology
- Uncertain market prospects except for wheat and rice though MSP is announced for the following 25 agricultural commodities:
    - Cereals – paddy, rice, wheat, jowar, bajra, maize, ragi, barley
    - Pulses – gram, tur, moong, urad, masur
    - Oilseeds – groundnut, soyabean (yellow/black), rapeseed and mustard, toria, sunflower seed, safflower, nigerseed
    - Sugarcane, cotton, jute, tobacco
- High indebtedness of farmers – the total debt of Punjab farmers for instance is estimated to be about Rs. 24,000 crores.

Similar conditions prevail in Haryana and Western UP. **Thus, the heartland of the green revolution is in grave trouble**. These areas need conservation farming which will help farm families to conserve and improve soil health, water quantity and quality and biodiversity. Some of the eco-technologies developed by the Punjab Agricultural University are: bed sowing of wheat saving 20 to 25% water, leaf colour chart saving 15% N application in rice, tensiometer based irrigation scheduling, zero tillage technology for wheat, and IPM in cotton saving 40% pesticides. Thus, there is vast scope both to promote Green Agriculture and to reduce the cost of production through enhanced factor productivity. **A course on Sustainability Science should be introduced in all Agricultural Universities. India will not be able to maintain a stable food security system, if the "fertile crescent" (i.e. Punjab, Haryana and Western UP) region is not saved through adequate support for conservation farming. Defending the gains already made in this region is an urgent task.**

An example of the need for support for Conservation Farming is provided by the situation in rice cultivation in the Punjab. At present, nearly 26 lakhs hectares are under rice in the Punjab. Much of the irrigation water used is ground water. The water table in the Central districts of the State producing rice and having 70% of the tubewells is receding at an alarming rate of 2 to 2.5 feet annually. At present about 30% of the tubewells have become submersible and it is estimated that during the next 10 years practically all the centrifugal pumps will become nonfunctional and will have to be converted into submersible pumps. It would therefore be advisable to restrict rice cultivation to 2 million hectares in the Punjab with a yield target of 5 tonnes per hectare. The remaining area can come under maize, pulses and oil seeds which are all at the moment in short supply. **Conservation Farming in the Punjab as well as in other intensive agriculture areas will involve a scientific programme of restructuring farming systems**. Such a restructuring is an urgent need in the interests of the long term livelihood security of Punjab farmers and food security of the nation.

Eternal vigilance is the price of stable agriculture. NCF had recommended in its Third Report the establishment of a science-based **National Biosecurity System. The sooner the NCF**

recommendation is implemented, the safer will be the country from the point of freedom from invasive alien species, which could cause potential harm to crop and animal husbandry, fisheries and forestry. The wheat crop for example, now faces threats from new strains of rusts like the 78S84 (similar to Yr 27 virulence) race of yellow rust Ug 99 strain of stem rust could also cause trouble. The steps taken to defend the gains already made should therefore include pest surveillance and management and gene deployment for checkmating the spread of pathogens. This is equally important in the case of poultry and animal enterprises.

In every State, the agricultural "bright spots" and "hot spots" will have to be mapped. The State should develop a strategy for enlarging the extrapolation domain of bright spots. Similarly, every State should develop a **Good Weather Code** to maximize the benefits of adequate moisture availability, a **Drought Code** to minimize the adverse impact of drought, and a **Flood Code** both to prevent excessive distress and damage, and to promote a post-flood production plan. In the desert areas of Rajasthan, the Good Weather Code should include provision for raising nurseries of appropriate plants, so that in years of excessive rainfall, an extensive tree planting and sand dune stabilization drive can be launched. This will help to strengthen the ecological infrastructure of the desert, and gradually convert the desert into an oasis. The Drought Code should include the adoption of crop life saving technologies and contingency plans to change the cropping pattern according to moisture availability. "**Be prepared**" – both to take advantage of a good monsoon and to reduce the impact of adverse seasons – should be our national motto in agriculture.

In the case of wheat, the following nine steps will help to improve wheat production during the 2006-07 **rabi** season:

- As a result of floods during August / September of 2006, Gujarat, South Rajasthan, Madhya Pradesh and Maharashtra, which account for nearly 7.0 million hectares of land sown to wheat, will need wheat varieties that will mature in 120 days. Farmers would like to sow more wheat due to current ruling prices and availability of good soil moisture. **Preparedness for timely supply of seeds is necessary, along with other inputs**.
- Adequate and timely availability of credit is essential, particularly due to the financial loss suffered during **kharif**.
- Wheat should be sown before 15 November in NW Plain Zone to get maximum yield and to escape from potential heat stress in March.
- Use of seed drill to capture soil moisture, adopt proper seeding depth and complete sowing in time should be recommended. **Resource conserving tillage agronomy should be subsidized to save water and improve yield.**
- Avoid sowing wheat during late October in the Punjab as such early sown wheat often suffers if temperatures during December are above normal.
- Balanced use of fertilizers including Zinc should be promoted and overuse of urea should be curtailed.
- Pre emergence weed control or a post emergence chemical weed control should be recommended over the Indo-Gangetic plain

- Varietal diversity should be ensured all over the wheat growing region and farmers should grow three of four different varieties to avoid genetic vulnerability to diseases.
- Wherever limited irrigation facility is available, atleast one irrigation around crown root stage / early tillering should be recommended followed by need based weeding and fertilization.

Rabi and **boro** rice production can be enhanced considerably by giving attention to balanced fertilization, particularly to the supply of the needed micronutrients like zinc, Boron and sulphur. Together with plant protection the enhancement of soil health will help to improve productivity atleast by an additional tonne per hectare. There are nearly 5 million ha under **rabi** and boro rice in the country and improved varieties are available for all the States where rice is cultivated between November and May. Striking progress in improving the yield of rainfed maize, soybean, sorghum, green gram, blackgram, pigeon pea, chickpea, finger millet (ragi), pearl millet (bajra), castor etc., can be achieved through balanced fertilization (NPK and the needed micronutrients). Seeds of improved varieties should be maintained in Village Seed Banks in rainfed areas, so that alternative cropping strategies can be introduced depending upon monsoon behaviour. Improved cultivars alone can enhance productivity by 10 to 50%. Varietal choice should be based on the likely moisture availability. The short duration chickpea variety **Shwetha** (ICCV2) has revolutionized chickpea production in Andhra Pradesh. The productivity increased from 470 kg per ha in 1993 to 1084 kg per ha in 2004. Area also increased seven fold. There are nearly 12 million ha of rice fallows in MP, Orissa, Jharkhand, Chattisgarh and West Bengal. In such rice fallow areas, chickpea can be grown by using residual soil moisture. Simple seed priming technologies like soaking seeds in water and micronutrient solution for 6 hours and drying in shade will help in establishing a good chickpea crop in rice fallows. In Madhya Pradesh 2 million ha remain fallow during the **kharif** season. Using broad bed and furrow, balanced nutrient management and short duration soybean cultivars like Samrat, farmers in the Vidisha district were able to take a crop of chickpea or wheat during **rabi** and thereby double their income. Many such simple steps in soil-water-crop management can lead to major advances in both crop output and farmers' income. This is the pathway to making farming economically viable.

Extending the Gains

Eastern India (eastern UP, Bihar, Chattisgarh, Orissa, West Bengal, Assam and NE States) have a large untapped production reservoir even with the technologies now available. In these areas, poor water management, rather than water availability, is the major constraint. **The Indo-gangetic plains offer scope for becoming the major bread basket of India through an appropriate mix of technology, services and public policies.** In many of these areas, the aquifer should be enriched during the S W Monsoon period, and extensive ground water use should be promoted during the October – April period. Given the right strategy, the Ganges Water Machine could become the main anchor for our food security system. **Bihar in particular is a sleeping giant in the field of agriculture**. The work of IARI in the Dharbhanga district and Sone

Command area has shown that the wheat yield can be increased substantially with good seeds and improved agronomic practices. The major bottleneck is however the absence of a Grain Purchasing Machinery which will provide the MSP to farmers.

Making New Gains

The immediate prospect for making new gains lies in the areas of post-harvest technology, agro-processing and value addition to primary produce. NCF has made several recommendations in this area in its first four reports. In the longer term, there is need for new yield and quality breakthroughs in major crops through genomics and gene pyramiding. For example, Super Wheats capable of yielding about 8 t/ha are now in the breeders' assembly line. Such wheats have a complex pedigree and illustrate the importance of genetic resources conservation and exchange.

Super Wheats are semi-dwarf with robust stem, broad leaves, large spikes with more number of grains per panicle and more grain weight. The Super Wheat architecture in the breeders' assembly line, both at CIMMYT and AICWIP, is derived from a blend of Tetrastichon (Yugoslavia), Agrotriticum (Canada), Tetraploid Polonicum (Poland) Gigas (Israel), Morocco wheat (Morocco) and semi-dwarf wheats currently grown in India

In India, we can produce 100 Million t of wheat by 2015, by the following two steps:

- Average yield of 4 t / ha from 25 million ha
- Harnessing the large untapped yield reservoir in eastern, central and western India
- Every State should develop a detailed agricultural strategy for their major farming zones and systems based on the 3 pronged approach outlined above.

It would be useful if a comprehensive **Rural-Urban Parity Index** is prepared at the State and district levels in order to help in the continuous refinement of the Bharat Nirman (New Deal for Rural India) Programme. If the evolution of rurban societies is planned carefully, there will be symbiotic linkages between the village and the town, each enriching the other in both economics and employment opportunities. **Work, water and energy are the key needs of rural India**. The VIth Five Year Plan (1980-85) gave over riding priority to these sectors, together with attention to technology, natural resources conservation, women's empowerment and enhancement, and remunerative marketing (for the first time, the Vith Five Year Plan of India introduced special Chapters on Environment and Development, Women and Development and a New Deal for the Self-employed). **This strategy resulted in agricultural growth rate (5.7%) exceeding for the first time, the overall GDP growth rate (5.5%)** during the plan period. Thus, given right priorities and strategies, we can accelerate progress both in agricultural growth and agrarian prosperity.

Redesignating the Ministry of Agriculture as **Ministry of Agriculture and Farmers' Welfare** will lead to positive results only if this is accompanied by a structural reorganization. **Leading farmers should be inducted at the senior level in the Ministry for specific tasks and specific periods**. This will help to change the "beneficiary" mindset in agricultural planning to one of regarding farmers as innovators, policy planners and life-givers. Agriculture will then become our nation's pride and "Jai Kisan" will acquire true meaning and significance.

The last among the UN Millennium Development Goals refers to global partnerships for achieving the goals. Seemingly impossible tasks can be achieved through mobilizing the power of partnership. Therefore, it is important that agricultural scientists all over the world work together in solving not only present day problems in the area of farming but also emerging problems particularly with reference to climate change. Partnership between the public and private sectors is important both in the generation of new technologies and their spread based on the principle of social inclusion. This is where conferences of this kind play a significant role in advancing the cause of food for all and for ever.

References

Kesavan, P.C., and Swaminathan M.S., From Green Revolution to Evergreen Revolution: Pathways and Terminologies, Current Science, Volume 90, Number 2, 25 July 2006, pp. 145–146

A Dialogue between M.S. Swaminathan and Daisaku Ikeda, Revolutions: to green the environment, to grow the human heart, EastWest Books (Madras) Pte. Ltd., Chennai, 2005.

Effects of Climate Change on Food Production

Martin L. Parry and M.S. Swaminathan

Parry, Martin L. and M.S. Swaminathan (1992). Effects of Climate Change on Food Production. *Confronting Climate Change: Risks, implications and responses.* Irwing, M. Montzer (ed.), Cambridge University press, pp. 113–115

Editors Introduction

Like hunger, the stresses that arise from rapid climate change will fall most heavily on the poorest, the most vulnerable, and those least able to adopt new technology. Food supply is perhaps the arena in which the potential crisis could prove most severe. As Martin Parry and M.S. Swaminathan show in this chapter, global climate change will increase the stress on agricultural systems, potentially decreasing yields at the very time when demand for food is growing dramatically.

This chapter, an overview of the most likely impacts of climate change on food production, emphasizes that current scientific knowledge is insufficient to allow us to predict the regional impacts of climate change on food production with much confidence. The news is not all bad, however. Increased concentrations of carbon dioxide have had a fertilization effect on some crops in the laboratory, when all other environmental conditions have been tightly controlled. Unfortunately, field conditions cannot always be so carefully controlled. And not all plants benefit equally from this effect. Wheat and barley have shown strongly positive effects; corn, a less positive response.

Parry and Swaminathan also note some strong and troubling trends in recent research results. The more pronounced warming in high latitude areas that is predicted by climate models will extend the Northern range of agricultural potential. But it may also decrease grain yields in the core of traditional agricultural areas. In the low-latitude countries, changes in temperature regimes will require farmers to alter crop-timing patterns that have been fine tuned over centuries of experience with traditional regional climates.

Perhaps more important for food production, climate model results also suggest that soil moisture will decrease in some already stressed areas as global temperatures rise. The areas most likely to see decreases in yield include North and West Africa, parts of Eastern Europe, North and Central China, Western Australia, and Eastern Brazil. In most cases, these are areas already strained by the pace of economic (and in some cases political) changes that make rapid, well-coordinated institutional responses unlikely.

But the most important impacts may be due to the least predictable aspects of the climate system - changes in the frequency, distribution, and severity of extreme weather events (See further Mitchell and Ericksen (1992), this volume). In a world warmed by the buildup of greenhouse gases, the number of days with temperatures above 30"C is likely to increase significantly while the number of days below 0°C is likely to decline. Such changes, along with changes in the frequency of floods, droughts, and wind storms, will necessitate major adjustments in the agricultural sector to prevent significant decreases in yield.

Martin Parry, Director of the Environmental Change Unit at the University of Oxford University, has been studying the effects of climate change on agriculture for more than a decade. His has been one of the few clear and stable voices urging caution, systematic analysis and careful preparation of adaptive response strategies to minimize the risks of rapid climate change. M.S. Swaminathan, former President of IUCN, is an eloquent spokesman for the shared requirements of environment and development. Swaminathan, as former director of the International institute for Rice Research in Manila, has made major scientific contributions to the development of practical techniques to enhance agricultural yields. Between them, these two leading scientists bring an unmatched depth of vision and compassion to this issue.

Parry and Swaminathan conclude that global food production can be sustained at levels that might occur in the absence of climate change, but only at a high cost. Vast new areas may have to be brought under cultivation. They caution that the gains in production in some traditional agricultural areas would be unlikely to balance possible large-scale reductions in these areas. Adequate response strategies may require changes in land use and tenure patterns, changes in management approach, and changes in crop and livestock husbandry. Successfully integrating these responses in ways which do not increase cross-border tensions within and between regions may produce challenges to political leadership that are every bit as difficult as the problem of breeding better crops for the new world ahead.

If we do not solve these food production problems, the results may well be devastating -for reasons that link this chapter with Nathan Keyfitz' work on population (1992; this volume). By the end of the next century, the human population of our planet is likely to be at least twice as large as it is today. But even at today's level of about 5.2 billion, the human race is hard pressed to keep itself well-fed. Current annual rates of food production (approximately 5.2×10^{15} Calories per year), could provide enough food for approximately 6 billion people, if all the food were shared equally and nearly all the caloric input came from grains. (Each person would receive about 2350 Calories per day, the UN recommended daily dietary input.) If the world's current food output were used to provide every person with a diet like that typical in South America - containing about 10% of the caloric input from animal products - only 4 billion people could be fed at today's level of production. If everyone were to eat on a European-style diet, with about 30% of the calories from animal products, only 2.5 billion people could be fed from the food produced worldwide today (World Resources Institute, 1988).

Of course, the world's food output is not distributed equally - neither within countries nor between countries. Kates et al. (1988) note that among 46 countries where food consumption levels were below those needed to support a healthy, productive life, 43 were in sub-Saharan Africa, Asia, Latin America or the Caribbean. A World Bank study (1986) of 87 developing countries, indicated that in 1980, 34% of the population or nearly 730 million people, got less than 90% of the FAD/WHO minimum daily requirements not enough calories to support an active working life. What's more distressing, 16% (nearly 340 million people) received less than the amount of food required each day to prevent stunted growth and major health problems.

Even in poor countries, hunger is unevenly distributed. The Fifth World Food Survey (FAD, 1985) noted that in Bangladesh, for example, urban dwellers are better fed than their rural cousins. In rural areas, landless peasants averaged about 1900 Calories per day while members of house hold owning more than 7 hectares received an average of almost 2400 Calories per day.

In 1980, nearly 70% of the world's arable land was under the plough. Dramatic increases in productivity would be necessary in both food production and in the management of food distribution systems, just to keep pace with expected population growth over the next century. In other words, the situation is worthy of concern even without the threat of climate change; Parry and Swaminathan, in this chapter, put the challenges of maintaining food production and the risks of rapid climate change, finally, in perspective to each other.

1. Introduction

The purpose of this chapter is to consider the likely effects of changes of climate on the ability of the world to feed its population. We consider, first, aspects of climate change that will be most important in their effect on agricultural potential. Secondly, we discuss the "direct" or fertilization effect of CO_2 on crop production. This is followed by a discussion of the possible shifts in potential production areas that could result from geographical shifts of climate, and the changes in yield levels that may occur in these regions. The chapter concludes with an overall assessment of the consequences of climate change for food production. We outline the responses that should be taken now, to minimize the negative effects and maximize the potentially positive benefits of climate change.

We conclude that changes of climate could provoke serious reductions in food production potential, particularly in regions that are at present vulnerable to climate variability The most vulnerable areas include traditional agricultural zones in the semi-arid and humid tropics. Many of these regions are characterized by low incomes, and have a limited ability to adapt through technological change. Sharing of knowledge and technological assistance with these regions by those in the advanced industrialized societies will be critically important. The co-development of agricultural technologies specifically adapted to local conditions will strengthen each nation's capabilities to respond and adapt to food shortages and new agricultural needs.

We note that the rate of future climate change will be critical in determining the sign and magnitude of climate change impacts. The costs will be made high if the changes are rapid, or if the perturbations in local weather are unstable and erratic. And, if the frequency, severity, and duration of extreme weather events increases the risks to farmers and farm economies will be amplified.

Fanners worldwide are generally conservative people. They prefer to continue what has worked well in the past, rather than to embrace radical change. If the patterns of regional climate change gradually, fanners and the institutions that support them will respond positively and adapt successfully. Some will diversify their holdings and their behavior. Others will switch to new crops and alternative cultivars. Thus, we expect that if changes occur slowly, and with an easily predictable pattern, losses will be minimized.

2. The Most Threatening Types of Climate Change

Although increases in average temperature worldwide are the most predictable consequence of greenhouse gas emissions, many other aspects of global and regional climate will also change. Some of these changes will be much more significant than others, with respect to agricultural output.

2.1. *Climatic extremes*

It is not clear whether changes in the variability of temperature will occur as a result of climate change. However, even if variability remains unaltered, an increase in average temperatures would result in the increased frequency of temperatures above particular thresholds. Changes in the frequency and distribution of precipitation are less predictable, but the Intergovernmental Panel on Climate Change (IPCC) concluded that the combination of elevated temperatures, and drought or flood, probably constitutes the greatest risk to agriculture (Parry and Duinker, 1990).

2.2. *Warming in high latitudes*

There is relatively strong agreement among general circulation model (GCM) predictions that greenhouse-gas induced warming will be greater at high latitudes (IPCC, 1990). This will reduce temperature constraints on high latitude agriculture, and increase the competition for land here, by bringing previously uncultivable land within the range that can support traditional cultivars (Parry and Duinker, 1990). Warming at low latitudes, although less pronounced, is also likely to have a significant impact on agriculture - to the extent that it alters availability of soil moisture or raises temperatures above the threshold of tolerance for key crop cultivars.

2.3. *Poleward advance of monsoon rainfall*

In a warmer world, the intertropical convergence zones and polar frontal zones may advance further poleward, as a result of an enhanced ocean-continent pressure gradient. Because air over the continental land masses will warm more rapidly than air over the oceans, the key zones of interaction between weather fronts are likely to advance poleward. If this were to occur, then total rainfall could increase in some areas of Asia, Africa and Australia that are now exposed to monsoon rainfall, but there is currently little agreement on which regions might be most affected (IPCC, 1990). Rainfall episodes could also be more intense and thus flooding and erosion could increase

2.4. *Reduced soil water availability*

Probably the most important consequences for agriculture could stem from higher potential evapotranspiration (the process by which water returns from plants and soils to the atmosphere). Enhanced evapotranspiration would be primarily a consequence of higher air and land surface temperatures. Even in the tropics, where temperature increases are expected to be smaller than elsewhere, and where precipitation might increase, the increased rate of loss of moisture from plants and soil could be considerable (Parry, 1990; Rind et al., 1989). Such losses may be somewhat reduced by greater air humidity and increased cloudiness during the rainy seasons, but could be pronounced in the dry seasons.

3. Potential Direct Effects of Greenhouse Gases on Crops and Livestock

3.1. *Effects of CO_2 increase on plant growth*

The impacts of elevated concentrations of carbon dioxide (CO_2) on crop growth and yield have been studied for nearly 100 years (Wittwer, 1985). The effects of other greenhouse gases - methane, dinitrogen oxides and chlorofluorocarbons - have yet to receive similar attention. Early experiments that increased the CO_2 concentration in greenhouses resulted in yield enhancement in lettuce, tomatoes, cucumbers and several ornamental plants. Kimball (1985) assembled 770 observations on the impact of higher levels of atmospheric CO_2 in crop yield, and found that the yield increase overall was of the order of 32% (if all other growth related factors were held constant).

Why does this take place? The most direct effect of high levels of CO_2 that has been verified in plants is an increase in leaf and canopy photosynthetic rates. It has been observed that an increase in photosynthesis may continue at concentrations up to four times the pre-industrial level of CO_2, or 1000 parts per million (Wittwer, 1985). Increased metabolic activity in plants may lead to conversion of up to 51% of the CO_2 that was transformed into carbohydrates being oxidized through photorespiration, and released again as CO_2. In the long term, photorespiration rates are likely to decrease with a rise in atmospheric CO_2 levels.

An increase in biomass production invariably occurs with CO_2 enrichment. This may not always come from an increase in net photosynthesis. Increasing CO_2 concentration leads to a partial closure of leaf stomates. This reduces leaf transpiration rates, and promotes water use efficiency. Increased water use efficiency is an important outcome of higher levels of CO_2. Some studies have shown that the yields of wheat varieties subjected to water stress were as high or even higher at high CO_2 levels, as compared to wheat grown with adequate irrigation but at normal CO_2 levels (Gifford, 1979, Havelka et al., 1984).

3.1.1. *C3 and C4 plants*

The metabolic processes of several hundred thousands of plant species have been described and analyzed by scientists. These plants are categorized by the particular pattern of photosynthetic processes that each plant uses to convert CO_2 and HP to carbohydrates. The two most important groups are called C3 and C4 plants. These terms reflect the number of carbon atoms characteristically involved in photosynthesis.

Of the many thousands of plants on the planet, global food security depends only on about 20 plant species. Of these, 16 have a C3 photosynthetic pathway. C3 plants include: rice, wheat, barley, oats, mung bean, cowpea, chickpea, pigeon pea, potato, sweet potato, cassava, sugar beet, banana and coconut. Most fruits and vegetables are C3 plants. In general, C3 crops exhibit a positive response to higher levels of CO_2, Cure (1985) has reported a yield increase of nearly 40% with CO_2 enrichment in wheat and barley. In C3 species, increases in leaf area, weight per unit area, leaf thickness, stem height, seed and fruit numbers and weight have been observed to increase at higher CO_2 levels. The carbon-nitrogen (C:N) ratio has been found to increase and the harvest index has generally been altered in favour of the economically important part of the plant.

Maize, by contrast, is a C4 plant. In maize, higher than normal levels of atmospheric CO_2 have been reported to induce greater water use efficiency (Rogers et al., 1983).

Though stomatal conductance declines with increasing CO_2 concentration in both C3 and C4 plants, C3 plants seem to benefit in dry matter production more from higher CO_2 level due to the following reasons:

(a) higher leaf expansion
(b) increase in the photosynthetic rate per unit leaf area and
(c) increase in water use efficiency.

Photorespiration rates may decrease with an increase in atmospheric CO_2,

3.1.2. *The limits of knowledge about CO_2 effects*

These results would tend to suggest that a higher CO_2 content in the atmosphere may be an advantage, particularly in areas where water may be a limiting factor. Unfortunately, most studies on direct CO_2 effects have involved a CO_2 concentration of 600 ppm. We need studies using levels ranging from 450 to 500 ppm.

More important, under field conditions, the anticipated gains may not be realized due to the complex nature of the linkages involved between the biological effects of photosynthetic efficiency and temperature, sunlight and moisture availability. Furthermore, the optimistic results of the laboratory studies presume that rainfall conditions and soil moisture availability will remain unchanged in a CO_2-enhanced world. This is not necessarily a good assumption. The combination of increased atmospheric CO_2 with, for example, reduced rainfall (or reduced soil moisture availability at critical periods in the growth cycle) could result in no change or in significant yield reductions. In some cases, the integrated impact of a rise in temperature and CO_2

concentration on the yield of wheat and rice may be negative (Sinha and Swaminathan, 1991). Under conditions in North India, every 0.5°C increase in temperature resulted in a drop in the duration of wheat crop by 7 days, which in turn reduced yield by 0.45 tonnes per hectare (t/ha). If mean temperatures rise by 1-2 °C, the adverse impact on yield will be even greater.

4. Potential Effects of Climate Change

There are two broad types of effect that changes of climate may have on agriculture: first, effects on the geographical limits to the regions where different types of crops and livestock can be produced; and secondly, effects on the potential yields of crops and livestock in these regions.

4.1. *Effects on geographical limits to agriculture*

4.1.1. *Changes in thermal limits to agriculture*

Increases in temperature can be expected to lengthen the growing season in areas where agricultural potential is currently limited by insufficient warmth - resulting in a poleward shift of thermal limits of agriculture. The consequent extension of potential will be most pronounced in the northern hemisphere, because of the greater extent here of temperate agriculture at higher latitudes.

There may, however, be important regional variations in our ability to exploit this shift. For example, the greater potential for exploitation of northern soils in Siberia than on the Canadian Shield may mean relatively greater increases in potential in northern Asia than in northern N. America (Parry, 1990).

A number of estimations have been made concerning the northward shift in productive potential in mid-latitude northern hemisphere countries. These relate to changes in the climatic limits for specific crops under a variety of climatic scenarios, and are therefore not readily compatible (Newman, 1980; Blasing and Solomon, 1983; Rosenzweig, 1985; Williams and Oakes, 1978; Parry and Carter, 1988; Parry, et al, 1989). They suggest, however, that a 1°C increase in mean annual temperature would tend to advance the thermal limit of cereal cropping in the mid-latitude northern hemisphere by about 150-200 km, and to raise the altitudinal limit to arable agriculture by about 150-200 m.

While warming may extend the margin of potential cropping and grazing in mid-latitude regions, it may reduce yield potential in the core areas of current production. This is because higher temperatures encourage more rapid maturation of plants and shorten the period of grain filling (Parry and Duinker, 1990). An important additional effect, especially in temperate mid-latitudes, is likely to be the reduction of winter chilling (vernalization). Many temperate crops (such as barley and oats) require a period of low temperatures in winter, to either initiate or accelerate the flowering process. Reduced vernalization results in low flower bud initiation and, ultimately, reduced yields. It has been estimated that a 1°C warming would reduce effective winter chilling by between 10 and 30%, thus contributing to a poleward shift of temperate crops (Salinger, 1989).

Increases in temperature are also likely to affect the crop calendar in low latitude regions, particularly where more than one crop is harvested each year. For example, in Sri Lanka and Thailand a 1°C warming would probably require a substantial re-arrangement of the current crop calendar, which is finely tuned to present climatic conditions (Kaida and Surarerks, 1984; Yoshino, 1984).

4.1.2. *Shifts of moisture limits to agriculture*

There is much less agreement between GCM-based projections concerning greenhouse gas-induced changes in precipitation, than there is about temperature - not only concerning changes of magnitude, but also of spatial pattern and distribution through the year. For this reason it is difficult to identify potential shifts in the moisture limits to agriculture. This is particularly so because relatively small changes in the seasonal distribution of rainfall can have disproportionately large effects on soil moisture, and thus on the viability of agriculture in tropical areas. This influence takes place largely through changes in growing period when moisture is sufficient, and thus through the timing of critical episodes such as planting.

However, recent surveys for the IPCC have made a preliminary identification of those regions where there is some agreement among the GCM experiments, concerning the regional implications for soil water availability if a doubled level of atmospheric CO_2 comes to pass (Parry, 1990; Parry and Duinker, 1990). It should be emphasized that coincidence of results for these regions is not statistically significant. The regions where soil water may be reduced are shown in Table 1.

Table 1. Regions where soil water may be reduced by global climate change.

	Decreases of soil water in Dec, Jan, Feb	Decreases of soil water in June, July, Aug
Africa	North-east Africa	North Africa
	Southern Africa	West Africa
Europe		Parts of Eastern Europe
Asia	Western Arabian Peninsula	North and central China
	Southeast Asia	Parts of Soviet Central Asia and Siberia
Australasia	Eastern Australia	Western Australia
N. America	Southern USA	Southern USA and Central America
S. America	Argentine Pampas	Eastern Brazil

4.1.3. *Regions affected by drought, heat stress and other extremes*

Probably among the most important impacts of global warming for agriculture, but about which least is known, are the possible changes in climatic extremes - such as the magnitude and frequency of drought, storms, heat waves and severe frosts (Rind et al., 1989; Mitchell and

Ericksen, 1992, this volume). Some modelling evidence suggests that hurricane intensities will increase with climatic warming (Emanuel, 1987). This has important implications for agriculture in low latitudes, particularly in coastal regions.

The effects of heat or cold stress on crops are often disproportionate to the magnitude of the stress itself, and crop yields often exhibit a nonlinear response to heat or cold stress. Thus, changes in the probability of extreme temperature events can be significant (Meams et al., 1984; Parry, 1976). In addition, even assuming no change in the standard deviation of temperature maxima and minima, we should note that the frequency of hot and cold days can be markedly altered by changes in mean monthly temperature. To illustrate, in a climate where CO_2 levels had doubled (a 2 x CO_2 climate), the number of days in which temperatures would fall below freezing would decrease from a current average of 39 to 20 in Atlanta, Georgia (USA), while the number of days above 90°F would increase from 17 to 53 (EPA, 1989).

The frequency and extent of area over which losses of agricultural output could result from heat stress, particularly in tropical regions, is therefore likely to increase significantly. Unfortunately, no studies have yet been made of this. However, the apparently small increases in mean annual temperatures in tropical regions (about 1 to 2 °C under a 2 x CO_2 climate) could sufficiently increase heat stress on temperate crops such as wheat, so that these are no longer suited to such areas. Important wheat-producing areas such as North India could be affected in this way (Parry and Duinker, 1990).

There is a distinct possibility that, as a result of high rates of evapotranspiration, some regions in the tropics and subtropics could be characterized by a higher frequency of drought, or a similar frequency of more intense drought, than at present. Current uncertainties about how regional patterns of rainfall will alter mean that no useful prediction of this can at present be made. However, it is clear in some regions that relatively small decreases in water availability can readily produce drought conditions. In India, for example, lower than-average rainfall in 1987 reduced food grains production from 152 to 134 million tonnes (Mt), lowering food buffer stocks from 23 to 9 mt. Changes in the risk and intensity of drought, especially in currently drought-prone regions, represent potentially the most serious impact of climatic change on agriculture both at the global and the regional level.

4.1.4. *Effects on the distribution of agricultural pests and diseases*

Studies suggest that temperature increases may extend the geographic range of some insect pests currently limited by temperature (EPA, 1989; Hill and Dymock, 1989; Porter et al., 1992). As with crops, such effects would probably be greatest at higher latitudes. The number of generations per year produced by multivoltine (i.e. multigenerational) pests would increase, with earlier establishment of pest populations in the growing season and increased abundance during more susceptible stages of growth.

An important unknown, however, is the effect that changes in precipitation amount and air humidity may have on the insect pests themselves - and on their predators, parasites and diseases.

Climate change may significantly influence interspecies interactions between pests and their predators and parasites.

Under a warmer climate at mid-latitudes there would be an increase in the over-wintering range and population density of a number of important agricultural pests, such as the potato leafhopper - which is a serious pest of soybeans and other crops in the USA (EPA, 1989). Assuming planting dates did not change, warmer temperatures would lead to invasions earlier in the growing season, and probably to greater damage to crops. ill the US Com Belt, increased damage to soybeans is also expected, due to earlier infestation by the com earworm.

Examination of the effect of climatic warming on the distribution of livestock diseases suggests that those at present limited to tropical countries, such as Rift Valley fever and African Swine fever, may spread into the midlatitudes. For example, the horn fly, which currently causes losses of $730.3 million in the US beef and dairy cattle industries, might extend its range under a warmer climate leading to reduced weight gain in beef cattle and a significant reduction in milk production (Drummond, 1987; EPA, 1989).

In cool temperate regions, where insect pests and diseases are not generally serious at present, damage is likely to increase under warmer conditions. In Iceland, for example, potato blight currently does little damage to potato crops, because it is limited by the low summer temperatures. However, under a 2 x CO_2 climate, those summer temperatures may be 4°C warmer than at present. As a result, crop losses to disease may increase by up to 15% (Bergthorsson et al., 1988).

Most agricultural diseases have greater potential to reach severe levels under warmer and more humid conditions (Beresford and Fullerton, 1989). Cereals would be more prone to diseases such as Septoria. In addition, increases in population levels of disease vectors may well lead to increased epidemics of the diseases they carry. To illustrate, increases in infestations of the Bird Cherry aphid (*Rhopalosiphum padi*) or Grain aphid (*Sitobian avenae*) could lead to increased incidence of Barley Yellow Dwarf Virus in cereals

4.2. *Effects on crop yields*

4.2.1. *In humid and semi-arid low latitude regions*

An analysis of more than 100 cooperative trials - conducted by the International Rice Research Institute (IRRI) under irrigated conditions in 22 countries over eight years indicates that rice yields are significantly affected by temperature and solar radiation. When the minimum temperature increases from 18 to 19°C, there is a decrease in yield of about 0.7 tònnes per hectare (t/ha). At a range of sites, varying in latitude from 31° N to 6°N, yields drop by 10–20% with a rise in temperature of 1-2°C during the November-December harvest period. Yields are most severely affected at higher latitudes.

Re-adjustment of the harvest period prior to the onset of higher temperatures would help, but this would be difficult where the crop season has to be tailored to the monsoon rainfall period.

Flowering response in crops to integrated factors needs study, since this is a major determinant of yields. Because yields are largely determined by flowering responses, it would be useful to study the integrated effects of increased temperatures, elevated CO_2 levels, and altered rainfall patterns on overall yields.

Considerable genetic variability occurs among genotypes of rice, wheat and other plants in relation to tolerance to several biotic and abiotic stresses. More studies at the micro-level are required, in order to understand the interrelationships between CO_2 concentration, temperature and precipitation. We also need integrated studies to understand the impact of other greenhouse gases on yield. It is possible to assemble genes from a wide range of land races through a properly planned breeding programme.

Another aspect of this situation which needs careful study is the impact of changes in CO_2 levels, temperature, precipitation, ultraviolet - B radiation on the incidence of pests. Most of the studies on the direct impact of CO2 have been carried out under conditions of optimum pest management. Pest and disease problems are more severe in tropical regions, due to multiple cropping and the availability of alternate hosts throughout the year (Table 2). Temperature influences the multiplication patterns of insect pests. For example, the seed potato industry in northern India is based on the fact that aphid transmitted virus diseases do not occur during November-December. Hence, the potato crop raised during this aphid-free season could be kept for seed purposes Weeds take a heavy toll on crop yield. The widely prevalent weeds that most severely affect C3 crop plants tend to incorporate the C4 metabolic pathway. Conversely, for a C4 plant such as maize, 19 of the 38 major weeds reported in the United States are C3 plants. Some of the serious weeds, insect pests and pathogens of today may become less important under altered climate conditions, and some of the less important ones may become more serious. Experience with high yielding varieties of rice in Asia shows that, when the micro-environment is altered through irrigation and fertilizer application, pests which were not important before, such as the brown plant hopper, become a serious concern.

Table 2. Crop disease in temperate and tropical regions.

	Number of diseases reported	
Crop	**Temperate**	**Tropical**
Rice	54	500–600
Maize	85	125
Citrus	50	248
Tomato	32	278
Beans	52	250–280

Source: Swaminathan, 1986

In the semi-arid tropics, farmers face enormous uncertainty about future yields. That uncertainty stems from our ignorance of future changes in precipitation - both its quantity and its distribution over time and space. For example, a series of studies conducted at the International Crops Research Institute for the Semi-Arid Tropics (ICRSAT) modelled the impact of climate on the yield of sorghum in two semi-arid, tropical regions of India. The two regions, otherwise, had contrasting climate: Anantapur is dry with a highly variable, erratic rainfall, while Hyderabad experiences a relatively dependable, 6-month rainy season. The study, using a sorghum growth model developed at Texas A&M University, found that yields increased when both rainfall and temperature increased. (In Anantapur, under those conditions, yields would more than triple.) On the other hand, if temperature increase alone, without any related changes in rainfall, would lead to more than 10% losses of grain and dry matter in both locales. This is due to increased transpiration and subsequent soil moisture stress.

In general, much depends upon whether increases in temperature are accompanied by increases in available moisture. Increases in temperature, - even the relatively small increases of about 1.5°C projected in lower latitudes for a doubling of greenhouse gases - would increase rates of evapotranspiration by 5 to 15%. If there were not compensating increases in rainfall, this would tend to reduce yields. In northern India, for example, temperature increases of even 0.5°C probably would reduce wheat yields due to heat stress, by about 10% if rainfall did not increase (Van Diepen et al., 1987).

The effects on maize in semi-arid areas may be similar. In central China, for example, maize yields are estimated to decrease on average by 3% per 1°C (Terjung et al., 1989) No impact studies based on climatic scenarios from global climate model (GCM) experiments have, at the time of writing, been made in Africa, primarily because there is little agreement between GCMs concerning changes in rainfall - in this region as in others, rainfall is the climatic variable that most determines variations in agricultural yield. Some preliminary work has been completed in South Africa, indicating that an increase in mean annual temperatures of 2 to 4 °C in Natal Province (consistent with the range of GCM projections for 2 x CO_2 climates) would increase rates of evapotranspiration by 5% to 15% and reduce plant productivity by about the same amount (Schulze, personal communication.).

In Kenya, a recent case study considered the effects of the driest 10% of years from the recent past. These suggest reductions of maize yields by 30%-70% (Akong'a et al., 1988). Any changes in the frequency of such years would substantially affect the average output of agriculture in the region, but we cannot at present estimate how these might change.

4.2.2. *In the midlatitude grainbelts*

One of the most serious threats to world food supply could result from decreased productive potential in the mid-latitude "grainbelt" (i.e. the grain-exporting regions of North America, Australia, Southern Africa, South America, and the southern regions of what was the USSR. In the USA increased temperatures and reduced crop water availability are estimated to lead to a decrease of yields of all the major unirrigated crops (EP A, 1989). The largest reductions are projected for the south and south-east. In the most northern areas of the US, however, where temperature is currently a constraint on growth, yields of unirrigated maize and soybeans could increase as higher temperatures increase the length of the available growing season.

When the direct effects of increased CO_2 are considered, yields may increase more generally in northern areas; but the production of most crops in the US will probably be reduced because of heat stress and diminished rainfall. The largest reductions are expected to be in sorghum (-20%), com (13%) and rice (-11%), with an estimated fall in net value of agricultural output of $33 billion. If this occurred, American consumers would face slightly higher prices, although supplies are estimated to meet current and projected demand. However, exports of agricultural commodities could decline by up to 70%, and this could have a substantial effect on the pattern of world food trade (EPA, 1989)

On the Canadian prairies, where growing season temperatures under a 2 x CO_2 equilibrium climate would be about 3.5 °C higher than today, uncertain changes in rainfall could decrease average potential yields by 10 to 30%. Spring wheat yields in Saskatchewan are estimated to fall by about 30% (Williams et al., 1988). Since Saskatchewan at present produces 18% of all the world's traded wheat, such a reduction could have global implications.

In Ontario, even sizeable increases in precipitation would be offset by increases in evapotranspiration - with consequent increased moisture stress on crops. Maize and soybean would thus become very risky in the southern part of the province. In the north, where maize and soybean cannot currently be grown commercially because of inadequate warmth, cultivation may become profitable but this is not expected to compensate for reduced potential further south.

In Australia, wheat production could increase, assuming increased summer rainfall, decreased winter rainfall, and a general warming of 3 °C. Increases are expected in all states except Western Australia, where more aridity might cause a significant reduction in output (Pittock, 1989). The major impact would probably be on the drier frontiers of arable cropping. For example, increases in rainfall in subtropical northern Australia could result in increased sorghum production at the expense of wheat. Increased heat stress might shift livestock farming and wool production southward with sheep possibly replacing arable fanning in some southern regions. Many areas currently under fruit production would no longer be suitable under a 3 °C warming, and would need to shift southwards or to higher elevations in order to maintain present levels of production.

All of these changes would also be affected by changes in the distribution of diseases and pests.

4.2.3. *In northern marginal regions*

Some of the most pronounced effects on agriculture would be likely to occur in high latitude regions, because GHG-induced warming is projected to be greatest here and because this warming could remove current thermal constraints on farming. Inappropriate terrain and soils are, however, likely to limit the increase in extent of the fanned area. Thus, in global terms, production increases would probably be small (Parry, 1990). A summary of available information is given below.

In Iceland, with mean annual temperatures increased by perhaps 4.0 DC, hay yields on improved pastures would increase by about two-thirds and herbage on unimproved rangelands by about a half (Bergthorsson, et al., 1988). The numbers of sheep that could be carried on the pastures would be raised by about 250%, and on the rangelands by two-thirds, if the average carcass weight of sheep and lambs is maintained as at present. At a guess, output of Icelandic agriculture could as much as double with a warming of 4°C.

In Finland, an increase in summer warmth by about a third and precipitation by about half (consistent with a 2 x CO_2 climate) could increase barley and spring wheat yields by about 10% in the south of the country, but slightly more in the north - due to relatively greater warming and lower present day yields (Kettunen et al., 1988).

The only other northern region for which an integrated impact assessment has been completed is in north European Russia. Here a 2.5 °C wanning, (consistent with a 2 x CO_2 climate), is estimated to decrease winter rye yields by about a quarter, due to faster growth and increased heat stress under the higher temperatures (Pitovranov et al., 1988). However, crops such as winter wheat and maize, which are currently low yielding because of the relatively short growing season in these regions, are better able to exploit the higher temperatures and exhibit yield increases.

Production could increase in regions currently near the low-temperature limit of grain growing. In the northern hemisphere, those regions would include the northern Prairies, Scandinavia, and north European Russia; in the southern hemisphere, southern New Zealand, and southern parts of Argentina and Chile could be affected. But it is reasonably clear that, because of the limited area unconstrained by inappropriate soils and terrain, increased high-latitude output will probably not compensate for reduced output at midlatitudes. The implications of this for global food supply and food security are considered later in this chapter.

4.3. *Effects on livestock carrying capacity*

Here, the information is extremely scant. The only model based study using GCM-derived scenarios of future climate has been completed in Iceland. There, increases in mean annual temperature of 4 °C and in rainfall of 15% are estimated to increase the carrying capacity of sheep on improved grassland by about two-and-a-half times, and on rough pasture by more than a half (Bergthorsson et al., 1988).

Few other regions, however, are currently as constrained by inadequate warmth as Iceland. One other beneficiary might be Patagonia, in the southern part of Argentina and Chile, where grass production and cattle grazing are limited by temperature rather than rainfall. Further north, in the southern pampas of Argentina, increases in rainfall would be needed to compensate for the higher rates of evapotranspiration (about 10%) that would stem from higher mean temperatures (2°C to 4 °C under the GCM 2 x CO_2 climates).

The productivity of the rangelands of Africa depends almost wholly on the amount and timing of rainfall. In Kenya, for example, forage yield in the driest 10% of years is reduced by 15% - to 60% - from its average (Akong'a, et al., 1986). In these dry years, the carrying capacity of livestock can fall by 10 to 40% and milk yields to zero.

Projected increases in summer rainfall in eastern Australia are expected to increase grass growth, but this is likely to be offset by the poorer nutritive value of tropical compared with temperate species. Loss of the Mediterranean-type climatic zones of Victoria and Western Australia - which are the current principal lamb and wool producing areas together with increased heat stress of both cattle and sheep, could mean that livestock productivity would decrease (Peannan, 1988)

5. Likely Effects on World Food Supply and Food Security

Although, on average, global food supply currently exceeds demand by about 10 to 20%, its year-to-year variation (which is about ±10%) can reduce supply in certain years to levels where it is barely sufficient to meet requirements. In addition, there are major regional variations in the balance between supply and demand, with perhaps a billion people (about 15% of the world's population) not having secure access to sufficient quantity or quality of food to lead fully productive lives.

For this reason, the working group on food security at the 1988 Toronto Conference on The Changing Atmosphere concluded that: "While averaged global food supplies may not be seriously threatened, unless appropriate action is taken to anticipate climate change and adapt to it, serious regional and year-to-year food shortages may result, with particular impact on vulnerable groups" (Parry and Sinha, 1988). Statements such as this are, however, based more on intuition than on knowledge derived from specific study of the possible impact of climate change on the food supply. No such study has yet been completed, although one is currently being conducted by the US Environmental Protection Agency and is due to report in 1992.

The information available at present is extremely limited. It has for example, been estimated that increased costs of food production due to climate change could reduce per capita global GNP by a few percentage points (Schelling, 1983). Others have argued that technological changes in agriculture will override any negative effects of climate changes and, at the global level, there is no compelling evidence that food supplies will be radically diminished (Crosson, 1989). Recent reviews have tended to conclude, however, that at a regional level, food security could be seriously threatened by climate change - particularly in less developed countries in the semi-arid and humid tropics (Sinha et al., 1988; Parry, 1990; Parry and Duinker, 1990).

Analyses conducted for the IPCC, designed to test the sensitivity of the world food system to changes of climate, indicate what magnitudes and rates of climatic change could possibly be absorbed without severe impact and, alternatively, what magnitudes and rates could seriously perturb the system (Parry, 1990; Parry and Duinker, 1990). These suggest that yield reductions of up to 20% in the major midlatitude grain exporting regions could be tolerated without a major interruption of global food supplies. However, the increase in food prices (perhaps, as much as 7% under a 10% yield reduction) could seriously influence the ability of food deficit countries to pay for food imports, eroding the amount of foreign currency available for promoting development of their non-agricultural sectors. It should be emphasized that these analyses are preliminary; more work is necessary before we have an adequate picture of the resilience of the world food system to climatic change.

On balance, the evidence is that food production at the global level can, in the face of estimated changes of climate, be sustained at levels that would occur without a change of climate, but the cost of achieving this is unclear. It could be very large. As we have seen, increases in productive potential at high mid-latitudes and high latitudes, while being of regional importance, are not likely to open up large new areas for production. The gains in productive potential would be unlikely to balance possible large-scale reductions in some major grain-exporting regions at mid-latitude. Moreover, there may well occur severe negative impacts of climate change on food supply at the regional level, particularly in regions of high present-day vulnerability, least able to adjust technically to such effects.

The average global increase in overall production costs could thus be small (perhaps a few per cent of world agricultural GDP). Much depends, however, on how beneficial are the so-called 'direct' effects of increased CO2 on crop yield. If plant productivity is substantially enhanced, and more moisture is available in some major production areas, then world productive potential of staple cereals could increase relative to demand, with food prices reduced as a result. If, on the contrary, there is little beneficial direct CO_2 effect, and climate changes are negative for agricultural potential in all or most of the major food-exporting areas, then the average costs of world agricultural production could increase significantly. These increased costs could amount to more than 10% of world agricultural GDP.

6. Response Strategies

Three broad types of adjustment strategies may be anticipated: changes in land use, changes in farm management, and changes in crop and livestock husbandry (Parry, 1990; Parry and Duinker, 1990).

6.1. *Changes in land use*

6.1.1. *Changes in farmed area*

Farmed area can be extended where environmental factors and economic incentives permit. Expansion may be most marked in Russia and northern Europe, where terrain and soils will permit further reclamation (Pitovranov, et al., 1988; Squire and Unsworth, 1988). But it may be more limited by inappropriate soils in much of Canada, with the exception of the Peace River region in northern Alberta, and parts of Ontario (Smit, 1987, 1989). There may also be potential for high-latitude reclamation of farmland in some of the valleys of central Alaska, in northern Japan, and in southern Argentina and New Zealand (Jager, 1988; Salinger et al., 1990).

Warming may also tend to induce an upward extension of farmed area in highland regions. For example, in the European Alps a 1 °C warming could raise climatic limits of cultivation by about 150 meters (Balteanu et al., 1987). However, these inclusions on the semi-natural environment may confront the extensive existing rangeland economies of mountain regions, which will come under pressure both from the upward advance of more intensive agriculture, and from afforestation

The broadscale changes in crop location imply a general poleward shift of present-day agricultural zones. This is likely to be most pronounced in mid- and high latitudes, partly because warming will be most marked here, but largely because latitudinal zoning is most evident in these regions, as a result of differences in available warmth for crop maturation.

The northern limit of maize production in the UK, for example, is at present located in the extreme north of the country. It would shift about 300 km northwards for each DC rise in mean annual temperature (Parry et al., 1989). In southern Europe, higher temperatures imply a more northerly location of present limits of citrus, olives, and vines (Imeson et al., 1987). In central North America, zones of fanning types are estimated to shift about 175 km northwards for each 1 °C of warming, resulting in reduced intensity of use in the south, but increased intensity in the north where soils and terrain permit. A sizeable area of output, currently located in the northern Great Plains of the USA, would relocate in the sourthem Canadian prairies. Other losses of Great Plains cropped acreage may be partially compensated for by increases in cultivated areas in the Great Lakes region (EPA, 1989).

Similar southward shifts of land have been suggested for the southern hemisphere, perhaps up to six degrees of latitude (670 km) under a 3-4 °C warming (Salinger et al., 1990). It should be emphasized, however, that these broadscale effects will be much influenced, at local levels, by regional variations in soils, by the competitiveness of different crops, and most importantly, by regional patterns of rainfall - none of which can be predicted at present.

6.1.2. *Changes in crop type*

In regions where there are substantial increases in the warmth of the growing season (and where output is currently limited by temperature, rather than rainfall), substitution of crops with higher thermal requirements - that would make fuller use of the extended and more intense growing season should allow higher yields. In Hokkaido (in Northern Japan), for example, the adoption of a late-maturing rice variety (at present grown in Central Japan) would enable greater advantage to be taken of the now-warmer climate, with yields increased by about 25% (Yoshino, et al., 1988).

Where moisture, rather than temperature, is the climatic constraint on output, or where increases in temperature could lead to higher rates of evapotranspiration - and thus to reduced levels of available moisture - there may occur a switch to crops with lower moisture requirements. Once again, the lack of information on likely changes in rainfall makes further speculation on this unprofitable, particularly at lower latitudes. However, there is some evidence that, at high mid-latitudes, a switch from spring to winter varieties of cereals would be one strategy for avoiding losses resulting from more frequent dry spells in the early summer. This might be the case in Scandinavia and on the Canadian prairies (Koster et al., 1987; Williams et al., 1988).

6.2. *Changes in management*

6.2.1. *Irrigation*

Substantial increases in the need for irrigation (and its costs) are likely, as it substitutes for moisture losses. The most detailed estimates yet available are for the USA, where irrigation requirements may increase by about 25% in the southern and 10% in the northern Great Plains under a 2 x CO_2 climate. Given the likely increased rate of groundwater depletion, this will probably lead to significantly higher costs of production, with consequent shifts to less water demanding uses in the most affected areas (EPA, 1989). Substantially increased irrigation needs are also projected for most of western and southern Europe (Imeson et al., 1987). In other places where available water will be reduced, although data are not available, it is probable that effects will be similar. Tighter water-management practices (Gleick, 1992, this volume) will be necessary to yield higher irrigation efficiency.

6.2.2. *Fertilizer use*

More use of fertilizers may be needed to maintain soil fertility where increases in leaching result from increased rainfall. In other regions, warming may increase productive potential to the extent that current levels of output can be achieved with substantially lower amounts of fertilizers. In Ireland, for example, fertilizer use could possibly be halved under a warming of 4°C, while maintaining present -day output (Bergthorsson et al., 1988). Much will depend on other factors: the extent that higher CO_2 levels make nutrients more limiting (thus requiring more use of fertilizers), or changes in future energy prices (and their effect on the cost of fertilizers).

6.2.3. *Control of pests and diseases*

The costs of these are likely to alter substantially, although it is quite impossible to specify them with any degree of detail. Possibly most important for global cereal production may be the costs of controlling the spread of subtropical weed species into current major cereal-producing regions (EPA, 1989).

6.2.4. *Soil drainage and control of erosion*

Adjustments in management are likely to be needed in tropical regions, particularly those characterized by monsoon rainfall, where there may be an overall increase in rainfall receipt and intensity. Recent assessments have indicated that, over the longer term, reduced soil fertility, increased salinity, and the costs of erosion control may more than offset the beneficial effects of a warmer climate, leading ultimately to reduced yields and higher production costs (Pitovranov et al., 1988).

6.2.5. *Changes in farm infrastructure*

Regional shifts of farming types and altered irrigation requirements imply major changes in capital equipment for farm layout, and in agricultural support services (credit, marketing, etc.). In the USA, it has been estimated that these will be substantial (EPA, 1989). Because of the large costs involved, only small, incremental adjustments may occur without changes in government policies

6.3. *Changes in crop and livestock husbandry*

In particular, there are likely to occur many alterations in the timing of farm operations such as tillage (ploughing, sowing, harvesting), fertilizing, and pest and weed control (spraying). The timing of these in the present farming calendar is frequently affected by climate. Particular aspects of husbandry are also likely to be affected - density of planting, the use of fallowing and mulching, and the extent of intercropping. A change of climate implies a retuning of these strategies to harmonize with the new set of climatic conditions

6.4. *Integrating the responses*

The overall response to the predicted changes in climate has to be in the form of an integrated package of long-term and short-term measures. The longer-term measures which can help to promote a balance between carbon emission and carbon absorption, including "green" energy policies and afforestation, are well known. In the short term, measures will have to be developed for each agro-ecological region, to help insulate the human and animal populations from the adverse impact of drought or floods, and impart greater stability to agricultural production.

Swaminathan (1982, 1986) has described in detail the ways in which these twin goals can be achieved. Contingency plans for planting alternate crops will have to be developed for drought and flood-prone areas. This will call for building seed reserves of alternate crops as well as alternate photo-insensitive short-duration varieties. In many areas in the tropics, seed reserves are essential for crop security, just as food grain reserves are important for food security.

The contingency plans should include optimizing production in the Most Favourable Areas (MFA) through improved input delivery systems and remunerative output pricing policies. MF As are generally those with assured irrigation facilities and with soil conditions which permit sustainable intensification. In the areas which are liable to be affected most seriously (MSA), effective public distribution systems and fodder and feed banks for animals will have to be developed. The idea is similar to that of building storm shelters along the coastal areas prone to cyclonic damage. Drinking water security for the human and animal populations is another area which needs location specific planning.

Ultimately, the most effective method of both avoidance and adaptation is public education and awareness. The Crop Weather Watch Groups recommended by Swaminathan (1982) at the village level could be a method of awareness generation as well as of micro-level monitoring of the interaction between climate factors and crop productivity.

The response strategies should include steps to diversify employment opportunities. Famines in India in the 19th Century were not famines of food, but of work. Where there is work, there is purchasing power; and where there is purchasing power, there is invariably food, except when communication is disrupted due to war or civil strife. Crop failures affect most severely societies which rely heavily on the primary sector for the security of their livelihoods. For example, a study by Resources for the Future found the greatest economic threat in the MINK (Missouri, Iowa, Nebraska and Kansas) region of the US, (where the percentage of farm income to total income is 3.4 times the percentage of farm income to total income found in the USA as a whole). Even this region may not suffer seriously due to changes in precipitation and temperature, since the share of the farm sector in the income of this region is only 3.87%. A diversification of employment opportunities can be brought about only through appropriate technologies and trade opportunities.

Unless developing countries are assisted technologically and economically to put in place appropriate response mechanisms, the sufferings of their people may increase as a result of adverse changes in climate caused by factors beyond their control.

Adaptive and anticipatory research strategies are important. The Centre for Research on Sustainable Agricultural and Rural Development, Madras (CSARD) has, for example, established a Genetic Resources Centre for Adaptation to Sea Level Rise in Tamil Nadu (India). At this centre, genotypes are being assembled to process “candidate genes” for conferring resistance to sea water intrusion and flooding

7. Conclusions

Certain steps that are necessary to avoid and/or to adapt to new climatic situations may also be important even if there is no global warming. These include improving energy efficiency, planting trees and the adoption of sustainable lifestyles. Without changes in current high-consumption lifestyles, it will be difficult to promote sustainable development.

While development action must continue, there is need for more inter-disciplinary and inter-institutional collaborative research on all aspects of climate change. The organization of an international research network on climate change and food security with the following aims would be useful in this context:

- Estimation and monitoring of the emission of greenhouse gases as a result of farm operations under different agro-ecological and technological conditions.
- Relative, as well as integrated, analysis of the agricultural impact of higher concentrations of CO_2 methane, dinitrogen oxides, and CFCs.
- Standardization of techniques - both for minimizing agriculture's contributions to the accumulation of greenhouse gases, and for withstanding additional biotic and abiotic stresses on crop and farm animals arising from climate change.

- Promotion of research and training in the field of restoration ecology, in the afforestation of degraded forests and in the sustainable management of tropical rain forests.
- Anticipatory research and development measures in coastal areas to avoid or minimize the adverse impact of potential changes in sea levels.
- Standardization of methods for deriving benefit from higher atmospheric CO_2 concentration.
- Standardization of post-harvest technologies for perishable agricultural commodities, based on non-CFC dependent refrigeration methods.
- Stimulating policy research designed to strengthen the public policy back-up both for avoiding adverse changes in climate and for adapting to new growing conditions.

To date, less than a dozen detailed regional studies have been completed that serve to assess the potential impact of climatic changes on agriculture. It should be a cause for concern that we do not, at present, know whether changes of climate are likely to increase the overall productive potential for global agriculture, or to decrease it. Neither do we know which regions are likely to gain or lose in terms of productive potential, although we have indicated what the regional level impacts might be.

Huge uncertainties remain, and the risks attached to such levels of ignorance are great. A comprehensive, international research effort is required, now, to redeem the situation. In the meantime, actions should be taken to increase the resilience of the food production system world-wide, but especially in the vulnerable areas of developing countries. Additional actions to reduce post-harvest losses, and to increase the efficiency of food processing and distribution systems makes sense today - whether or not the world is on the edge of a major global climate change.

References

Akong'a, J., T.E. Downing, N.T. Konijn, D.N. Mungai, H.R. Muturi and H.L. Potter (1988). The effects of climatic variations on agriculture in Central and Eastern Kenya. *The Impact of Climatic Variations on Agriculture, Volume 2, Assessments in Semi-Arid Regions*, Parry, M.L., Carter, T.R, and Konijn, N.T., eds., Kluwer, Dordrecht, The Netherlands.

Balteanu, D., P. Ozenda, M. Huhn, H. Kerschner, W. Tranquillini and S. Bortenschlager (1987). Impact analysis of climatic change in the central European mountain ranges, G, *European Workshop on Interrelated Bioclimatic and Land Use Changes,* Noordwijkerhout, The Netherlands, October 1987.

Beresford, R.M. and RA. Fullerton (1989). Effects of climate change on plant diseases. Submission to Climate Impacts Working Group, May 1989, 6pp.

Bergthorsson, P., H. Bjornsson, O. Dyrmundsson, B. Gudmundsson, A. Helgadottir and J.V. Jonmundsson (1988). The effects of climatic variations on agriculture in Iceland. In *The Impact of*

Climatic Variations on Agriculture, 1 Cool Temperate and Cold Regions, M.L. Parry, T.R Carter and N.T. Konijn, eds., Kluwer, Dordrecht, The Netherlands.

Blasing, T.J. and A.M. Solomon (1983). *Response of North American Corn Belt to Climatic Warming*, prepared for the US Department of Energy, Office of Energy Research, Carbon Dioxide Research Division, Washington, DC, DOE/N88-004.

Crosson, P. (1989). Greenhouse warming and climate change: why should we care? *Food Policy*, **14**(2), 107–118.

Cure, J.D. (1985). Carbon dioxide doubling responses: a crop survey. *Direct Effects of Increasing Carbon Dioxide on Vegetation*, B.R. Strain and J.D. Cure, (eds.), US DOE/ER-0238, Washington, USA, pp. 100–116.

Drummond, R.O. (1987). Economic aspects of ectoparasites of cattle in North America. Symposium, *The Economic Impact of Parasitism in Cattle*, XXIII World Veterinary Congress, Montreal, pp. 9–24.

Emanuel, K.A. (1987). The dependence of hurricane intensity on climate: mathematical simulation of the effects of tropical sea surface temperatures. *Nature*, 326, 483-485.

Environmental Protection Agency (EPA), (1989). *The Potential Effects of Global Climate Change on the United States*, Report to Congress, Washington, DC.

Gifford, RM. (1979) *Austral. J. Plant Physiol.*, **6**, 367–378. Havelka, U.D., V.A. Wittenbach and M.G. Boyle, (1984). *Crop Sci.* **24**, 1146–1150.

Hill, M.G. and J.J. Dymock (1989). *Impact of Climate Change: Agricultural/Horticultural Systems*. DSIR Entomology Division, submission to New Zealand Climate Change Programme, Department of Scientific and Industrial Research, New Zealand, 16pp.

Imeson, A., H. Dumont and S. Sekliziotis (1987). *Impact Analysis of Climate Change in the Mediterranean Region,* F, *European Workshop on Interrelated Bioclimatic and Land Use Changes*, Noordwijkerhout, The Netherlands.

Intergovernmental Panel on Climate Change (IPCC), (1990). *The IPCC Scientific Assessment of Climate Change: Policymakers Summary*, Geneva and Nairobi: WMO and UNEP.

Jager, J. (1988). *Developing Policies for Responding to Climate Change*, Report of the Workshops in Villedi, Austria and Bellaggio, Hal, Stockholm Environment Institute, Stockholm, Sweden.

Kaida, Y. and V. Surarerks (1984). Climate and agricultural land use in Thailand. In *Climate and Agricultural Land Use in Monsoon Asia.* M.M. Yoshino, (ed.), University of Tokyo Press Tokyo, pp. 231–253.

Kettunen, L., J. Mukula, V. Pohjonen, O. Rantanen and U. Varjo (1988). The effects of climatic variations on agriculture in Finland. *The Impact of Climate Variations on Agriculture: 1, Assessments in Cool Temperature and Cold Regions.* M.L. Parry, T.R. Carter and N.T. Konijn, eds. Kluwer, Dordrecht, The Netherlands, pp. 511–614.

Kimball, B.A. (1985). *Direct Effects of Increasing Carbondioxide on Vegetation*, B.A. Strain and J.D. Cure, eds., National Technical Information Service, US Dept. Commerce. Springfield, Virginia, pp. 185–204.

Koster, E.A., E. Dahl, H. Lundberg, R. Heinonen, L. Koutaniemi, B.W.R. Torssell, R. Heino, J.E. Landmark, L. Stronquist, T. Ingelog, L. Jonasson and E. Eriksson (1987). *Impact Analysis of Climatic Change in the Fennoscandian Part of the Boreal and Sub-Arctic Zone,* D. European Workshop on International Bioclimatic and Land Use Changes, noordwijkerhout, the Netherlands, October 1987.

Mearns, L.O., R.W. Katz and S.H. Schneider (1984). Extreme high temperature events: changes in their probabilities with changes in mean temperatures. *Journal of Climatic and Applied Meteorology*, **23**, 1601–1613.

Newman, J.E. (1980). Climate change impacts on the growing season of the North American Corn Belt, *Biometeorology* **7**(2), 128–142.

Parry, M.L. (1976). *Climatic Change, Agriculture and Settlement.* Dawson, Folkestone, England.

Parry, M.L. (1990). *Climate Change and World Agriculture* Earthscan London, 165 pp.

Parry, M.L. and T.R. Carter (1988). The assessments of the effects of climatic variations on agriculture: aims, methods and summary of results. *The Impact of Climatic Variations on Agriculture, 1, Assessments in Cool Temperate and Cold Regions*, M.L. Parry, T.R. Carter and N.T. Konijn, (eds.), Kluwer, Dordrecht, The Netherlands.

Parry, M.L. and P.N. Duinker (1990). *The Potential Effects of Climatic Change on Agricultur. In Intergovernmental Panel on Climate Change* (1990) the IPCC Impacts Assessment. WMO and UNEP, Geneva, Switzerland, pp. 2-1–2-45.

Parry, M.L. and S.K. Sinha (1988). Food Security, Working Group Report. In Conference Proceedings, *The Changing Atmosphere: Implications for Global Security.* Toronto, Canada, 27-30 June, WMO-No, 170, World Meteorological Organisation, pp. 321-3.

Parry, M.L., T.R. Carter and J.H. Porter (1989). The greenhouse effect and the future of UK agriculture. *Journal of the Royal Agricultural Society of England*, pp. 120–131.

Pearman, G.I. (1988). *Greenhouse: Planning for Climate Change* Melbourne, Australia: CSIRO.

Pitovranov, S.E., V. Iakimets, V.E. Kislev and O.D. Sirotenko (1988). The effects of climatic variations on agriculture in the subarctic zone of the USSR. *The Impact of Climatic Variations on Agriculture: Volume I, Assessments in Cool, Temperate and Cold Regions*, M.L. Parry, T.R. Carter and N.T. Konijn, (eds.), Kluwer, Dordrecht, The Netherlands.

Pittock, A.B. (1989). *Potential Impacts of Climatic Change on Agriculture, Forestry and Land Use*, personal communication, pp. 5.

Porter, J.H., M.L. Parry and T.R. Carter (1992). The Potential Effects of Climatic Change on Agricultural Insect Pests. *Agricultural and Forest Meteorology*, in press.

Rind, D., R. Goldberg and R. Ruedy (1989). Change in climate variability in the 21st Century. *Climatic Change*, **14**, 5–37.

Rogers, H.H., J.F. Thomas and G.E. Bingham (1983). Response of Agronomic and Forest Species to Elevated Atmospheric Carbon-Dioxide. *Science*, **220**, 428–429.

Rosenzweig, C. (1985). Potential CO_2-Induced Climate Effects on North American Wheat-Producing Regions. *Climatic Change*, **7**, 367–389.

Salinger, M.J. (1989).. *The Effects of Greenhouse Gas Warming on Forestry and Agriculture.* Draft report for WMO Commission of Agrometeorology, 20 pp.

Salinger, M.J., W.M. Williams, J.M. Williams and R.J. Martin (eds.), (1987, 1988, 1990). *Carbon Dioxide and Climate Change: Impacts on Agriculture.* New Zealand: New Zealand Meteorological Service; DSIR Grasslands Division; MAFTech, 1990.

Schelling, T. (1983). Climate Change: Implications for Welfare and Policy. *Changing Climate: Report of the Carbon Dioxide Assessment Committee*, Washington, DC: National Academy of Sciences.

Schulze, R.E. (1989). *Hydrological Responses to Long-Term Climatic Change*, Personal Communication, 9 pp.

Sinha, S.K. and M.S. Swaminathan (1991). Deforestation, Climate Change and Sustainable Nutrition Security. *Climatic Change.*

Sinha, S.K., N.H. Rao and M.S. Swaminathan (1988). Food Security in the Changing Global Climate. Conference Proceedings, *The Changing Atmosphere: Implications for Global Security,*

Toronto, Canada, 27–30 June, 1988, WMO-No. 170, World Meteorological Organisation, 1988, pp. 167–192.

Smit, B. (1987). Implications of Climatic Change for Agriculture in Ontario. *Climate Change Digest,* CCD 87-02, Downsview, Ontario, Environment Canada.

Smit, B. (1989). Climate Warming and Canada's Comparative Position in Agriculture. *Climate Change Digest*, CCD89-01. Downsview, Ontario, Environment Canada.

Smit, B., M. Brklacich, R.B. Stewart, R. McBride, M. Brown and D. Bond (1989). Sensitivity of crop yields and land resource potential to climate change in Ontario. *Climatic Change*, **14**, 153–74.

Squire, G.R. and M.H. Unsworth (1988). Effects of CO_2 and Climatic Change on Agriculture, *Contract Report to the Department of the Environment.* Sutton Bonnington, UK, Department of Physiology and Environmental Science, University of Nottingham.

Swaminathan, M.S. (1982). *Science and Integrated Rural Development.* Concept Publishing Company, New Delhi. 354 pp.

Swaminathan, M.S. (1986). Building national and global food security systems. In *Global Aspects of Food Production,* M.S. Swaminathan and S.K. Sinha, (eds.), Tycooly Press, pp. 417–449.

Terjung, W.H., H.-Y. Ji, J.T. Hayes, P.A. O'Rourke and P.E. Todhunter (1989). Actual and potential yield for rainfed and irrigated maize in China. *International Journal of Biometeorology*, **28**, 115–135.

Van Diepen, C.A., H. Van Keulen, F.W.T. Penning de Vries, I.G.A.M. Noy and J. Goudriaan (1987). Simulated variability of wheat and rice yields in current weather conditions and in future weather when ambient CO_2 has doubled. In *Simulation Reports CABO-TT*, 14., Wageningen, The Netherlands: University of Wageningen.

Williams, G.D.V. and W.T. Oakes (1978). Climatic resources for maturing barley and wheat in Canada. In *Essays on Meteorology and Climatology*: in Honour of Richard W. Longley, Studies in Geography Mono.3., 367–385, Haye, K.D. and E.R. Reinelt, (eds.), University of Alberta, Edmonton.

Williams, G.D.V., R.A. Fautley, K.H. Jones, R.B. Steward and E.E. Wheaton (1988). Estimating effects of climatic change on agriculture in Sasketchewan, Canada. In *The Impact of Climatic Variations on Agriculture: 1, Assessments in Cool Temperate and Cold Regions*, M.L. Parry, T.R. Carter and N.T. Konijn, (eds.), pp. 219–379, Kluwer, Dordrecht, The Netherlands.

Wittwer, S.H. (1985). Carbon Dioxide Levels in the Biosphere Effects on Plant Productivity. *Crit. Rev. Plant. Sci.*, **2**(3), 171–198.

Yoshino, M.M. (1984). Ecoclimatic Systems and Aagricultural Land Use in Monsoon Asia. *Climate and Agricultural Land Use in Monsoon Asia,* M.M. Yoshino, ed., University of Tokyo Press, Tokyo, pp. 81–108.

Yoshino, M., T. Horie, H. Seino, H. Tsujii, T. Uchijima and Z. Uchijima (1988). The Effects of Climatic Variations on Agriculture in Japan. *The Impact of Climatic Variations on Agriculture, 1, Assessments in Cool Temperate and Cold Regions.* Dordrect, The Netherlands, Kluwer, pp. 725–868.

Sustainable Food Security in Africa: Lessons from India's Green Revolution

M.S. Swaminathan
M. S. Swaminathan Research Foundation, Chennai, India

Swaminathan, M.S. (2003). Sustainable Food Security in Africa: Lessons from India's Green Revolution. *South African Journal of International Affairs*. **10**(1): 11–26

1. Introduction

Forty years ago, food experts had predicted serious food famines in Asia and Africa. In their famous book, *Famine 1975*, Paul and William Paddock (1967) characterized India as a hopeless case and predicted that it could not be saved from hunger and famine. Today, India has about 60 million tonnes of grain stocks, which include the quantities required for maintaining a minimum food security reserve and for supplying the vast public distribution network of fair price shops. China, Vietnam, Bangladesh, Pakistan, Indonesia and several other Asian countries, also regarded as nations unable to feed themselves, have all made spectacular progress.

Ironically, when *Famine – 1975* was published Africa was not considered to have a food problem. However, sub-Saharan Africa, formerly a food-exporting region, became a net food importer in the 1970s. Currently, the rate of growth of net food imports is still rising.

Since India was in the same position 35 years ago, it would be useful to recount how the scenario changed from "a begging bowl to a bread basket" situation. The severe famine in Bengal in 1942-43 resulted in nearly three million deaths and prompted Jawaharlal Nehru, the first Prime Minister of Independent India, to remark, "Everything else can wait, but not agriculture". This priority was reflected in land reform expansion of irrigation facilities, and greater support for research and the production of inputs, particularly seeds and fertilisers.

Fertiliser trials with wheat and rice in the 1950s revealed that the varieties then cultivated lodge (fall over) when about 20 kg of nitrogen is applied per hectare. As well as making the crop difficult to harvest, lodging also makes it difficult to provide irrigation during grain development, when the crop benefits most from water availability. Thus, with these varieties, it proved difficult to get an enhanced response to additional nutrients and water.

India's search for crops that could respond to higher levels of nutrition started in 1952, when a programme was initiated for crossing fertiliser-responsive *japonica* rice with standard *indica* strains. The goal was to select lines with the ability to use about 100 kg of nitrogen, producing about five tonnes of rice, per hectare. Several genetic problems including semi-sterility arose, rendering the search difficult. With the advent in the early 1960s of semi-dwarf, non-lodging varieties, initially from Taiwan and subsequently from the International Rice Research Institute (IRRI) in the Philippines interest in *indica-japonica* crosses waned, and semi-dwarfed indica rices provided the material for breeding high-yielding varieties.

When I joined the Indian Agricultural Research Institute (IARI) in late 1954, I started a research programme for developing non-lodging and fertilizer responsive varieties of wheat, on the lines of the work earlier initiated in rice. At that time the research strategy adopted had three components. First, crosses were made between the semi-dwarf, stiff straw *compactum* and *sphaerococcum* sub-species of T.*aestivum* with cultivated bread wheat. Second, attempts were made to induce *erectoides* mutants in commercial wheat varieties through the use of radiations and chemical mutagens. Third, studies on the potential for increasing straw stiffness through different chemical treatments were initiated. Unfortunately, in all these three approaches, short and stiff straw was always associated with short panicles with fewer grains. The reason why straw stiffness became such an essential prerequisite for favourable response to water and fertilizer was the tendency, among the then cultivated tall wheat varieties, to lodge when fertilizer was applied. Also, as earlier mentioned, such lodging made it difficult to give an irrigation during the grain development phase, when the crop will benefit much from water availability. Thus, with the earlier tall varieties, it was difficult to get economic response to the application of mineral fertilizers and adequate irrigation water. In wheat, yields stagnated at less than one tonne per ha. This is why the breeding of non-lodging varieties was accorded such high priority during the fifties, when India had taken to the path of expanding the area under irrigation and manufacturing mineral fertilizers (Swaminathan, 1993).

During the late fifties, scientific publications on the work done under the leadership of Dr Orville Vogel in Washington State of the United States of America, on the transfer of dwarfing genes from the Japanese Norin-10 wheat variety to North American winter wheats, started appearing. When requested, Dr Vogel was kind enough to send seeds of Gaines, a semi-dwarf winter wheat variety with red grains. He further indicated that Dr N E Borlaug in Mexico had semi-dwarf varieties in a spring wheat background, which will grow better under the short day conditions prevailing in India during winter months.

In March 1962, a few dwarf spring wheat strains entered by Dr Borlaug in the International Wheat Rust Nursery organised by the U S Department of Agriculture were grown in the fields of the Indian Agricultural Research Institute (IARI). Their phenotype was most impressive. They had reduced height and long panicles, unlike the earlier hybrids between *aestivum* and *compactum* and *sphaerococcum* and the induced *erectoides* mutants which had short height and small panicles. In 1963, Dr Borlaug sent a wide range of semi-dwarf material which provided the initial germplasm for stimulating an accelerated advance in wheat productivity and production.

In 1964, a National Demonstration Programme was started in farmers' fields, both to verify the results obtained in research plots and to introduce farmers to the new opportunities opened up by semi-dwarf varieties for improving very considerably the productivity of wheat. When small farmers, who with the help of scientists organised the National Demonstration Programme, harvested over five tonnes of wheat per hectare, its impact on the minds of other farmers was electric. The clamour for seeds began and the area under high yielding varieties of wheat rose from four ha in 1963-64 to over four million ha in 1971-72. A small Government programme became a mass movement.

Wheat production in India rose from 10 million tonnes in 1964 to 17 million tonnes in 1968. In 1968, the Government of India released a stamp titled "Wheat Revolution" to commemorate the significant progress then being made in enhancing the production of wheat (Fig.1,2 &3). Similar results were obtained with semi-dwarf varieties of rice (Fig.4). In 1968, Dr William Gaud of the United States coined the term "Green Revolution" to stress that what was happening in the wheat and rice fields of Asia was revolutionary and not just evolutionary progress.

2. Agriculture in the 21st Century

At the dawn of the 21st century, we can look back with pride and satisfaction on the revolution which farm men and women brought about in our agricultural history during the 20th century. Writing about the role of Indian farm families in initiating the Wheat Revolution, I wrote in the Illustrated Weekly of India in 1969 and I quote:

"Brimming with enthusiasm, hard-working, skilled and determined, the Punjab farmer has been the backbone of the revolution. Revolutions are usually associated with the young, but in this revolution, age has been no obstacle to participation. Farmers, young and old, educated and uneducated, have easily taken to the new agronomy. It has been heart-warming to see young college graduates, retired officials, ex-armymen, illiterate peasants and small farmers queuing up to get the new seeds. At least in the Punjab, the divorce between intellect and labour, which has been the bane of Indian agriculture, is vanishing."

While we can and should rejoice about the past achievements of farmers, scientists, extension workers and policy makers, there is no room for complacency. We will face several new problems, of which the following are important.

- First, increasing population leads to increased demand for food and reduced per capita availability of arable land and irrigation water.
- Second, improved purchasing power and increased urbanisation lead to higher per capita food grain requirements due to an increased consumption of animal products.
- Third, marine fish production is tending to become stagnant and coastal aquaculture has resulted in ecological and social problems.
- Four, there is increasing damage to the ecological foundations of agriculture, such as land, water, forests, biodiversity and the atmosphere and there are distinct possibilities for adverse changes in climate and sea level.
- Five, while dramatic new technological developments are taking place, particularly in the field of biotechnology, their environmental, food safety and social implications are yet to be fully understood.
- Finally, gross capital formation in agriculture is tending to decline in both public and private sectors during the present decade. The rate of growth in rural non-farm employment has been poor

Since land and water will be shrinking resources for agriculture, there is no option in the future except to produce more food and other agricultural commodities from less per capita arable land and irrigation water. In other words, the need for more food has to be met through higher yields per units of land, water, energy and time. It would therefore be useful to examine how science can be mobilised for raising further the ceiling to biological productivity without associated ecological harm. It will be appropriate to refer to the emerging scientific progress on the farms as an "*ever-green revolution*", to emphasise that the productivity advance is sustainable overtime since it is rooted in the principles of ecology, economics, social and gender equity and employment generation.

3. The African Challenge

The tasks ahead, are formidable. In the mid-1980s, sub-Saharan Africa was the only major region of the world whose population growth rates were still increasing while per capita food production was declining. The rate of rural to urban migration has been high, with cities growing at around 6 percent per annum in the 1960s and 1970s (World Bank, 1981, p.179). As has been stressed in several reports of the United Nations Environemnt Programme (UNEP), rising human populations, with expanded food needs, put great pressure on the stabilizing elements in ecosystem structures such as perennial vegetation, stable water table, adequate stream flow and fertile soil profiles. When drought strikes, such overworked land is unable to protect itself. Prolonged drought is not the only cause of desertification, however. Deforestation, overgrazing, overcultivation, poor irrigation and unscientific land-use practices are also important contributors. Desertification has continued to spread at the same rate as in 1977, when a UN conference held at Nairobi developed a plan to combat it, and now threatens 40 percent of the continent's land surface.

The fragility of soils, diversity of crops and pests, variability of climate and scarcity of water characteristic of many countries south of the Sahara all make it imperative to develop ecologically sound technologies adapted to local socio-economic conditions. A conservative estimate indicates that 75 percent of the food grown and eatern in Africa is produced predominantly by women on small landholdings of less than 2 hectares. Unfortunately, so far, most extension and development projects have bypassed women.

Experts agree that hunger is primarily a rural phenomenon in Africa. However, malnutrition is also rising in urban areas because of the unplanned migration of the rural poor to towns and cities. The relation between production and consumption often is quite direct, much more so in most of Africa than in Asia, where food production affects the poor primarily through its effects on rural employment and prices. We therefore need a strategy that will deal with the problems of production, trade and consumption in an integrated manner.

The first step in the strategy is to develop an agricultural balance sheet in each country, indicating national strengths and weaknesses. The weaknesses of the production base in many African countries have been pointed out in numerous seminars, symposia and publications. For example, 44 percent of Africa's land is desert, arid or semi-arid, 18 percent has low fertility and 22 percent has shallow depth and excess water and several health problems arise from the incidence of vector-borne diseases such as schistosomiasis and malaria. Compounding these problems is the spread of HIV/AIDS.

Only six countries account for 70 percent of all irrigated land in Africa. As would be expected, the percentage of temporary and permanent cropland under irrigation varies tremendously from one African nation to another.

With the exception of Sudan, Madagascar and Nigeria, irrigated agriculture plays a minor role in the economies of sub-Saharan countries, whose water potential is very limited. For example, countries like Angola, Central African Republic, Gabon and Zambia had less than 0.3 percent of cropland under irrigation in 1982. Of the total area of modern irrigation developed in sub-Saharan Africa upto the eighties – 2.64 million hectares – Sudan alone, with 1.7 million hectares, accounts for nearly two thirds. Similarly, of the total area of small-scale or traditional irrigation in the region – 2.38 million hectares – Madagascar and Nigeria (with 800,000 hectares each) account for slightly over two thirds.

Deforestation and over-grazing are serious problems in many parts of Africa. Soil and water conservation and management and afforestation programs clearly deserve the highest priority. Integrated programmes of water harvesting, watershed development and wasteland reclamation are urgently needed.

In many respects, the Indian agricultural scene is different from that of Africa. Monsoon India has a reasonably good rainfall. The scope for groundwater use is considerable, and crops like rice and wheat occupy large areas. The soil in general is fertile, and the rural infrastructure is developed. Marketing is better organized, and a balance between producer and consumer policies has been achieved.

4. Food Production Strategies: Lessons from Indian Experience

To promote food production on a sustainable basis, we need mutually supportive packages of technology, services and public policies. I will deal with each of them briefly, based on the experience gained in India.

4.1. *Package of Technology*

Farming systems and practices have evolved over centuries, and what are usually referred to as traditional faring systems have evolved through observation and experience to suit specific agroecological conditions. Such systems, however, are generally not intended to raise crops or farm animals for market. When the process of modernization sets in, the principal catalyst is opportunity for remunerative marketing. When food crops are grown for market, the distinction between them and traditional cash crops disappears. For example, in several parts of India, wheat or rice is the main cash crop for numerous farming families with small holdings.

Market-oriented agriculture succeeds only with a high degree of efficiency both at the production and post-harvest stages. The goal is to achieve the highest yield possible per unit of land, water, time and labour. This is where improved technologies become essential.

During the first 20 years after India's independence, production advances were achieved largely by increasing the area under cultivation. With the pressure of a growing population, this pathway of increasing production could only lead to a dead end. Improved productivity and intensified cropping became necessary to produce the food needed for a growing population.

One of the first tasks, therefore was to build a scientific infrastructure to stimulate and sustain rapid agricultural advance. Agriculture is, by and large, a location-specific vocation; hence, a dynamic national research system is a must for sustaining a dynamic production program.

India is fortunate to have an excellent network of agricultural research institutes and universities. At the national level, the Indian Council of Agricultural Research (ICAR) supports and coordinates scientific research, training and extension education in crop husbandry, horticulture, animal husbandry, fisheries and agroforestry. ICAR is a unique organization in that it has concurrent responsibility for both research and education. It supports over 40 Agricultural and Veterinary Universities. Unfortunately, in most developing countries the responsibilities for research, extension and education are divided among different agencies and ministries.

An important element in developing strong research organizations is personnel policies that are designed to attract and retain dedicated and high-quality staff. ICAR introduced an agricultural research service (ARS) in 1974 to encourage specialization and to provide job security and opportunities for promotion, even without the occurrence of vacancies. A National Academy of Agricultural Research Management was set up in Hyderabad to provide opportunities for inservice training to the members of ARS as well as to promote a scientific culture within the organization. A National Academy of Agricultural Sciences also exits, which provides from time to time policy guidance in crucial areas, such as land and water conservtion and management.

Research programs in India are carried out in a large number of central and state research institutes and national bureaus such as the National Bureau of Soil Survey and Land Use Planning and the National Bureau of Plant Genetic Resources. Linking the agricultural universities and central institutes are the All-India Coordinated Research Projects. These projects bring together scientists working in different institutions and disciplines in a symbiotic partnership. **The individual strengths of the different research institutes may vary, but the collective strengths of All-India Coordinated Research Projects are considerable.**

Once improved technologies are developed by scientists, the experimental findings are verified in farmers' fields before they are recommended for widespread adoption. ICAR has developed the following methods of demonstrating and verifying new experimental findings:

(1) The "lab to land" and "land to lab" programmes, which aims at extending new experimental findings to small farmers;
(2) National demonstrations, which are designed to show that small farmers can achieve high yields; and

(3) Whole-village or watershed operational research projects, which can help identify the major constraints responsible for the gap between potential and actual farm yield and income on the basis of different agro-climatic zones.

Since illiteracy is still widespread in rural India, ICAR has organized a network of Farm Science Centres (Krishi Vigyan Kendra) all over the country for disseminating the latest technical skills through a learning-by-doing scheme. Such a **technicracy** (i.e. technical literacy) programme has helped to reach the unreached in terms of skill and knowledge empowerment.

Agricultural research institutions and universities have identified themselves with the farmers, who often spend many hours visiting the experimental fields and discussing with scientists problems of mutual interest.

Technology development has been based not only on individual factors of production but also on farming systems as a whole. In summary, ICAR has been able to achieve the following:

(1) A close integration of research, training and extension education;
(2) Personnel policies that attract and retain high-quality staffs and that promote life-long specialization in specific areas;
(3) A national grid of cooperative experiments bringing together scientists from different disciplines in a working partnership to achieve specific research goals;
(4) Close linkages between research and development agencies; and
(5) Farmers as partners in the testing and refinement of technologies.
(6) Linking production, processing and marketing in the form of an integrated system.

Let me give two examples of how the research system responded to a specific challenge. After India's independence in 1947, fertilizer factories were established and production began, both in the private and public sectors. Fertilizer use in farmers' wheat fields was started in the 1950s, using low doses of about 20 kilograms of nitrogen per hectare. Studies showed that there was no economic response to this low level of nitrogen application. Hence, in the early 1960s, research on the breeding of varieties that could respond well to water and fertilizer application was begun. The semidwarf wheat varieties from Mexico, introduced through the help of Dr N E Borlaug, provided suitable strains that could respond well to good soil fertility and water management.

A National Demonstration Program was started in 1964-65 to illustrate new opportunities for improving productivity by using management-responsive varieties. A small government program was soon converted into a mass movement by farmers, and, as a result, the area planted to semidwarf varieties increased from about 4 hectares in 1964 to about 4 million hectares in 1971. From a peak production of about 12 million tonnes yielding varieties in 1964-65, wheat production has now reached a level of 80 million tonnes.

India's progress in wheat production is well-known. However, it is important to realize that such rapid progress was possible only because of the existence of a dynamic national research system with the capacity to derive maximum benefit from international research centers like the International Centre for Maize and Wheat Improvement (CIMMYT) in Mexico.

The second example is of recent developments in rice production, which accounts for about 40 percent of the total food grain production in India. Like grain production, rice production was stagnant until the mid-1960s, when new genetic strains were introduced that were able to respond to good soil fertility and water management. Rice is predominantly a crop of the southwest monsoon period (May to October) when problems of pest and water control make it difficult to increase yield. Fortunately, the All-India Coordinated Rice Research Project of ICAR, helped to provide location-specific varieties and technologies for different parts of the country. Because of new technologies, non-traditional rice areas, like the Punjab, and non-traditional seasons became important in rice production. The government's provision of adequate energy to pump water from tubewells has enabled the Punjab to maintain high rice yields even in severe drought.

To sustain yields at high and stable levels, effective monitoring and early warning systems are needed, particularly with regard to soil conditions and plant and animal health. In agriculture, environmental conditions are dynamic and constant vigilance is necessary. To cite an example, there was virtually no economic response to phosphorus application in the Punjab in the 1960s. In the early 1970s, however, phosphorus application because essential to obtain substantial returns from applied nitrogen. In a few years, zinc deficiency became widespread. Thus, when farming practices shift from low to high productivity levels, new problems of plant and soil health may arise, making a location-specific research capability vital.

The national goal of agricultural research is the improvement of the productivity, profitability, stability and sustainability of a country's major farming systems. Sustainability has to be viewed from the ecological as well as the social and economic angles. We cannot allow depreciation of basic agricultural assets like land, water, flora and fauna. While ecological sustainability is fundamental to the future of agriculture, economic sustainability is vital to generate the interest of farmers in increased production. At the same time, social equity both in gender and economic terms, is important to avoid social disruption. An effective national agricultural research system should look at all aspects of the spectrum. Improving productivity without detriment to the long-term production potential of the soil and creating economic viability that results in a satisfactory "take-home income" for farmers are both essential. To succeed in achieving the triple goals of ecological, economic and social sustainability, biological and social scientists work together as a team.

4.2. *Package of Services*

Applied research is no different from "ivory tower" research if the resulting new technologies are not transferred to the farmers' fields. In most countries of South and Southeast Asia, where the average size of a farm holding is usually less than one hectare, the cooperation and assistance of atleast a million farming families is needed to produce an extra million tonnes as grain. No production target can be achieved without first asking the question, "**why should farmers produce more, and how will they do it**?".

If this question is asked at the outset of a development project, priorities can be established to ensure that services that enable all farmers to take advantage of new technologies are provided. In India, both government and private agencies have been active in providing inputs like seeds, fertilizers, irrigation water, pesticides and, very importantly, credit. Government policies in the development of effective input supply services have evolved over a period of time. Public sector companies like the National Seeds Corporation and credit institutions like the National Bank for Agriculture and Rural Development have become very important instruments in India for enabling small farmers to adopt new technologies. Credit cards, known as "Kisan Credit Cards" have been issued to millions of resource poor farmers.

The mass media, particularly radio and local language newspapers, have been extremely important in the dissemination of agricultural information. Television increasingly is becoming a powerful communication tool. Extension services have grown in capacity and effectiveness.

The major goal of the input supply system is to render new technology accessible to all farmers. Ensuring equality of access to appropriate technology should be the foundation of agricultural extension and development planning. Knowledge and skill transfer systems and methods of providing inputs needed to convert knowledge into field accomplishment must be synchronized in space and time. Input delivery systems must be tailored to specific sociocultural conditions.

In India, several innovative systems of input delivery to illiterate farmers have been developed by both governmental and non-governmental agencies. For example, credit fairs have been organized to eliminate red tape and corruption and to ensure that credit is utilized for the intended purpose. The goals of a credit fair are the following:

(1) Identify a well-designed production or input supply program such as the construction of tubewells, farm ponds or other minor irrigation elements, or the supply of fertilizers, seeds and other production inputs.
(2) Specify the criteria for credit eligibility, including the amount of credit and conditions under which it will be made available.
(3) Organize a tie-in between credit supply and the supply of inputs for which the credit is intended.
(4) Announce three to four weeks before the date of the fair the details of the program, including the location and conditions of eligibility.
(5) Organize teams of fair managers consisting of representatives from the banks and input supply agencies. They should have a short training program to prepare them to supervise the smooth operation of credit and input supply. In addition to the regular extension personnel, volunteers from agricultural universities and institutions could be mobilized to ensure that the program is carried out efficiently and effectively.

In 1975, the National Commission on Agriculture of India issued a comprehensive report, reviewing the progress in Indian agriculture since 1927 and suggesting methods of meeting the future challenges. It recommended the organization of country-wide farmers' service societies to provide farmers with credit and the inputs for which the credit is intended, all under the same roof

(Government of India, 1976, pp. 254-83). Such "single window" services enable farm women and men save time and get all the needed services at onc place and in one visit. Farm women, in particular, are particularly "time poor" owing to the multiple burden on their time, arising from home keeping, child rearing and economic activities.

In recent years, the availability of key inputs in remote areas has been monitored rigorously. A checklist of all essential activities is verified at the village level during a specified two-week period, usually several weeks before the date of sowing. This system provides early warning of potential critical input supply bottlenecks so they can be removed before the onset of the sowing season.

4.3. *Package of Government Policies*

Even if good technologies are available and the input supply system is efficient, farmers will be able to derive benefit from them only if some basic steps are taken by government. The first area of action is in the realm of land reform. Farmers need to have a long term stake in the land to make them invest in the infrastructure essential for sustained productivity. Security of tenure, land ownership patterns and the size of the farmholdings need particular attention. Under conditions in Africa, livestock reform is also important, so that the cattle population is within the population supporting capacity of the ecosystem.

In the past, farmers based land-use decisions largely on the needs of both the family and the immediate neighborhood. With the modernization of agriculture, farmers produce food grains and other commodities not only for themselves, but, more importantly, for the market. When this transition takes place, opportunities for producer-oriented and remunerative marketing become an important factor in sustaining and stimulating farmers' interest in modern technology. The smaller the farm, the greater, is the need for marketable surplus, so that some cash income becomes available.

Input-output pricing policies become crucial. The turning point in Indian agriculture took place in 1964, when the government set up a Prices Commission to recommend minimum prices for food grains in order to provide an incentive to farmers and a Food Corporation to assure that farmers received those prices. The government has honoured its commitment to farmers by purchasing all surplus grain offered to government agencies until today. For this reason, the government now has fairly substantial grain reserves, now exceeding 50 million tonnes.

4.4. *Symphonic and Knowledge-Intensive Production Systems*

The term "symphonic agriculture" designates the evolutionary stage reached in the development of sustainable agricultural production systems when all of the components of an agricultural action plan become mutually reinforcing. When synergetic packages of technology, services and public policies are developed and introduced in a mutually supportive manner, agricultural progress is rapid. Further, government investment policies ultimately determine the fate of rural professions. Without adequate rural infrastructure such as roads, warehouses, electricity and facilities for education and health care, rural areas will not attract technically qualified people. In fact, the migration of educated and well-trained people from villages to towns is the most serious form of brain drain in most developing countries. In India, as in most developing countries, youth below the age of 21 constitute the majority of the population. Thus, the future of agriculture and of rural professions will depend greatly on the government's ability to make agriculture both economically and intellectually attractive. This calls for greater effort in developing and popularising knowledge-intensive production systems. At the same time, the minimum essential living facilities and the opportunity to acquire consumer goods must be provided in rural areas.

Essentially the message from the Indian experience is the following

(1) Develop a strong national research capability in agriculture that can help to provide optimum returns from the land, water, livestock and capital resources.
(2) Develop methods of transferring know-how and skills to farmers and provide them with the inputs necessary for converting technological advances into production gains.
(3) Introduce agrarian reform, rural development and communication concepts, input-out pricing policies and other programs that can stimulate and sustain the growth of market-oriented farming.
(4) Stimulate consumption by the rural and urban poor through various measures including a food-for-work and employment guarantee projects.
(5) Consider agriculture not just as a means of producing food for the urban population but also as a powerful instrument for increasing income and employment. Such an approach is essential to enhance the purchasing power of the rural population. This calls for a labor-friendly approach in technology development and transfer.

These components of an agricultural action plan are the basis for a sustainable agricultural production system.

4.5. *Sustainable Nutrition Security*

Africa is a continent with a wide diversity of agricultural conditions and accomplishments. At one end of the spectrum are countries like Egypt, which has a production program, Mabrouk-4, that aims to produce 4 tonnes of rice per acre (10 tonnes per hectare) (International Rice Research Institute, 1986, p.84). At the other end of the spectrum, several countries in the Sahel region produce scarcely 500 kilograms per hectare of sorghum or millet. In contrast, American farmers obtain a yield of 8 to 10 tonnes per hectare.

For Africa to develop sustainable nutrition security will require that each country develop an integrated program designed for its own particular conditions and requirements and supported by government policies. The term "nutrition security" is used in this context (rather than "food security" or "sustainable agricultural production") because it helps to focus attention on all the components essential for giving every humanbeing an opportunity to express his or her innate genetic potential. The following are the major components of a sustainable national nutrition security system for Africa (Swaminathan, 1984, p.35):

(1) *Food Security*, resulting in adequacy and stability of food supplies and economic and physical access to food, as described by FAO.
(2) *Drinking water security*, resulting in a reduction in intestinal infection and in the incidence of such diseases as diarrhea and hepatitis.
(3) *Minimum income security*, giving to all human beings the requisite purchasing power for balanced nutrition. Depending upon needs and possibilities, steps to ensure minimum income security may take the form of food-for-work and food-for-development programs for unskilled and skilled labour; crop and livestock insurance for small farmers and sharecroppers; midday meal and supplementary feeding programs for children; and various other social security measures for economically underprivileged farmers and consumers. The rights to food and to work must become correlated rights. Developing countries cannot afford the advanced social security measures adopted by developed countries. Nevertheless, they have to realize that freedom from hunger can be achieved only by winning the battle against unemployment.
(4) *Nutrition intervention*, leading to the protection of old and infirm persons, preschool children and pregnant and nursing mothers from undernutrition and malnutrition. Nutrition intervention programs may also take the form of oral administration of vitamin A and other special programs to cater to specific needs.
(5) *Nutrition education*, culminating in a widespread understanding that every family and every local community can and should solve the problems of malnutrition affecting them through a malady-remedy analysis.
(6) *Population stabilization*, leading to a proper balance between population growth and resource availability, which is essential for improving the quality of life, including nutrition. Hence, population education and stabilization programs must become an integral part of a national nutrition security action plan.

5. Conclusion

Agriculture is the backbone of the ecological and livelihood security systems of most developing countries. Also, agricultural progress is the best safety net against hunger and deprivation since it touches the lives of a majority of the population. Arable land and irrigation water are however shrinking resources for agriculture. Therefore, there is need to raise, productivity in perpetuity without ecological or social harm, a pathway which I termed over 10 years ago as "ever-green revolution". Unlike the green revolution technologies, which are commodity centred, the ever-green revolution technology is based on the conservation and enhancement of natural resources (Fig.5). I will like to conclude with a statement made by E O Wilson (2002) in relation to my concept of ever-green revolution,

"The problem before us is how to feed billions of new mouths over the next several decades and save the rest of life at the same time, without being trapped in a Faustian bargain that threatens freedom and security. No one knows the exact solution to this dilemma. ***The benefit must come from an evergreen revolution****. The aim of this new thrust is to lift food production well above the level obtained by the green revolution of the 1960s, using technology and regulatory policy more advanced and even safer than those now in existence."*

Synergy between technology and public policy was responsible for India's agricultural progress. Investments in science and technology, education, extension, irrigation, fertilizer, rural roads and markets paid rich dividends. The same factors have played a part in accelerating agricultural progress in many regions of Africa. Marco Quinones, Regional Director for Africa of the Sasakawa – Global 2000 programme, brings out the need for greater attention to input and outpricing in the following words, "As long as farm gate prices for fertiliser are so high and grain prices so low", "many smallholders will be unable to fully embrace high-yield technology".

References

1. Swaminathan, M. S. (ed.) (1993). Wheat Revolution: A dialogue. Macmillan India Ltd, Madras 164 pp.
2. Swaminathan, M. S. (1996). Sustainable Agriculture: Towards an Ever-green Revolution. Konark Publishers, New Delhi. 232 pp.
3. Swaminathan, M. S. (2002). From Rio de Janeiro to Johannesburg – Action Today and not just promises for tomorrow. EastWest Books (Madras) Pte. Ltd, 224 pp.
4. Wilson, E. O. (2002). The Future of Life. Alfred A Knopf, New York, 229 pp.

Sustainable Food and Water Security

M.S. Swaminathan
UNESCO Chair in Ecotechnology

Swaminathan, M.S. (2003). Sustainable Food and Water Security. *Ankuram – Environment Quarterly*. **1**(1): 23–33

1. Introduction

It is widely agreed that the breathing spell provided by the green revolution for achieving a balance between population growth and food production will soon get exhausted, unless we take steps to foster an ever-green revolution based on principles of ecology, gender and social equity, economics, employment generation and energy conservation. Will the 21st century be one of hope or despair on the food front? Will there be enough water for human, agricultural, industrial and ecosystem needs? I would like to deal with these critical questions in the following four parts:

- Food security
- Water security
- Climate management
- Safeguarding the ecological foundations of sustainable agriculture: agrobiodiversity

The present global trends in the areas of preventing adverse changes in climate and sea level and in the protection of the ecological foundations for sustainable agriculture are not encouraging. However, there is still change for achieving the goals of "food and drinking water for all" in the coming century, because of the uncommon opportunities opened up by science and technology (Swaminathan, 1996). This paper describes some of the emerging challenges and opportunities.

2. Food Security

The concept of food security has been undergoing refinement during the last 50 years. Immediately after World War II, food security meant building emergency grain reserves and ensuring the physical availability of food in the market. After the onset of the green revolution in the late sixties, it became obvious that economic access to food is equally important for ensuring food security at the household level. During the eighties, it became evident that the gender dimension of food security should receive attention, in view of the growing feminisation of poverty and agriculture. This was highlighted at the World Conference on Women held at Beijing in 1995. The principle of social access, with reference to women and marginalised communities was hence added to the concept of food security. Finally, after the UN Conference on Environment and Development held at Rio-de-Janeiro in 1992, there has been an increasing understanding of the role of environmental factors in food security. The ecological foundations essential for sustained agricultural progress are increasingly under stress due to human activities. Agenda 21 of UNCED addresses these concerns. Without safe drinking water and environmental hygiene, the biological absorption and retention of food will be poor. Thus, environmental access to food becomes important.

Based on the above considerations, the Science Academies Summit held in July 1996 at the M S Swaminathan Research Foundation, Madras, India, in preparation for the World Food Summit convened by FAO in Rome in November 1996, proposed the following comprehensive definition of food security.

Policies and technologies for Sustainable Food Security should ensure:

- that every individual has the *physical, economic, social and environmental access* to a balanced diet that includes the necessary macro- and micro-nutrients, safe drinking water, sanitation, environmental hygiene, primary health care and education so as to lead a healthy and productive life.
- that food originates from efficient and *environmentally benign production technologies* that conserve and enhance the natural resource base of crops, animal husbandry, forestry, inland and marine fisheries.

The term "green revolution" was coined in 1968 by Dr William Gaud of the US Agency for International Development to highlight the opportunities opened up by the semi-dwarf varieties of wheat and rice to increase production through higher productivity. 1968 was also the year when the Indian Government issued a special postage stamp to commemorate India's wheat revolution (Swaminathan, 1993). Early in January 1968, I made the following statement in my Presidential address to the Agricultural Sciences section of the Indian Science Congress (Swaminathan, 1968).

> "*Exploitive agriculture offers great possibilities if carried out in a scientific way, but poses great dangers if carried out with only an immediate profit motive. The emerging exploitive farming community in India should become aware of this. Intensive cultivation of land without conservation of soil fertility and soil structure would lead, ultimately, to the springing up of deserts. Irrigation without arrangements for drainage would result in soils getting alkaline or saline. Indiscriminate use of pesticides, fungicides and herbicides could cause adverse changes in biological balance as well as lead to an increase in the incidence of cancer and other diseases, through the toxic residues present in the grains or other edible parts. Unscientific tapping of underground water will lead to the rapid exhaustion of this wonderful capital resource left to us through ages of natural farming. The rapid replacement of numerous locally adapted varieties with one or two high-yielding strains in large contiguous areas would result in the spread of serious diseases capable of wiping out entire crops. Therefore the initiation of exploitive agriculture without a proper understanding of the various consequences of every one of the changes introduced into traditional agriculture, and without first building up a proper scientific and training base to sustain it, may only lead us, in the long run, into an era of agricultural disaster rather than one of agricultural prosperity.*"

Thus in the same year when the term green revolution first came into use, I had stressed the need for concurrent attention to productivity improvement and the conservation of the ecological foundations essential for sustainable advances in agricultural productivity. I pointed out that such an integrated approach is the pathway to an ever-green revolution or sustained progress in enhancing yield per ha. (Swaminathan, 1996).

Since the term sustainable agriculture is now widely used, it may be appropriate to recall what Varro, a Roman land owner said in first century B.C.

> *"Agriculture is a science which teaches us what crops should be planted in each kind of soil, and what operations are to be carried on, in order that the land may produce the highest yields in perpetuity."*

Varro was probably the earliest to define what we now call "sustainable agriculture".

In the coming century, we will have to produce more food and other agricultural commodities under conditions of diminishing per capita arable land and irrigation water resources and expanding biotic and abiotic stresses. However, it is obvious that the pathway for productivity enhancement has to be different from that associated with the "green revolution". It is equally important that agriculture should help developing countries not only to produce enough food for the growing population but should also lead to the generation of more income and opportunities for skilled employment. How can we develop agriculture into an effective instrument for creating more income, jobs and food?

Integrated Farming Systems (IFS)

These triple goals can be achieved if we develop and disseminate ecologically, economically and socially sustainable integrated farming systems (IFS). The seven pillars of IFS are the following:

i. *Soil Health Care*

This is fundamental to sustainable intensification, IFS fosters the inclusion of stem nodulating legumes like *Sesbania rostrata,* incorporation of *Azolla*, blue green algae and other sources of symbiotic and non-symbiotic nitrogen fixation and promotion of cereal-legume rotation in the farming system. In addition, vermiculture composting and organic recycling constitute an essential components of IFS. IFS farmers are trained to maintain a Soil Health Card to monitor the impact of farming systems on the physical, chemical and microbiological components of soil fertility. Also, in many tropical countries, crop-livestock integrated farming systems help to conserve soil fertility.

ii. *Water harvesting and management*

IFS farm families include in their agronomic practices measures to harvest and conserve rain water, so that it can be used in a conjuctive manner with other sources of water. Where water is the major constraint, technologies which can help to optimise income and jobs from every litre of water are chosen and adopted. Maximum emphasis is placed on on-farm water use efficiency and on the use of techniques such as drip irrigation, which help to optimise the benefits from the available water.

iii. *Crop and Pest Management*

Integrated Nutrient Supply (INS) and Integrated Pest Management (IPM) systems form important components of IFS. The precise composition of the INS and IPM systems will depend on the components of a farming system as well as on the agro-ecological and soil conditions of the area. Computer aided extension systems will provide farm families with timely and precise information on all aspects of land, water, pest and post-harvest management.

iv. *Energy management*

Energy is an important and essential input. Besides the energy efficient systems of land, water and pest management described earlier, every effort will have to be made to harness biogas, biomass, solar and wind energies to the maximum extent possible. Solar and wind energy can be used in hybrid combinations with biogas for farm activities like pumping water and drying grains and other agricultural produce.

v. *Post harvest management*

IFS farmers will not only adopt the best available threshing, storage and processing measures, but will also try to produce value-added products from every part of the plant or animal. Post harvest technology assumes particular importance in the case of perishable commodities like fruits, vegetable, milk, meat, egg, fish and other animal products and processed food. A mismatch between production and post-harvest technologies affects adversely both producers and consumers. Growing urbanisation leads to a diversification of food habits. Therefore there will be increasing demand for animal products like milk, cheese, eggs as well as for fruits, vegetables and processed food. Agro-processing industries can be promoted on the basis of an assessment of consumer demand. Such food processing industries should be promoted in villages in order to increase employment opportunities for rural youth. In addition, they can help to mitigate micronutrient deficiencies in the diet. A well planned urban green belt movement around towns and cities will help to improve both human and animal nutrition and the environment.

Investment in sanitary and phytosanitary measures is important for providing quality food both for domestic consumers and for export. To assist the spread of IIFS, Governments should make a major investment in storage, roads, transportation and on sanitary and phytosanitary measures. ISO 9000 and ISO 14000 standards of environmental management should be popularised.

vi. *Choice of the Crop and animal components of farming systems*

In IFS, it is important to give very careful consideration to the composition of the farming system. Soil conditions, water availability, agro-climatic features, home needs and above all, marketing opportunities will have to determine the choice of crops, crop varieties, farm animals and aquaculture systems. Small and large ruminants will have a particular advantage among farm animals since they can live largely on crop biomass. Backyard poultry farming can help to provide supplementary income and nutrition.

vii. *Information, skill, organisation, management and marketing empowerment*

IFS is based on the principle of precision farming. Hence, for its success, IFS system needs a meaningful and effective knowledge and skill empowerment system. Decentralised production systems will have to be supported by a few key centralised services, such as the supply of credit, seeds, biopesticides, and animal disease diagnostics. Ideally, an Information Shop will have to be set up by trained local youth in order to give farm families timely information on meteorological, management and marketing factors. Organisation and Management are key elements and depending on the area and farming system, steps will have to be taken to provide to small producers the advantages of scale in processing and marketing.

IFS is best developed through participatory research between scientists and farm families. This will help to ensure economic viability, environmental sustainability and social and gender equity in IFS villages. The starting point is to learn from families who have already developed successful IFS procedures.

It should be emphasised that IFS will succeed only if it is a human-centered rather than a mere technology-driven programme. The essence of IFS is the symbiotic partnership between farming families and their natural resource endowments of land, water, forests, flora, fauna and sunlight. Without appropriate public policy support in areas like land reform, security of tenure, credit supply, rural infrastructure, input and output pricing and marketing, small farm families will find it difficult to adopt IFS.

Further developments in this regard include intensive research and testing, in the United States of America, intensive research and testing of Low External Input Sustainable Agricultural Practices (LEISA) are in progress. Under a project supported by the European Union, a computer model, GOAL (General Optical Allocation of Land Use) has been developed, which calculates optimal land use in the Community of the 12 member states (Rabbinge and Van Latesteijn, 1992). The objectives incorporated in the GOAL model are the following:

Class of objectives		Objective
Agricultural	1	maximize soil productivity
	2	minimize costs of agricultural production
Socioeconomic	3	maximize total employment in agriculture
	4	minimize regional decrease in employment in agriculture
Environmental	5	minimize input of nutrients per unit of acreage
	6	minimize input of nutrients per unit of product
	7	minimize input of pesticides per unit of acreage
	8	minimize input of pesticides per unit of product

Results of recent studies reveal that most land use trends indicate that there will be a dramatic decrease in farm land. About one third of the present area under cultivation will be sufficient once productivity in the EC reaches the optimum. Similarly studies at the International Water Management Institute (IIMI) have shown that nearly 50% of the additional irrigation water required by 2025 can be met by improved efficiency of irrigation.

Meeting the Challenge

How can we meet these challenges?

Our first aim should be to eliminate the widely prevalent yield gap (i.e. gap between potential and actual yields in farmers' fields) with the best currently available technologies through appropriate packages of technology, services and public policies. As we approach the new century and millennium we are experiencing three major revolutions in science and technology, which will influence agricultural technology is a fundamental manner. It will therefore be appropriate to make a brief reference to them.

i) the gene revolution - which provides a molecular understanding of the genetic basis of living organisms, as well as the ability to use this understanding to develop new processes and products for agriculture, the environment, and for human and animal health.
ii) the information and communications revolution - which allows a very rapid growth is the systematic assimilation and dissemination of relevant and timely information, as well as a dramatically improved ability to access the universe of knowledge and communicate through low cost electronic networks, and
iii) the ecotechnology revolution - which promotes the blending of the best in traditional knowledge and technology with modern science.

In principle, these three types of advances - when coupled with improvements in management science and governance - greatly increase the power of a scientific approach to genetic improvement, agronomics, the integrated management of natural resources and ecosystems, and the management of local and regional development policies. However, these scientific revolutions seem to be proceeding at an ever increasing pace, with most of the action occurring in a few places in industrialised nations. Developing countries should lose no further time in harnessing these technologies for achieving the twin goals of natural resources conservation and food security.

The Gene Revolution

The past ten years have seen dramatic advances in our understanding of how biological organisms function at the molecular level, as well as in our abilities to analyze, understand, and manipulate DNA molecules, the biological material from which the genes in all organisms are made. The entire process has been accelerated by the Human Genome Project, which has poured substantial resources into the development of new technologies for working with human genes. The same technologies are directly applicable to all other organisms, including plants. Thus, a new scientific discipline of genomics has arisen. This discipline has contributed to powerful new approaches that can be used in agriculture as well as in medicine and has helped to promote the biotechnology industry.

Several large corporations in Europe and the United States have made major investments in adapting these technologies to produce new plant varieties of agricultural importance for large-scale commercial agriculture. The same technologies have equally important potential applications for addressing food security in the developing world.

The key technological developments in this area are:

a. *genomics*: the molecular characterization of species.
b. *bioinformatics*: data banks and data processing for genomic analysis
c. *transformation*: introduction of individual genes conferring potentially useful traits into plants, trees, livestock, and fish species.
d. *molecular breeding*: identification and evaluation of useful traits by use of marker assisted selection, which greatly speeds up traditional breeding processes
e. *diagnostics*: identification of pathogens by molecular characterization
f. *vaccine technology*: use of modern immunology to develop recombinant DNA vaccines for improved disease control against lethal diseases of animal and fish.

Developing countries must exploit these techniques and develop partnerships with advanced research institutions for this purpose. There are widespread public concerns about the potential adverse impact of genetically modified organisms (GMOs) on human health and the environment. Some of these concerns are genuine. In order to take advantage of recombinant DNA technologies without associated harm to human or ecological health, it is important that every country has in place suitable institutional structures and regulations for biosafety, bioethics and biosurveillance. A recent statement issued by the Royal Society of London on "Genetically modified plants for food use" provides guidelines for the safe handling of biotechnological applications (Royal Society, 1998).

The Information Technology Revolution

New communication and computing technologies will have profound implications in everyday research activities.

a. Access to the Internet will soon be universal, and it can provide unrestricted low-cost access to information, as well as highly interactive distance learning. The Internet will not only facilitate interactions among researchers, but also greatly improve their ability to communicate effectively with the potential users of their research knowledge.
b. Computing makes it possible to process large-capacity databases (libraries, remote sensing and GIS data, gene banks) and to construct simulation models with possible applications in ecosystem modeling, preparation of contingency plans to suit different weather probabilities and market variables.
c. The software industry is continuously providing new tools that increase research productivity and create new opportunities for understanding complex systems of growing conditions.
d. Remote sensing and other space satellite outputs are providing detailed geographic information useful for land and natural resources management.

All countries should take advantage of these opportunities.

The Ecotechnology Revolution

Knowledge is a continuum. There is much to learn from the past in terms of the ecological and social sustainability of technologies. At the same time, new developments have opened up uncommon opportunities for developing technologies which can lead to high productivity without any adverse impact on the national resources base. Blending traditional and frontier technologies leads to the birth of ecotechnologies with combined strength in the following areas.

- Economics
- Ecology
- Equity
- Employment
- Energy

For example, in the area of water harvesting and sustainable use, there are many lessons to be learnt from the past. In the Thar desert area of Rajasthan, India drinking water is available even in areas with 100 mm annual rainfall, largely because women are continuing to harvest water in structures called kunds. In contrast, drinking water is scarce during summer months in some parts of the North east India with an annual rainfall of 15000 mm. Thus, there is need to conserve traditional wisdom and practices, which are often tending to become extinct (Anil Agarwal and Sunita Narain, 1996). The decision of the World Intellectual Property Organization (WIPO) to explore the intellectual property needs, rights and expectations of holders of traditional knowledge, innovations and culture is hence an important step in widening the concept of intellectual property. Principles of ethics and equity demand that this invaluable component of IPR gets included when the TRIPS agreement (Trade-related Intellectual Property Rights) of the World Trade Organization comes up for review in 1999. FAO has been a pioneer in the recognition of the contributions of farm families in genetic resources conservation and enhancement by promoting the concept of "Farmers Rights". Like WIPO, UPOV (Union for the Protection of New Varieties of Crops) should also undertake the task of preparing an integrated concept of breeders' and farmers' rights.

Precision Agriculture

Precision agriculture involves a systems approach to experimental design and agronomic practices. It needs inter-disciplinary research drawing on expertise in a range of subject areas such as agronomy, plant science, genetics, soil science, entomology, meteorology, weed science, plant physiology, plant pathology, ecology and economics (National Research Council, 1997). Agricultural extension workers using information technology will play an increasingly important role in crop production and natural resource management. The curricula of Agricultural Schools and Colleges need to be modified to make precision agriculture the road to an ever-green revolution. Precision agriculture is particularly valuable for increasing opportunities for skilled employment in the farm sector. For example, computer software development, equipment fabrication and sales, custom hiring of software and farm equipment, local production of biofertilizers, biopesticides and drip irrigation equipment and consultancy services can all provide new opportunities for unskilled workers to become skilled.

Precision farming methods have to be based on scientific land and water use planning. It will need concurrent attention to natural capital stocks and nature's services. Examples of stocks include: soils and soil nutrients, biodiversity, water, minerals, forests and oceans. Examples of natures' services include: water cycles, nutrient cycles, carbon sequestration and waste recycling, Agro-forestry and other sustainable systems of land management need to be popularised in areas experiencing varying degrees of desertification. Costanzo et al (1997) have assessed the value of 17 ecosystem services and flows for 16 biomes (desert, tundra, ice/rock excluded) at $ 33 trillion per year. By comparison, global GNP is now $ 18 trillion per year. Food production services of crop lands is $ 0.13 trillion per year. Thus, the need to conserve ecosystem services and flows under varying agro-ecological conditions through precision farming practices can hardly be over-emphasised.

In the emerging Knowledge intensive agricultural era, international cooperation is vital for taking the benefits of new technologies to those who have so far been bypassed by new knowledge and techniques. Women farmers and farm labour need particular attention in any agricultural extension and development programme designed to reach the unreached. The gender dimension needs to be internalised in all research, educational and development programmes intended to promote natural resources conservation based advances in agricultural productivity and production.

Sustainable agriculture in the 21st century will be based on the appropriate use of biotechnology, information technology and ecotechnology. Practical achievements in bringing about the desired paradigm shift will depend upon public policy support and political action, Regulation through legislation, social mobilization through local level community organizations and education through the mass media and information shops will all be needed to meet the dual demands for food and ecological security. It would be useful to consider how these concepts can be applied to water, which will be a key constraint to food and health security in the coming millennium.

2. Water Security

2.1. *Physical access*

Humanity now uses over one-half of the total accessible freshwater runoff. Projected growth rates in irrigated area are significantly lower than in the recent past. For the world as a whole, irrigated area is projected to grow at 0.6 percent per year, as compared to 1.5% during 1982-93. Current global water use is around 4500 cubic kilometers. However, freshwater is distributed unevenly across the globe. Countries with freshwater resources in the range of 1000-1500 cubic meters per capita per year face water stress, particularly in drought years. Agriculture is by far the biggest user of water, accounting for more than 70 per cent of water withdrawals worldwide and more than 90 per cent water withdrawals in several low income developing countries. Population rich and land hungry countries like India and China, have no option except to produce more food and other farm commodities from less per capita arable land and irrigation water availability in the coming millennium. The largest increases in irrigated area during the coming decades are expected in India and China. However, even in India, the projected 1995 to 2020 rate of growth in irrigated area of 1.2 per cent per year is well below the rate of 2.0 per cent per year achieved during 1982 to 1993. Much of the additional irrigation water comes from groundwater and this source is being increasingly exploited in an unsustainable manner. Major irrigation projects are running into serious environmental and social problems.

The adequate availability and equitable and efficient use of irrigation water are essential for converting the green revolution into an evergreen revolution. A study in eight Asian countries revealed that of the additional 117 million tons of rice produced between 1965 and 1988, 34 million tons can be directly attributed to irrigation. Even the remaining attributed to new varieties and fertilizer would not have been possible without water. Irrigation water has helped to increase not only crop productivity but cropping intensity, thereby leading to an increase in yield per day. In the tropics and sub-tropics, where there is abundant sunshine during most of the year, multiple cropping techniques have spread fast, both because of irrigation projects and the availability of photo-intensitive crop varieties. This has facilitated crop-livestock integration, through the introduction of fodder crops in the rotation.

Some of the urgent steps needed in the area of irrigation water conservation and sustainable use are:

- Harvest and conserve rain water
- Promote conjuctive water use of river, rain, ground, sea and sewage water, in appropriate combinations.
- Prevent unsustainable exploitation of the aquifer
- Ensure efficiency, economy and equity in water use through cooperative management of watersheds and command areas.
- Regulate the expansion of water markets and water lords
- Introduce proactive measures to avoid water conflicts

2.2. *Economic access*

Government pricing policy, particularly with reference to electricity for pumping ground water, often results in inefficient and unsustainable exploitation of precious water resources. On the other hand, private ownership of ground water resources leads to the emergence of water lords and water markets. A national policy relating to access to water for all should be designed in such a manner that first, irrigation does not result in long term harm to soil health and secondly, gender and social equity in sharing the available water is ensured. Participatory management of irrigation water resources, including systems of rotational distribution of water, will help to foster the equitable and efficient use of water. Pricing policies should signal the inter-generational equity aspects of water use.

Irrigation has been an important factor in poverty alleviation in several developing countries. When water is available, there is not only crop security but also opportunities for multiple cropping, mixed cropping and agro-forestry systems of land use. Irrigation has a multiplier effect on employment both at the production and post-harvest phases of agriculture. It makes Integrated Intensive Farming Systems (IIFS) possible. Irrigation water thus enhances purchasing power.

2.3. *Environmental access*

This again is a multi-faceted issue. Sewage and industrial water recycling, including their incorporation in conjuctive water use systems will confer both public health and economic benefits. Ecological problems associated with unscientific water use such as water logging, salinisation and soil erosion are well known. Serious nutritional problems such as arsenic poisoning, due to the tapping of the deeper layers of the aquifer, have been reported in Bangladesh and the State of West Bengal in India.

Environmental and social problems are also serious in several major multi-purpose irrigation projects. For example, China's Three Gorges Dam over the Yangtze river will help to halt floods in the South of China and bring irrigation water to the northern part of the country. However, environmentalists have serious concerns. Similarly, the Narmada project in India, designed to provide drinking water to 40 million people and irrigate 1.8 million ha. will lead to the displacement of nearly 250,000 people and to inundation of 117,000 hectares of land. It is obvious that every nation will have to weigh carefully the pros and cons of large multipurpose water projects and choose the ones which confer maximum social benefit with the least ecological harm. There is need for large numbers of professionals trained in Green Audit procedures with reference to irrigation projects. Opposing unsustainable development alone is not enough; there must be equal emphasis on proposing sustainable options. This is going to be a major challenge in the next century with reference to irrigation projects.

2.4. *Social access*

This again has several dimensions. Gender inequity is most serious in the case of drinking water, since women are invariably entrusted with the responsibility of fetching water for domestic use. Destruction of forests leading to the disruption of hydrologic cycles has affected adversely the nutrition and livelihood security of women.

Social cohesion and cooperation in the harvesting, storage and use of water will help enormously to strengthen irrigation water security. Anil Agarwal and Sunita Narain (1997) have illustrated the power of social action, characteristic of the past but fast vanishing now, with the following examples.

i. *Jaisalmer*

A district in the Thar desert of Rajasthan, India. Annual rainfall is 100 mm. During the drought year of 1987, the Government's piped water supply ran dry. But there was enough water for the people who stuck to their rainwater harvesting structures called kunds (small water conservation structures).

ii. *Cherrapunji, Meghalaya*

A village in the northeast with annual rainfall of 15,000 mm. Yet this village suffers from water storage during summer months.

Peoples' participation in water harvesting, conservation and efficient use will foster both sustainable food and drinking water security. However, there will be no cooperation in water saving unless there is equity in water sharing. Irrigation Water delivery systems and on-farm management of water can be made more efficient, if community-centred systems of water management are promoted. Globally, more than 50% of the cultivated area will continue to depend on rainfall and hence, rainwater management is vital for sustainable food security. In the past, famines were invariably associated with the failure of rainfall. Irrigation systems have helped to reduce variability in production from year to year and have provided insulation against total crop failure in years of drought.

Competing Demands for Water

Currently, water use goes to four major sectors - domestic needs including drinking water, agriculture, industry and ecosystem conservation. In most calculations, the need for water to maintain ecosystems, particularly those rich in biodiversity, is not taken into consideration. For example, many mangrove forest ecosystems, which occur in the estuaries of major rivers, are adversely affected when the flow of fresh water goes down. Salinity then goes up and not all Mangrove species can withstand a high degree of salinity. Indiscriminate deforestation disrupts hydrologic cycles and increases the frequency of floods and drought. Hence, water allocation policies should include adequate provision for safeguarding the integrity of critical ecosystems. Disruption of hydrologic cycles hastens the process of desertification, as is evident from the ravines of the sub-Himalayan zone. There is need to standardise methods for calculating the water requirements for the conservation and sustainable management of forests and natural ecosystems, so that this need can get integrated in sectoral assessments.

The need for policy making and implementation structures to deal with water allocation and use issues in a holistic manner is becoming urgent in many countries. This will be clear from the recent decision of the International Irrigation Management Institute (IIMI) in Sri Lanka to change its name to "International Water Management Institute (IMI)" in order to enable it to look at water problems in its totality. River basins are highly integrated hydrological systems with the same water flowing and recycling through the agricultural, domestic, industrial and environmental sectors.

The following aspects will need integrated attention while developing a holistic approach to water management.

- Demand: While global demand projections are useful, what matters to individuals is the local availability of water. The need for water for household use and for agriculture, industry and ecosystem conservation will have to be assessed both locally and nationally. The qualitative aspects of water should receive equal attention.

- Supplies: Different methods of enhancing water availability will have to be promoted at the local and regional level. An appropriate mix of major, medium and mini-irrigation projects will have to be fostered. Conjunctive use of different sources of water such as rain, river, ground, sea and waste water will have to be promoted, as is being done effectively in countries like Israel. Computerised systems of water management and delivery need popularisation. Water Information Shops can be started in areas characterised by severe water scarcity.
- Management: Efficient systems of water management, including equity in distribution and the control of pollution will have to receive attention. Seasonal fluctuations in demand will have to be addressed through suitable management protocols.
- Conflict resolution: Conflicts are likely to grow at the national and regional levels on sharing water. At the local level, conflicting inter-sectoral demands, of the kind described later, will have to be resolved. Suitable institutional structures will have to be developed

for a proactive resolution of emerging conflicts. The Water Court operating in the city of Valencia in Spain since many centuries is a good example of a local initiative in resolving conflicts amicably. In the new millennium, conflicts are likely to arise between the need of water for human use and that needed for irrigation. There are also possibilities of conflicts between countries, if changes in precipitation and temperature occur as a result of climate change induced by green house gas emissions. The Framework Convention on Climate provides a mechanism for cooperation among countries in preventing adverse changes in rainfall, temperature and sea level.

- Technology development and dissemination: This has to receive high priority, since technologies are now available to harvest every drop of water and use it economically and efficiently. Local level water users' associations can help to save and share water based on the principles of equity and efficiency.
- Public awareness, social mobilisation and information empowerment: This is an area of great importance, particularly in countries where a majority of farmers operate small holdings. In India and China, for example, the average size of holding is less than one hectare. Group cooperation will be essential under conditions of small and fragmented holdings for both water harvesting and efficient use. Without equity in water sharing, cooperation in water saving will not be forthcoming.
- Resources: The requisite managerial, institutional and financial resources will have to be mobilised for achieving the above objectives. Institutional structures will be needed for demand forecasting and management and for advice on efficient water use.
- New Partnerships: Coalitions of all concerned - scientists and engineers, political leaders, mass media, civil society, farm families, private sector industry, womens' organizations and bilateral and multilateral donors - will have to be formed to tackle water problems on an end to end basis.
- Population: With every increase in population, there is a decline in per capita availability of water, a situation which can be altered only by new technologies such as solar desalination of sea water. For example, water availability per capita in India was over 5000 cubic meters (m^3) per annum in 1950. It now stands at hardly more than 2000 cubic meters per capita. By year 2025, per capita availability is projected at only 1500 cubic meters (Fig.1). Such quantitative data alone are not adequate to get a real picture of the water availability status. Pollution affects water quality both in rivers and ground water. Also, there are gross inequalities between basins and geographic regions. Water markets, if they are organised in a non-exploitative manner, can help to meet the minimum household and agricultural needs.

Managing Competing Demands

Sandra Postel (World Watch Paper 132, September 1996) has described the problems faced in dividing waters for different uses. In a more recent paper (December 1997), Mark W Rosegrant of the International Food Policy Research Institute and Claudia Ringler of the International Irrigation Management Institute, have dealt with the global impacts of water reallocation from agriculture on food production. Their calculations indicate that the projected reductions in agricultural water availability will be substantial by 2020. The reduction can be as much as 24 percent in China and 21 percent in India. Their model also suggests that reallocation of water out of agriculture can have a dramatic impact on global food markets. In developing countries, yield growth for all cereals will slow from 1.20 percent annually to 1.07 percent per year during the period 1993 to 2020. The area decline during the same period will be from 0.29 to 0.23 percent. Rice will suffer most, since it needs larger quantities of irrigation water. Consequently, the average price of rice is projected to increase by 68 percent between 1993 and 2020.

In addition to direct impacts on agricultural production, water transfers can negatively affect business activities, fishing and hydropower generation. Under conditions of scarcity, water markets grow. They can be of benefit to those who do not own a well or other source of water, if they function in a regulated environment, where making profit out of water scarcity becomes unethical. Rosegrant and Ringler recommend policy reforms such as the establishment of secure water rights to users, the decentralization and privatization of water management functions to appropriate levels, pricing reform, markets in tradable property rights, and the introduction of appropriate water saving technologies. While developing public policies for specific agro-ecological and socio-economic conditions, the trade-offs among various policy options will have to be carefully considered.

Often, local solutions will have to be found to manage water scarcity. Contingency plans and alternate cropping strategies will have to be developed for different water availability situations. Crop life saving irrigation methods will help to optimise yield under conditions of water scarcity. Most of these methods will require the active cooperation of all the families residing in a watershed. Management procedures relating to inter-sectoral availability of water should keep in view the needs of women. It is women who are mostly in charge of fetching and managing water at the household level. Hence, the gender dimension should be internalised in all technological and policy issues relating to water.

The Next Steps

Trends in water consumption indicate that demand for water for household and industrial uses in developing countries could double as a proportion of total water demand in the next 25 years . Scope for water supply expansion will at the same time be limited because development of irrigation and urban water supplies is becoming increasingly expensive, and often involve high costs in terms of environmental degradation and human resettlement. Without fundamental reform of water management, the rapid growth in urban water demand will require large transfers of water from irrigated agriculture, thereby threatening food security. Hence, water supply and demand should be managed in an integrated fashion, considering simultaneously all uses and sources.

How can we accomplish this objective? At the global level several initiatives like organisation of a Global Water Partnership and a World Water Council have been taken in recent years. At a meeting held at Valencia in Spain in July 98 under my Chairmanship, the participants recommended the establishment of an International Water Centre for undertaking research, analysis, appraisal, information dissemination, training and consultancy activities. Such a Centre could also specialise in water laws and help in resolving water disputes through analysis and information. It was felt that we urgently need a new age instrument for promoting sustainable water security in the world. The major features of the proposed **Valencia International Centre for Cooperation in Water Management** are as follows:

The mission of this new Centre shall be **to contribute to the anticipation, prevention and resolution of water conflicts**. The Centre will strive to achieve its mission through the compilation and dissemination of information, applied research, partnerships, training and other assistance. The Valencia Centre shall in particular address cross-jurisdictional and cross-sectoral issues at all levels --local, national, regional and international. It shall concentrate on the following tasks during the first five years.

- mapping "hot spots" locations with reference to potential water conflicts.
- offering advice and assistance on impartial and informed mediation techniques and services;
- linking to data bases on traditional and modern methods of water conservation and use;
- promoting awareness among decision-makers, professionals and the public that water is a social, economic and ecological commodity.
- forging broad-based coalitions of all concerned with efficient water management and prevention of conflicts within and between different jurisdictions;
- developing on-line data bases on conflict resolution experts and institutions who can assist in the resolution of water disputes and establishing an inventory of case studies as a knowledge and reference tool;
- undertaking participatory research with the involvement of the principal stakeholders;
- organising training programmes and policy dialogues to bolster conservation and efficient use of all sources of water;
- spreading information on the management of drought and water scarcities.

During 1999-2000, the Centre should concentrate on building expertise in applying modern information and communication technologies to the tasks at hand. Depending on needs, it should organise a global network of institutions engaged in water issues linking and making accessible their respective databases and on-line expertise.

The Centre should compile and publish an "atlas of hot spots" from the point of view of threats to human livelihood and food security arising from acute water shortages. It should also assemble information in the effective management of water during severe droughts and on successful water basin transfers of water. Jointly with the Global Environment Facility (GEF) and other institutions, it may launch a programme for science-based management of international rivers. It should also seek to work with the Secretariats of the Climate Change and Desertification conventions on water issues and other relevant programmes and entities, such as the Global Water Partnership and the World Water Council.

Charity begins at home – while global mechanisms and institutions are important, it is essential that every country sets up institutional structures to deal with national and local level problems. International Conflicts over water are often precipitated by a failure to meet local, provincial, national and regional water demands for household, agriculture and industrial uses as well as for environmental / ecosystem services. The various sources of water and sectoral needs are indicated in Fig. 2.

An institutional structure for dealing with the multiple dimensions of water management at the country level is proposed in Fig 3. I have suggested the title "National Water Trust" instead of the commonly used term National Water Authority, to emphasise that water management should be carried out in the trusteeship mode. We must consider ourselves as trustees of water sources and not as owners. This will help to instill the feeling that management of water should not only be in the interests of the present generation but also of the generations yet to be born. Thus, the guiding principle for the work of the National Water Trust will be intra- and inter-generated equity.

Provincial and local level units of the National Water Trust can be organised, according to needs. The National Water Trust will serve as the hub of a grid of institutions including policy making bodies. It will thus serve as the flagship of a national system for sustainable water security.

Biotechnology and Water Security: Opportunities

Among the technologies those are currently available, biotechnology has enormous promises and implications for food security, human nutrition and water quality management. Development of transgenic plants, microbes and animals during the last few years have opened-up new prospects for developing stress tolerant genetic material that is likely to contribute towards addressing problems of water scarcity, increased salinization in the agricultural sector, increased contamination and ground water and solving several other associated problems in waste management.

Drought and Salinity Stress

Drought and salinity stress, sometimes in combination, are expressed by a series of morphological, physiological, metabolic and molecular changes in plants, which adversely affect the plant growth and productivity. Numerous studies conducted on biochemical and molecular aspects point out that the ability of plants to tolerate water stress is determined by multiple biochemical pathways that facilitate retention and/or acquisition of water, protect chloroplast functions and maintain ion homeostasis. Esstential pathways include those that lead to synthesis of osmotically active metabolites and specific proteins that control ion and water flux, suport scavenging of oxygen radicals or may act as chaperons. The ability of plants to detoxify radicals under conditions of water deficit is probably most crucial requirement. many stress-tolerant species accumulate methylated metabolities, which play a crucial role as osmoprotectants and as radical scavengers. Their synthesis is correlated with stress induced enhancement of photorespiration. Contrary to that of the biotic stress, abiotic stress is under polygenic control. This would therefore require transfer of multiple gene systems and accumulation of engineered mechanism trough breeding that would result in increased tolerance of plant systems to water stress. The strategy for engineering tolerance to water stress should include:

- Control over expression, activity and ion discrimination properties of potassium transport system.
- Capabilities for enhanced sodium / proton antiport in plasma membrane of root cells and in the tonoplast of mesophyll or other sink cells
- Osmatic adjustment as a result of metabolic accumulation of complex photo-inhibitory damage.
- Protection of sensitive metabolic reactions by protecting either individual enzymes, protein complexes or membrane structures by increased capacity for hydroxyl-radical scavenging.

A combination of following approaches would be required to achieve the goal of increased tolerance to water stress.

- germplasm collection of tolerant genotypes to be used as source of new genes
- conventional breeding and selection methods
- novel molecular and biotechnological methodologies to identify stress-related genes and use them as probes for selection of tolerant genotypes and for generation of transgenic plants
- mapping and generation of molecular markers to be used in breeding systems, functional genomics studies, QTL analysis and screening
- generation of tolerant genotypes by genetic engineering (transgenics) and somatic cell genetic (haploidization, mutation breeding) methods
- rapid seed and clonal propagation of desired genotypes

The research agenda of molecular and traditional breeding should include all known physiological and molecular mechanisms of stress adaptation, such as; escape and avoidance; osmoprotection; membrane transport and ion-uptake; structural water binding proteins; ABA signal transducers; additional mechanism etc. It is therefore, desirable to undertake **mission-oriented world-wide** research partnership on drought and salinity research.

There is, hence, no time to relax. A Technical Panel which at MSSRF in February 99 has recommended the launching of a "**Global Biotechnology Research Partnershi for Water Security**". This network is becoming an absolute necessity in the context of the present trends and the future needs.

Globally, we are faced with three major problems associated with the water security namely; Salinity, Drought and Water Quality. Many of the physiological and biochemical response mechanisms in relation to salinity and drought stress are of similar nature. Hence at present it would be ideal to take both these stresses in combination. It is, therefore, proposed to include two mini-networks addressing these specific areas. Each of this mini-network, in addition to exploring possibilities of continued use of the conventional technologies, would broadly have three major focus.

Basic Biochemical and Molecular Studies

- Understanding the plant respnse mechanism to water stress
- Identification of compounds which are expressed under water stress and deciphering biochemical pathways
- Screening of wild relatives and cultivars for stress tolerant genotypes through genomic studies (QTLs)
- Developing probes and markers related to stress tolerance
- Studies related to the biochemical mechanisms including enzymatic pathways in aerobic / anaerobic bioremedial conditions

Identification and isolation of novel genes including identification of new source material

- Isolation of known stress tolerant genes and comparative studies on their efficiency in transgenic systems
- Identification and isolation of novel genes
- Screening of unexploited genetic material for stress tolerance, for example *Acacia tortolis, Prosopis julifloral* (for drought resistance); mangroves (*porteresia coarctata, Salicornia* sp., *Sueda* sp., *Avicennia marina* etc., for salinity tolerance) and microbes (*Pseudomonas*; for biological remediation.
- Understanding the genetic basis for enhancing the capabilities of detectors "biosensors" (to include whole cell, enzymes or sequences), degraders and sequestors of heavy metals.

Molecular Breeding and Development of Transgenics for Crop Productivity and Biological Remediations

- Development of delivery systems for targeted species including transformation studies resulting in transgenics with novel genes.
- Environmental impact analysis of transgenic organisms and related studies for field applications.
- Marker assisted selection and breeding.

Such a network would therefore involve participation and partnershp among

- Applied researchers working in the problem areas
- Researchers involved in molecular biology / genetics
- Researchers engaged in developing transgenic organisms

Each network could be co-ordinated by a co-ordinating centre. These co-ordinating centres would identify collaborators / partners for undertaking focused research to address the practical problems. Exchange of information, expertise and results among the collaborators would be through electronic media and regular interative meetings. Partnership development in such collaborations are envisioned through cost-free collaborations, cost-sharing collaborations and general services to be met by the projects. These three centres would be functioning under the umbrella of the **Global Biotechnology Research Network for Water Security**, that in turn will be co-ordinated by an appropriate organisation / institution. These centers should ideally be linked to the Genetic Enhancement Centers to provide practical pre-breeding / novel material to the grass-root level breeders / researchers for developing location specific adaptable genotypes / technologies for addressing the water stress and water quality problems. The coordinating center would be advised by a Technical Advisory Group for which the members could be drawn depending on the expertise and experience in the area of research. It would also have a donor consortium. The panel also recommends that water commission may set up two expert panels on specific themes on water stress and water quality to prepare state-of-art reports on the trends of research, available technologies, key institutions and likely partners for the research partnership. They could also be entrusted with the task of developing a framework for the Global Biotechnology Research Partnership. It also recommends that the global partnership should develop a knowledge system for collection and dissemination of information on these two key areas. The overall structure envisaged is in the line of some of the existing international networks in the area of biotechnology, e.g. Rice Biotechnology Programme supported by the Rockefeller Foundation. It is also recommended that the period of assured funding for such a network should be about 10 years so that practical benefits could be realised at the end.

Water Security: Bridge to a Millennium of Hope

1998 marks the bi-centenary of Thomas Malthus's essay on population. In 1798, Malthus warned "the period when the number of men surpass the means of subsistence has long since arrived". When Malthus wrote his essay, the global population was less than the current population of India alone, namely 970 million. There is adequate food in the world today at current levels of purchasing power. Hunger is presently more related to economic access than physical access. Irrigation water availability and management have played a pivotal role in keeping Malthusian predictions at bay. Our ability to maintain a satisfactory balance between population and food production will depend both upon population policies and food production strategies. Every nation will have to develop an appropriate mix of water supply augmentation and demand management through socially and ecologically constructed policies and technologies. Comprehensive policy reform that promotes efficient use of existing water supplies will be needed in most countries. The National Water Trust proposed in this paper would be of help in integrated planning and decision making.

Above all, cooperation between countries sharing the same river or water source will become increasingly important. This is also true for States within a country having a federal constitution. Differing perceptions on the value of artificial rainmaking is a case in point. National and international mechanisms for proactive action on potential water conflicts will have to be put in place. International cooperation is also essential for avoiding adverse changes in climate, particularly precipitation and sea levels, arising from the accumulation of greenhouse gases in the atmosphere. A major step in this direction has been taken at the Conference of Parties to the Framework Convention on Climate Change held at Kyoto in December, 97. Meanwhile, it will be prudent for all countries to be prepared for different weather probabilities based on computer simulation models. Both avoidance and mitigation strategies should be developed. It will be appropriate in this context to recall the words of Mahatma Gandhi.

"*Nature provides for everybody's need but not for everybody's greed*".

As we approach the new millennium, there are again apprehensions that an imbalance between human numbers and food needs will result in large scale famines in many developing countries, and may necessitate countries like China and India to resort to extensive food imports (Lester Brown, 1995). The following are some of the major factors underlying such concerns.

- Steady decline in per capita availability of irrigation water and arable land, as a result of a continuing rise in population as well as the diversion of prime farm land for non-farm uses.
- Increase in food demand to meet the needs of the growing population, to overcome prevailing undernutrition among 800 million children, women and men and to meet the additional food needs arising from greater purchasing power and increasing urbanisation and the consequent higher demand for animal products.
- Stagnation or even decline in marine food production.
- Stagnation in technological change.
- Fatigue of the Green Revolution caused by environmental, economic and social factors.
- Potential changes in climate, resulting in alterations in precipitation, temperature and sea level, as well as the possibility of increased ultraviolet-B radiation.

The above factors represent real challenges to both scientists and policy makers. There is no room for complacency. The FAO's World Summit of 1996 set a target for reducing the number of persons going to bed hungry by half by 2015 and several experts have expressed doubts whether even this extremely modest target can be achieved. In addition to protein-calorie undernutrition, FAO estimates that nearly 2 billion persons suffer from iron deficiency. In addition, deficiencies of iodine, vitamin A and other micronutrients are widespread. Such "hidden hunger" affects health and productivity. Further, one third of the children born in South Asia and the Sahelian region of Africa are characterized by low birth weight due to nutritional anemia in pregnant women. There is evidence that low birth weight children are handicapped in brain development. This can be considered to tbe cruelest form of inequity, since the new millennium will be a

"Knowledge Millennium", where information, knowledge and intellectual property will determine the pace and direction of economic growth and human well-being. As stressed at the UN Social Summit held at Copenhagen in 1995, the persistence of extreme poverty and deprivation in the midst of islands of prosperity will only lead to social disintegration and numerous mini conflicts. Balanced diet and safe drinking water are the first among the hierarchical needs of human beings and the international community and national governments must make every effort to ensure that all members of the human family have an opportunity for a productive and healthy life.

Today, in most parts of the world, undernutrition and malnutrition result from inadequate economic access to food and lack of clean drinking water, and not because of lack of food in the market. There is therefore need to address more intensive methods of harnessing science and technology for increasing income and employment opportunities. In most developing countries, the contribution of agriculture to GNP is going down, but there is no commensurate drop in the role of agriculture in providing opportunities for jobs and livelihoods. Therefore, the onus of providing more food, jobs and income falls on the farm sector. Job-led economic growth is the need of the hour.

While the challenges are great, progress in science and technology has opened up uncommon opprtunities for a food secure world. Advances in the fields of biotechnology and information, space, renewable energy and management technologies has been spectacular in recent years. At the same time, there is a growing realization that science for sustainable development must be rooted in the principles of ecology, economics, gender equity and ethics. This will involve blending traditional technologies and ecological prudence with frontier science, leading to the development of eco-technologies.

4. Impact of Climate Change

Among the different components of potential changes in climate, changes in precipitation will have the most serious consequences in areas where there is likelihood of either recurrent drought or floods. It is therefore important to promote anticipatory research for developing technologies and public policies which on the one hand, can help to mitigate the adverse consequences of droughts and floods, and on the other, help to maximise the benefits of favourable temperature and rainfall. Computer simulation models provide guidelines for such anticipatory research. For example, MSSRF has established a **Genetic Resource Centre for Adaptation to Sea Level Rise** in a Mangrove forest near Chidambaram in Tamil Nadu, India. At this Centre, a gene pool is being assembled for the purpose of breeding crop varieties tolerant to sea water intrusion. Thanks to rapid progress in genome mapping and molecular breeding, we can now design crops to suit different growing conditions. The development of designer crops to meet the challenge of climate change should receive high priority.

In addition to tapping frontier science, there is an equal need to preserve and use traditional wisdom. For example, the traditional water harvesting and saving techniques adopted in the past in India were very effective in insulating human and animal populations from acute water scarcity. Anil Agarwal and Sunita Narain (1996) have pointed out that in the desert region of Jaisalmer in India, with an annual rainfall of 100 mm, there is enough drinking water for the people even in severe drought years because of their habit of storing water in traditional rainwater harvesting structures called **Kunds**. In contrast, Cherrapunji, a village in the northeast of India, with an annual rainfall of 15000 mm. suffers from water shortage during summer months, since there has been forest denudation and since the local population have had no tradition of water harvesting and conservation.

During 1997, two major climate related events have attracted global professional, political and public attention. These are **El Nino** and the Kyto Protocol for reducing green house emissions. Tornadoes in Florida and California during 1998 have resulted in extensive media attention to the **El Nino** phenomenon in the United States of America. However, in the case of **El Nino**, we do not know what anticipatory steps we should take. Computer simulation models based on past **El Nino** experiences can however be prepared.

While in the past we used to deal only with the impact of climate on agriculture, we are now discussing also the potential **impact of climate change** on food production. Changes in temperature and precipitation can exert a profound influence on crop production. An increase in ultraviolet-B radiation as a result of damage to the ozone shield will also affect agriculture adversely. Unfortunately, there is little public discussion on such issues in most developing countries. Countries in the European Union prepared a proposal, in preparation of the December, '97 Kyoto meeting of the Conference of Parties to the Framework Convention on Climate, for achieving a 15% cut in the emission of greenhouse gases by the year 2010. This would require a drop of 800 million tonnes of CO2 emissions. Achieving this goal would cost between US dollars 15 to 21 billion or between 0.2 to 0.4 per cent of European GDP in the year 2010. Some of the steps involved in this transition are:

- UK's shift from coal to gas-fired stations will result in a 6% drop in greenhouse gas emissions by the year 2000.
- Closure of absolete factories in East Germany would result in a 12% cut in emission.
- Modernization of power stations and switching to cleaner fuels will result in cutting CO2 emissions by nearly 300 million tonnes.
- Increasing the contribution of renewable energy to about 10% of electricity generation by 2015. The real hope is wind power, using recent Danish technology which involves installing wind mills at sea, in shallow waters upto 45 feet deep.
- Introduction of fuel-efficient cars and transportation would lead to cutting down CO2 emissions by nearly 180 million tonnes.

Thus, the European strategy for preventing unfavourable changes in climate involve concurrent attention to the energy and transportation sectors. Developing countries should develop a nationally debated and accepted strategy for achieving a balance between carbon emissions and absorption. Prevention of deforestation and promotion of greening will help to increase carbon sequestration. While the industralised countries are largely responsible for the present situation where human activities are beginning to influence climate, it will be poor nations and the poor in all nations who will suffer most from adverse changes in climate. Effective action at home and emphasis abroad on a "Polluter pays" principle should be the two pronged strategy of developing nations in dealing with climate change issues.

Climate Management and Sustainable Food Security

1979 was a severe drought year in India. As Secretary for Agriculture, I then developed a detailed strategy for monsoon management. The three major components of this strategy are the following:

First, establish in every district a **Crop-Weather Watch Group** consisting of climatologists, farm scientists, representatives of farmers' and womens' organisations, concerned officers of government and representatives of financial institutions and mass media. The tasks to be addressed by such a Group include monitoring monsoon progression, development of contingency plans and alternative cropping strategies to suit different weather probabilities, building seed reserves of alternate crops and intensifying efforts in the area of water harvesting and minor irrigation. The aim is both to maximise the beneficial impact of a good monsoon on agricultural productivity and minimise the adverse impact of aberrant rainfall through efficient water saving and use, crop life saving practices and contingency land use plans.

Second, demarcate in each district the most favourable areas (MFA) from the point of view of intensifying agricultural production through appropriate public policies and investment, particularly in minor irrigation and water management. MFA areas are those where the moisture retention capacity of the soil is high and where irrigation facilities are either available or can be created. Compensatory production programmes designed to offset to the extent possible crop losses in the drought or flood affected areas, can be introduced in MFA areas.

Third, develop strategies for introducing effective relief and rehabilitation measures in the areas most seriously affected (MSA) either by drought or floods. In chronically drought prone areas, such measures should include earmarking community land for establishing cattle camps to save the lives of farm animals and identifying aquifers which can be conserved as "ground water sanctuaries" for being tapped for the supply of drinking water only when absolutely essential.

The above three-pronged strategy will help to minimise both human suffering and crop losses when monsoon behaviour is abnormal and results in drought or floods. A decentralised Rural Godown Scheme, also introduced in 1979, was designed to promote the decentralised storage of harvested produce, so as to prevent distress sales by farmers when the harvest is good and panic purchase by consumers when crop losses are high.

Like Peace Corps volunteers, governments should sponsor the training of atleast one male and one female volunteer in every village in the science and art of climate management. Such trained members can be designated "**Climate Managers**". Wherever possible, an appropriate technical institution should provide such village level Climate Managers information derived from computer simulation models, so as to help them to **be prepared** to handle both adequate and aberrant rainfall.

The need for a micro-level understanding and management of temperature and precipitation is evident from the fact that though rainfall may often be normal in national terms, the total food grain production may go down due to climatic variations at the local level. Micro-level management promotes the use of precision-farming techniques involving plant-scale agronomy rather than just area based approaches. Plant scale agronomy is knowledge and information intensive and affords opportunities for making farming intellectually stimulating, in addition to being economically rewarding. Precision farming methods will be very helpful in facing the challenges arising from climate change.

International Network for Anticipatory Research to Meet the Potential Impact of Climate Change on Food Security

The coming into force of a global convention on climate and the signing of a protocol for the protection of the ozone shield are indicators of the concerns of the international community on the potential adverse impact of increased green house gas emissions and damage to the ozone layer. The Inter-Governmental Panel on Climate Change (IPCC) has been monitoring global trends in climate change. The Global Environment Facility (GEF) has also included climate change as one of the four major areas of concern to GEF.

Computer simulation models of the potential impact of changes in temperature, precipitation and sea level are now available in many industrilized countries. Data on enhanced ultraviolet-B radiation on crop and farm animal productivity are also becoming available. The mandate of national and international agricultural research centres includes attention to the **stability** of crop and animal production. Many such Centres also hold large *ex situ* collections of germplasm. Hence, they are in a position of comparative advantage in helping to initiate anticipatory research to avoid and mitigate potential adverse changes in climate and sea level. Several International Agricultural Centres like the International Rice Research Institute have also long experience in studying the relationship between climate and crop yields.

Experience with field trials with transgenics in USA has shown that there is a distinct genotype × environment interaction with reference to the expression of Bt toxin carrying transgenic cotton. It is likely that genetically modified crops (GMOs) will be widely grown in the coming millennium in developing countries also.

Therefore it will be useful to organise an International Network for fostering anticipatory research to meet the potential impact of climate change on food security. Such a Network should be serviced by a Coordinating Unit which could advise on priorities for screening germplasm for tolerance to climatic changes. Genotypes, which can be used in breeding strains for tolerance to heat, coastal salinity, floods, etc., can then be identified.

An International Network, comprising IARCs, interested National Agricultural Research Systems (NARS) and Advanced Research Institutes could be established within the Consultative Group on International Agricultural Research (CGIAR) system, serviced by a coordinating unit.

Such proactive action will be an important step in helping NARS to optimize the benefits of favourable growing conditions and to minimize the adverse impact of unfavourable climatic changes. The cost of servicing the Network could be kept low through electronic information linkages with appropriate advanced research institutions and meteorlogical departments.

The Coordinating Centre could also advice IARCs on the progress being made in short and medium term weather forecasting and its implications for agronomic management of farming systems. Weather forecasting is an area where considerable progress ie being made and IARCs should take the lead in developing strategies for enhancing the stability of crop production based on the effective use of weather forecasts. Thus, a small initiative in this area could confer multiple benefits in achieving the goal of coupling productivity advance with stability of production.

In the past, food production was described as a "gamble of rainfall" in many countries. Today, we are in the fortunate position to harness new scientific tools to minimise the "gamble" component of agriculture. A marriage between modern science and traditional wisdom will help to ensure the food security is sustained under varying climatic scenarios. At the same time the global community should work in a concerted manner to avoid large human-induced changes in climate. Both avoidance and mitigation strategies should be developing at the local, national, regional and global levels, so that people everywhere will have a happy common future.

5. Safeguarding the Ecological Foundations of Sustainable Agriculture: Agrobiodiversity

The concept of Farmers' Rights developed in the forum of the International Commission on Plant Genetic Resources of FAO (now renamed as Commission on Genetic Resources for Food and Agriculture) has undergone considerable refinement during the last 10 years. As proposed originally, Farmers' Rights was a method of acknowledging the invaluable contributions of farm women and men to the conservation and improvement of plant genetic resources. Enhancement takes the form of selection for characters such as agro-ecological adaptation, resistance to biotic and abiotic stresses and improved culinary qualities, as well as knowledge addition through information on desirable traits. Knowledge addition is of particular value in the case of medicinal plants. Several books like the U S National Academy of Sciences publications on "The Lost Crops of the Incas" and "the Lost crops of Africa", bring out clearly the crucial role played by tribal and rural families in selecting and conserving plants of significance to the food and health security of human and farm animal populations (NRC, 1989 & 96).

Before the advent of well structured government sponsored methods of *in situ* conservation and *ex situ* preservation, the dominant method of conservation was *in situ on farm* by local communities. It is this system of preservation that has resulted in numerous folk varieties and rich intra-specific variability. The over 100,000 rice strains preserved cryogenically in Gene Banks like the one at the International Rice Research Institute (IRRI) in the Philippines, are the products of the in situ on-farm conservation traditions of farm families. *Ex situ* and *in situ* conservation methods are widely supported from public funds, since they are regarded as "public good" activities. Unfortunately, a very important component of the genetic conservation continuum, namely *in situ* on-farm conservation is yet to be recognised as an activity of supreme public interest. Tribal and rural families are thus conserving genetic variability for public good at personal cost. It is this inequity inherent in the current recognition and reward systems that the concept of Farmers' Rights seeks to end.

Convention on Biological Diversity and Farmers' Rights

CBD is a significant landmark among international agreements, since it incorporates for the first time the principles of ethics and equity in both access to genetic wealth and in sharing benefits. Article 15(1) of the CBD recognises the sovereign rights of nations over the genetic resources occurring in their respective countries. Article 15(5) stipulates the need for prior informed consent in the use of genetic resources, while article 15(4) proposes that access should be on mutually agreed terms. Several articles of CBD stress the need to recognise and conserve traditional knowledge and wisdom. CBD also draws attention to the critical role of women in the conservation and improvement of genetic resources.

CBD is legally binding, while the FAO undertaking on Plant Genetic Resources is not. The FAO Commission on Genetic Resources for Food and Agriculture has therefore been discussing methods of getting a suitable protocol incorporated in CBD which will capture the principles and commitments underlying FAO's International Undertaking in Genetic Resources. The method of getting such a protocol prepared and incorporated in CBD was discussed at the Fifth extraordinary Session of the Commission on Genetic Resources for Food and Agriculture held at Rome from 8-12 June, 1998. Article 12 of the draft undertaking discussed at this meeting deals with Farmers' Rights. This Article starts with the following statement.

"Parties recognise the enormous contribution that farmers of all regions of the world, particularly those in the Centres of origin and crop diversity, have made and will continue to make for the conservation and development of plant genetic resources which constitute the basis of food and agriculture production throughout the world". Different options are under consideration of this Commission for converting rights into financial rewards. The major difficulty arises from the fact that the contributions are often made by entire communities and therefore cannot be attributed to individuals. Therefore, procedures are needed to recognise and reward Community Contributions to genetic resources conservation and selection. In order to make progress, the Commission has requested FAO to carry out an analytical financial study on possible formulas for the sharing of benefits based on different benefit-indicators establishing the relative total amounts and relative contributions corresponding to each country and region.

TRIPS and Farmers' Rights

An important element of the World Trade Agreement which came into force in 1994 is the Trade Related Aspects of Intellectual Property Rights (TRIPS). According to article 27.2(b) of the TRIPS agreement, "plants and animals other than microorganisms" may be excluded from patentability. However, member countries are required to provide "for the protection of plant varieties by patents or by an effective *sui generis* system". Though not mentioned in the World Trade Agreement, it is generally assumed that an *effective sui generis* system will be one which is in conformity with the 1978 Act of the International Convention for the Protection of New Varieties of Plants (UPOV Convention). From April 24, 1998 the 1991 Act of UPOV has come into force. The 1991 Act provides countries with the option of restricting Farmers "Plant back" rights. It strengthens breeders' privileges and has the potential to reduce farmers' options.

The Member States of the UPOV Convention now total 37. Several other countries including India are currently preparing legislation in response to the obligations under the World Trade Agreement. The 1991 UPOV Act will impose restrictions on "seed savers". Following this Act, it will be necessary to draw a clear distinction between the following two categories of farmers.

Farmer-Cultivator — These are farmers who cultivate crops for the market or for self-consumption. In their case, rights refer to freedom to save seeds for raising crops and for entering into a limited exchange or sale in their neighbourhood.

Farmer Conservers — This is the group to which FAO's concept of Farmers' Rights applies. They practice *in situ* on-farm conservation and are value-adders in term of information and selection.

A US patent No 5723765, taken by Delta and Pine Land Co., and the United States of America represented by the Secretary of Agriculture on March 3, 1998 titled "control of plant gene expression" has aroused considerable concern in developing countries. This genetic technique is to enable seed companies to ensure that farmers do not raise the F2 population from F1 seeds. The technique, now christened in popular writing as "terminator technology", was the topic for discussion at the Conference of Parties to CBD at Bratislava in May,1998. The Working Group on Agricultural Biodiversity in Bratislava concluded that this technology needs careful study from the point of view of its impact on the conservation and sustainable use of biological diversity. The Group hence urged Parties, Governments as well as civil society and public and private institutions "to consider the precautionary approach in its applications".

The terminator technology will further reduce the rights of farmer-cultivators. They will have no option except to buy seed from the Seed Company every year. Genetic homogeneity in crops will then become more widespread, enhancing genetic vulnerability to pests and diseases. Hence, this aspect of Farmers' Rights needs careful consideration not only from the point of view of ethics and equity, but even more importantly from the view point of maintaining genetic diversity at the field level. Large areas may be covered by the same genotype and on-going efforts directed towards the revitalisation of the on-farm conservation traditions of farm families and the breeding of location specific varieties through participatory breeding methods will suffer.

Converting Farmers' Rights from rhetoric to reality

The first step in this direction is the introduction of legislation in all countries which will concurrently deal with breeders' and farmers' rights. Such an Act which can be titled "Plant Variety Protection and Farmers' Rights Act" should clearly distinguish between the rights of farmer-cultivators and farmer-conservers (Swaminathan, 1995, 1996). As far as farmer-cultivators are concerned, each country in accordance with national needs and socio-economic conditions will have to take decisions on issues like plant-back rights and the use of "terminator technology", For example, in India which has nearly 100 million operational holdings, denial of plant back rights or the use of the terminator mechanism will be disastrous, since over 80% of farmers plant their own seeds.

The area which is causing problem in international fora is the operationalisation of the rights of farmer-conservers. Here again, the difficulties to be overcome are in the following areas.

- How to recognise community contributions to genetic resources conservation and value addition through selection and information?
- If a particular gene critical to the success of a new variety has come from different areas, how to determine priority in recognition and reward?
- While it is feasible to develop a system of recognition of informal innovations in genetic resources conservation and improvement at the national level, how can this be handled at the international level?

In this regard three issues need to be considered.

- First, National Plant Variety Protection and Farmers' Rights Acts should provide for raising a Community Gene Fund through procedures like levying a 1% cess on the sale of all agricultural commodities. In the Act, there should be a mandatory provision requiring plant breeders to provide with as much precision as possible information on the source/area from which the genes critical to the success of the variety for which protection is sought, came. Where there are doubts, DNA finger printing methods can be used. Based on such information, the concerned National Authority can broadly identify the communities which have conserved / selected the donor parents. Based on the extent of spread of the new variety, funds can be provided to the Community for use in strengthening facilities for on-farm conservation or for such other purposes the community may decide. There are several examples in India and elsewhere (See Swaminathan, 1997) where benefit sharing practices have been adopted by research institutions / companies on the basis of a voluntary code of conduct.
- Second, CBD proposes bilateral agreements as the pathway for sharing biodiversity among countries. This procedure provides an opportunity for including respect and recognition for the contributions of farmer-conservers in the agreement for sharing benefits. This will have to be done on a case by case basis, except where generic bilateral agreements are entered into. In the case of agreements dealing with exchange in a generic manner, suitable provisions can be made for the recognition of both farmers and breeders.

- The pedigrees of modern plant varieties contain parents drawn from several countries. A National Act or a bilateral agreement can deal with the material occurring within a nation or between two nations. Hence, a multilateral system of recognition of Farmers' Rights needs to be built into the multilateral system of exchange of genetic resources (MUSE) currently being considered by the FAO Commission.

We have several decades of experience in implementing Breeders' Rights. This has led to four revisions of the UPOV Act. In the case of Farmers' Rights also, we should have the will to learn and to wait. There is no quick-fix to complex problems Breeders and farmers are allies in the fight against hunger and food insecurity. Their rights should therefore be mutually reinforcing and not projected to be antagonistic.

The TRIPS agreement will come up for review next year. We should enter the new millennium with a TRIPS Plus, where the Plus involves integrating the dimensions of equity and ethics in the patenting and *sui generis* systems of rewarding innovations. If this is not done, conflicts rather than cooperation will characterise efforts in the implementation of CBD.

For example, the Governing Council of the Organization of African Unity (OAU) at a recent meeting has approved the adoption of a model Act by Member States, which would lead to African countries not recognising any patent on a drug made from natural products found in Africa unless it acknowledges the ownership and contribution of the relevant community to the new product. Bioprospecting can lead to symbiotic biopartnerships rather than to biopiracy only if the rights of Farmer-conservers are recognised through an integrated package of legislative measures and voluntary codes of conduct. The provisions of TRIPS are due for review during 1999-2000. During this review, it should be ensured that the principles of ethics and equity embodied in CBD are incorporated into TRIPS.

6. Towards a Century of Hope

FAO, the International Food Policy Research Institute and the World Watch Institute have made projections regarding the likely food scenarios in the years 2010, 2020 and 2030 respectively. According to IFPRI, global demands for cereals is projected to increase between now and 2020 by 41 per cent and for meat by 63 per cent. The "food gap" i.e. the difference between production and demand for food could more than double in the developing world in the next 25 years. The World Watch Institute predicts that India may have to import more than 40 million tonnes of food grains by the year 2030. It has been predicted that China will also have to import more than 200 million tonnes by 2030. Where will all this food come from?

It is clear that the further intensification of food production in industrialised countries will be ecologically disastrous. On the other hand, inability to achieve the intensification, diversification and value-addition of farming systems and farm products will be socially disastrous in developing countries like India, where agriculture is the main source of rural income and employment. It is socially, politically and economically essential that India realises its potential for becoming a major supplier of farm products in the world. How can this be achieved?

First, we must develop a symbiotic partnership between the private sector industry and resource poor farmers in bridging the gap between potential and actual yields and in adding economic value to the time of agriculture labour. Some of the pathways through which such partnerships for sustainable food security can be promoed are described below.

- **Contract farming rather than corporate farming**: This will involve firm buy-back arrangements of crops and commodities produced for a company. Such an arrangement can also be made in the case of tree species by the paper and pulp industries.
- **Wasteland development**: This is an urgent task. Upgrading degraded land requires technology and capital. Industry-farmer linkages can help enterprises requiring wood as raw material.
- **Water harvesting and Pulses and oilseeds Villages**: In dry farming areas, industry-farmer partnerships will help to promote scientific water harvesting and the use of conserved water f or producing high value – low water requiring crops like pulses and oilseeds. This has become clear from the FICCI – SPIC – MSSRF and MSSRF – IOB projects in the low rainfall areas of Tamil Nadu.
- **Urban Green Belt Movement**: A green-belt movement can be fostered by industry linking the rural producer and the urban consumer in a symbiotic manner.
- **Seed Villages**: With the spread of hybrid strains even in self-pollinated crops like rice, there are many opportunities for business-farm familis collaboration in seed production and technology
- **Horticulture and animal husbandry**: Industry-producer linkages will be of particular value in perishable commodities, since these need capital and management intensive facilities at the post-harvest stage, such as cold storages and refrigerated transportation and processing technologies.
- **Agri-business enterprises**: The Small Farmer Agribusiness Programme designed by MSSRF at the instance of the Government of India needs for its success the active participation of industry in promoting value-addition to primary products and in linking products with markets.
- **Decentralised production supported by key centralised services**: There is need for such arrangements in the production and marketing of a wide range of processed foods, horticultural and animal products and goods produced in micro-enterprises. The All-Women Biotechnology Park being established at Chennai by the Government of Tamil Nadu with support from the Department of Biotechnology, Government of India, is an example.
- **Manufacture of farm implements**: This is another area where large industries can foster small scale manufacturing units in villages.
- **Farmers' Service Centres**: There is need for the growth of Farmers' service centres particularly in dry farming areas, for custom hiring of farm machinery and implements.

Such mutually beneficial industry-resource poor families linkages will help to strengthen the livelihood security of the economically under privileged sections of the Society and thereby help to promote economic access to food.

Second, we must end the prevailing mismatch between production and post-harvest technologies. Value-addition to primary products as for example spices, cereals, pulses, oilseeds, tuber crops and vegetables should receive high priority. Wheat and rice afford opportunities for the manufacture of a wide range of processed foods such as flakes, noodles, breads etc. In the case of rice, every part of the biomass – straw, bran and husk could be used for producing a wide range of valuable products. For realising the full potential in value addition, we should also tailor crop varieties to end uses. This is particularly important in *durum* (macroni) and bread wheats (biscuits and leavened bread) as well as in vegetable crops like tomato and potato. **Breeding Crop Varieties for the processed foods industry** is yet to receive the attention it deserves. New kinds of semi-processed foods involving nutritious crops like millets and pulses need to be produced.

Designer Crops

We should also design new crops to fulfill the needs of human health as well as food processing. For example, there is considerable variability in crops like rice and wheat in the content of iron, zinc and sulfur. There is also variability in tannin content and activity (tannins are phenolic components in plants that precipitate minerals and make them unavailable for absorption in the gut). For example, in rice varieties, the content of iron, zinc, sulfur and tannins has been found to vary from 40 to 77 ppm, 24 to 42 ppm, 1.69 to 2.79 ppm and 10 to 41 ppm respectively. Tannins not only reduce the nutritional value of foods with regard to iron and zinc; they also impede digestibility of proteins. It is possible now through the use of molecular breeding techniques to design new crop varieties which could cater to the health foods industry. India has a comparative advantage in providing to the global markets health and organic foods, because of its rich traditions in ecological farming.

In 1966, a programme for the development of a high-yielding basmati rice variety was initiated at the Indian Agricultural Research Institute, New Delhi. After over 20 years of carefully planned breeding and selection work, the world's first high yielding dwarf basmati variety **Pusa Basmati 1** was released. Pusa Basmati yields 1.0 to 1.50 tonnes more than traditional Basmati varieties. India is now exporting 0.5 to 0.6 million tonnes of high quality basmati rice valued at around Rs 1500 crores. The share of Pusa Basmati in such export is about 35 per cent.

By filling the gap between potential and actual yields and by adopting low external input sustainable agriculture and aquaculture methods, India cannot only meet the food needs of its growing population but can also help to meet the food gap in both developing and industrialised countries. Industrialised countries, whose agriculture is driven by capital, technology and subsidies should help developing nations to increase their agricultural trade. Unfortunately, the reverse is happening as exemplified by the patent taken in the United States of America for rice varieties belonging to the **Basmati** category. Basmati rice has been cultivated for centuries in North West India and the Punjab part of Pakistan and is a major source of agricultural export earning for India and Pakistan. Unless patenting procedures are rooted in the principles of ethics and equity, agriculture, which is the foundation for a healthy and productive life for all, will become a source of conflict, rather than a source of partnership for achieving freedom from hunger.

The growing problems of unemployment, poverty, feminisation of agriculture and unplanned migration of the rural landless to towns and cities resulting in the growth of urban slums can be solved only through an agricultural renaissance. The opportunities for this are great. But to avail of these opportunities, it is essential that both Governments and industry make a major investment in rural infrastructure, post-harvest technology, sanitary and phytosanitary measures and the information and skill empowerment of rural families. The future lies in precision farming based on the conservation of soil and water and the adoption of ecotechnologies developed through an appropriate blend of frontier technologies like biotechnology, information and space technologies, renewable energy and management technology with traditional wisdom and methods. Agriculture should not be regarded only as a profession for feeding the urban population, since it is the only pathway which can ensure both the livelihood security of a majority of our population and the ecological security of the countryside.

Given appropriate public policies and a greater investment in science and technology, and rural infrastructure, the nearly 500 million farming families who toil in sun and rain to feed the world will help produce adequate agricultural commodities to ensure a healthy and productive life for all children, women and men of our planet.

References

1. Anil Agarwal and Sunita Narain (1996). Dying Wisdom, Centre for Science and Environment, New Delhi. 404 pp.
2. Lester Brown (1995). Who will feed China? Wake-up call for a small planet. W W Norton Company, New York 163 pp.
3. Precision Agriculture in the 21st Century: Geospatial and Information Technologies in Crop Management. National Academy Press. Washington, D.C., 149 pp.
4. National Research Council, USA (1989). Lost crops of the Incas: Little known plants of the Andes with promise for worldwide cultivation. National Academy Press, Washington D.C., 415 pp.
5. National Research Council, USA (1996). Lost crops of Africa. Vol. I. Grains, National Academy Press, Washington D.C., 383 pp.
6. National Research Council, USA (1997). Precision Agriculture in the 21st Century. National Academy Press, Washington, D.C., 149 pp.
7. Rabbinge, R and H C Van Latesteijn (1992). Long term options for Land use in European Community. Agricultural Systems **40**, 195–210.
8. The Royal Society, London (1998). Genetically modified plants for food use. London, September, **98**, 16 pp.

9. Swaminathan, M.S. (1968). The age of algeny, genetic destruction of yield barriers and agricultural transformation. Presidential Address, Agricultural Science Section, Fifty fifth Indian Science Congress, January 1968. *Proceedings Indian Science Congress*, Varanasi.
10. Swaminathan, M.S. (1987). Abnormal Monsoons and Economic Consequences. The Indian experience. Chapter 6 in Monsoons. (ed.) J. S. Fein and P. L. Stephens. John Wiley & Sons, Inc. pp. 121–133.
11. Swaminathan, M.S. (ed.) (1993). *Wheat Revolution: A dialogue.* Macmillan India Ltd., Madras 164 pp.
12. Swaminathan (1994). *Uncommon Opportunities: An agenda for Peace and equitable development by International Commission on Peace and Food.* Zed books, London. 210 pp.
13. Swaminathan, M.S. (ed.) (1995). *Farmers' Rights and Plant Genetic Resources: A dialogue.* Macmillan India Ltd., Madras. 440pp.
14. Swaminathan, M.S. (ed.) (1996). *Agro-biodiversity and Farmers' Rights.* Konark Publishers Pvt Ltd. Delhi. 303 pp.
15. Swaminathan, M.S. (1996). *Sustainable Agriculture: Towards an Evergreen Revolution.* Konark Publishers Pte Ltd, Delhi. pp 232.
16. Swaminathan, M.S. (1997). Implementing the benefit sharing provisions of the Convention on biological diversity: Challenges and opportunities. *Plant Genetic Resources Newsletter*, No. 112. 19–27.

SECTION II

SCIENCE AND FOOD SECURITY — HOW SCIENCE IS USED TO GENERATE EFFICIENT AND OPTIMAL AGRICULTURAL OUTPUTS

Section II

Science and Food Security

How science is used to generate efficient and optimal agricultural outputs

Commentary by M.S. Swaminathan

The green revolution of the nineteen sixties provided a breathing spell for achieving an adjustment between population growth and food production. For the first time, the rate of growth in food production exceeded the growth rate of population. This became possible due to the introduction of crop varieties with a plant architecture and physiological rhythm which enabled the plant to utilize effectively the available soil nutrients and irrigation water. However in recent years, there has been a stagnation in productivity in the heartland of the green revolution in India, viz., Punjab, Haryana and Western Uttar Pradesh. This is largely because of the ecological problems associated with intensive agriculture caused by the excessive use of fertilizers and pesticides and unsustainable exploitation of soil and ground water. It has became clear that to sustain agricultural advance and food security, there is need for mainstreaming ecological considerations in technology development and dissemination. This is why the concept of ever-green revolution involving advances in productivity in perpetuity without ecological harm was proposed and popularized. The papers in this section deal with our understanding of the genetics and cytogenetics of improving crop productivity. Information on the genetic make-up of human population is also helpful.

SCIENCE AND SUSTAINABLE FOOD SECURITY

31st Jawaharlal Nehru Memorial Lecture

by

Prof. M.S. Swaminathan

13 November 1999
New Delhi, INDIA

SCIENCE AND SUSTAINABLE FOOD SECURITY

I. INTRODUCTION

The Bengal famine of 1942-43 provided the backdrop to our independence. Gandhiji said in Noakhali in 1946, "God is bread to those who are hungry" and that it is a fundamental duty of free India to ensure that every individual is enabled **to earn** his/her daily bread. This led Jawaharlal Nehru to pronounce in 1948 that "everything else can wait, but not agriculture."

From 1950-51 onwards, when the first Five Year Plan was developed, the different Five Year Plans have included strategies for promoting accelerated agricultural growth through mutually reinforcing packages of technologies, services and public policies. Policies for land reform such as ceiling on land holdings and security of tenure were introduced. Investment in irrigation received high priority. Technology development and dissemination, training, credit, rural infrastructure and home and external trade in agricultural commodities also received detailed attention. In spite of all these efforts, the rate of growth in agricultural production was not adequate until the beginning of the seventies to meet the needs of the fast growing population. Several eminent foreign experts concluded in the sixties that India has no option except to become the home of serious and recurrent famines.

It was in the midst of such an atmosphere of gloom that the then Prime Minister, Indira Gandhi, released a special stamp in July 1968 titled "the Wheat Revolution". The stamp carried a portrait of the Library building of the Indian Agricultural Research Institute (IARI), New Delhi, to indicate that this revolution was spearheaded by scientific research.

What did scientists do to merit this recognition? They helped to fill the missing gap in the package programme introduced in the early sixties under the Intensive Agriculture District Programme (IADP). The package programme had an important missing ingredient, namely genetic strains which can respond well to the rest of the package consisting of irrigation water and fertilizer. This missing link was provided through the High-Yielding Varieties Programme (HYVP) initiated in 1996-97. The various political and administrative decisions taken during the sixties which helped to transfer scientific progress into production in farmers' fields have been described by Shri C Subramaniam, who served with great vision as the Food and Agriculture Minister during the period 1964-67.[1]

Today, we are a leading Nation in the world in the production of wheat, rice, vegetables, fruits and milk. The milk revolution was the product of a management revolution, which led to small producers acquiring the power of scale in the areas of processing and marketing. When Thomas Malthus predicted in 1798 an emerging adverse imbalance between human numbers and the human capacity to produce food, the global population was about 960 million. Today, India's population alone is one billion. Synergy between science and public policy and farmer's receptiveness to new technologies have so far helped us to ensure that Malthusian fears do not come true.

In spite of satisfactory progress in food production, widespread food insecurity prevails at the level of individual households. It is now clear that sustainable food security involves concurrent attention to the following three components:

- **Availability** in the market, which is a function of production.

---

1. C Subramaniam, 1995. Hand of Destiny. Vol. 2 The Green Revolution. Bharatiya Vidya Bhavan, Mumbai.

- **Access** to food, which is related to purchasing power and hence to jobs or livelihoods.
- **Absorption** of food in the body, which is determined by access to safe drinking water, environmental hygiene and primary health care.

Poverty is now the primary cause of household food insecurity. The household is also not a homogenous unit, with women and girl children tending to suffer more from endemic hunger. Economic inequity is widening nationally and globally. According to UN data, the proportion of the planet's wealth enjoyed by one fifth of the world's population living in richer countries increased massively between 1960 and 1998. UNDP's Human Development Report 1999 points out that OECD countries, with 19% of the global population, have 71% of global trade in goods and services, 58% of foreign direct investment and 91% of all Internal users. The world's 200 richest people more than doubled their net worth between 1994-98, to more than $1 trillion. At the same time, 1.3 billion people – two thirds of whom are women – live without access to balanced diet, clean water, sanitation, primary health care and basic education. The gap between the rich and poor is widening both between and within nations and globalization is creating new threats to the livelihood and security of men and women living in poverty.

India's position in Human Development was 132 among 174 nations, according to UNDP's Human Development Report of 1999. HDI represents an integrated index involving criteria such as life expectancy at birth, adult literacy, school enrollment, and per capita income. India's human resource is further characterised by great gender inequity. India's position in Gender Development Index was 112 in 1999. Canada and Norway occupy the number 1 and 2 positions in the world both in human development and gender equity.

Our experience in agriculture has shown that synergy among science, public policy and social mobilisation can help us to leapfrog in accomplishments.

I believe that unless we make rapid progress in both human development and gender equity, we will not be able to harness the fruits of both traditional wisdom and frontier science for building a sustainable food security system in the country during the 21st century. Agriculture in the new century will be knowledge intensive and will be based on ecotechnologies which are not only environment friendly and economically viable but are also socially equitable both in gender and economic terms. Unless agriculture becomes both economically rewarding and intellectually satisfying, it will not be possible to attract and retain youth in farming professions in the years ahead. I would therefore like to deal with the implications of Jawaharlal Nehru's statement made 50 years ago that "agriculture cannot wait" for both scientists and political leaders in the first decade of the coming century.

II. AGRICULTURE IN THE 21ST CENTURY

As we say good bye to this century we can look back with pride and satisfaction on the revolution which our farm men and women have brought about in our agricultural history. Writing about the role of our farm families in initiating the Wheat Revolution in India, I wrote in the Illustrated Weekly of India in 1969 and I quote:

> "Brimming with enthusiasm, hard-working, skilled and determined, the Punjab farmer has been the backbone of the revolution. Revolutions are usually associated with the young, but in this revolution, age has been no obstacle to participation. Farmers, young and old, educated and uneducated, have easily taken to the new agronomy. It has been heart-warming to see young college graduates, retired officials, ex-armymen, illiterate peasants and small farmers queuing up to get the new seeds. At least in the Punjab, the divorce between intellect and labour, which has been the bane of our agriculture, is vanishing".

While we can and should rejoice about the past achievements of our farmers, scientists, extension workers and policy makers, there is no room for complacency. We will face several new problems and I would like to mention some of them:

- First, increasing population leads to increased demand for food and reduced per capita availability of arable land and irrigation water.

- Second, improved purchasing power and increased urbanisation lead to higher per capita food grain requirements due to an increased consumption of animal products.

- Third, marine fish production is tending to become stagnant and coastal aquaculture has resulted in ecological and social problems.

- Four, there is increasing damage to the ecological foundations of agriculture, such as land, water, forests, biodiversity and the atmosphere and there are distinct possibilities for adverse changes in climate and sea level.

- Five, while dramatic new technological developments are taking place, particularly in the field of biotechnology, their environmental, food safety and social implications are yet to be fully understood.

- Finally, Gross Capital formation in agriculture is tending to decline in both public and private sectors during the present decade. The rate of growth in rural non-farm employment has been poor.

Since land and water will be shrinking resources for agriculture, there is no option in the future except to produce more food and other agricultural commodities from less per capita arable land and irrigation water. In other words, the need for more food has to be met through higher yields per units of land, water, energy and time. It would therefore be useful to examine how science can be mobilised for raising further the ceiling to biological productivity without associated ecological harm. It will be appropriate to refer to the emerging scientific progress on the farms as an "*ever-green revolution*", to emphasise that the productivity advance is sustainable overtime since it is rooted in the principles of ecology, economics, social and gender equity and employment generation.

The green revolution has so far helped to keep the rate of growth in food production above population growth rate. The green revolution was, however, the result of "public good" research, supported by public funds. The technologies of the emerging gene revolution in contrast, are spearheaded by proprietary science and can come under monopolistic control. How can we take the fruits of the gene revolution to the unreached?

III. MEETING THE CHALLENGES AHEAD

A. The Gene Revolution

It was in 1900 that Mendel's laws of genetics were rediscovered. Gregor Mendel had published his work on inheritance patterns in pea in 1865 but it took 35 years to grasp their significance. Since 1900, we have witnessed steady progress in our understanding of the genetic make-up of all living organisms ranging from microbes to man. A major step in human control over genetic traits was taken in the nineteen twenties when H J Muller and L J Stadler discovered that radiations can induce mutations in animals and plants respectively.

In the nineteen thirties and forties, several new methods of chromosome and gene manipulation were discovered, such as the use of colchicine to achieve a doubling in chromosome number, commercial exploitation of hybrid vigour in maize and other crops, use of chemicals like Nitrogen Mustard and ethyl methane sulphonate to induce mutations and techniques like tissue culture and embryo rescue to get viable hybrids from distantly related species. In 1953 came the discovery of the double helix structure of DNA (deoxyribose nucleic acid), the chemical substance of heredity, by James Watson and Francis Crick. This triggered explosive progress in every field of genetics. As we approach the end of the 20th century, we see a rapid transition from Mendelian to molecular genetics with reference to applications in agriculture, medicine and industry. This brief capsule of genetic progress from 1900 to 1999 should be adequate to stress that knowledge and discovery represent a continuum, with each generation taking our understanding of the complex web of life to a higher level. It would hence be wrong to worship or discard experimental tools or scientific innovations because they are either old or new.

Just as it took 35 years for biologists to understand the significance of Mendel's work, it may take a couple of decades more to understand fully the benefits and risks associated with genetically modified foods. It will be prudent to apply the precautionary principle in areas relating to human health and environmental safety. As one who started his research in plant genetics 50 years ago, I must confess that I never dreamt that we can gain so much control over the very blueprints of life.

The past ten years have seen dramatic advances in our understanding of how biological organisms function at the molecular level, as well as in our ability to analyze, understand, and manipulate DNA molecules, the biological material from which the genes in all organisms are made. The entire process has been accelerated by the Human Genome Project, which has poured substantial resources into the development of new technologies for working with human genes. The same technologies are directly applicable to all other organisms, including plants. Thus, a new scientific discipline of genomics has arisen. This discipline has contributed to powerful new approaches in agriculture and medicine and has helped to promote the biotechnology industry.

Several large corporations in Europe and the United States have made major investments in adapting these technologies to produce new plant varieties of agricultural importance for large-scale commercial agriculture. The same technologies have equally important potential applications for addressing food security in the developing world.

Let me illustrate from our work the power of genetic modification to do immense good to agriculture and food security. It is now clear that the 21st century may witness changes in temperature, precipitation, sea level and ultraviolet β radiation, as a result of global warming. Such changes in climate are expected to adversely affect India and Sub-saharan Africa. All human induced calamities affect adversely the poor nations and the poor among all nations the most. This led us to initiate an anticipatory research programme to breed salt tolerant varieties of rice and other crop plants in coastal areas, in order to be prepared for sea water intrusion into farmland as a result of a rise in sea level. The donor of salt tolerance was a mangrove species belonging to the family Rhizophoraceae. Transferring genes for tolerance to salinity from mangrove tree species to rice or tobacco is an impossible task without recourse to recombinant DNA experiments. Thus, the immense benefits that can accrue from genomics and molecular breeding are clear.

What then are the principal concerns? In industrialised countries, the major concerns relate to the impact of GMOs on human health and the environment. These food and environmental safety concerns have been well documented and are widely known. The professionals, public and political leaders of developing countries are equally concerned about the food and environmental safety aspects of GMOs. The viewpoints of countries in the North on the ethical and social issues relating to GM crops have been dealt with in detail in a report published by the Nuffield Council on Bioethics in May 1999. What are the additional issues which agitate the minds of the public and professionals in the developing world?

The first issue relates to biosafety. Why are large biotechnology companies averse to the labelling of GM foods? In spite of over three years of intensive discussion in meetings sponsored by the Secretariat of the Convention on Biological Diversity, the negotiations broke down at Cartagena, Colombia, in February 1999. Thus, there is as yet no internationally agreed biosafety protocol, as called for under Article 19 of CBD. The absence of such a protocol will hurt the private sector the most.

Besides biosafety concerns, the following three issues agitate the minds of Indian public and professionals:

First, India is a land of small farm holdings. There are now 106 million operational holdings in the country, about 75% of which are one hectare or less in size. India has 25% of the global farming community and farming provides livelihoods to nearly 66% of the population, now numbering a billion. There is concern that expansion of proprietary science and shrinking of "public good" research supported from public funds may lead to a situation where the technologies of the future remain in the hands of a few transnational corporations. Only resource-rich farmers may have access to them, thereby widening further the already wide rich-poor divide. This could accelerate the process of social disintegration.

Second, the monopolistic control over crop varieties could lead to a situation where large areas are covered by very few genetic strains or hybrids. It is well known that genetic homogeneity enhances genetic vulnerability to biotic and abiotic stresses. Biotechnology companies are hence recommending resistance management strategies, as for example growing 30 to 40% non-Bt maize with Bt maize. Unfortunately, it is now clear that a single gene could confer resistance to several Bt strains including strains with which the insect has had no prior contact. What will happen to the livelihoods of farm men and women operating small holdings with institutional credit and with no crop insurance cover, if GM maize, soyabean, rice, potato or other crops are affected by serious diseases as a result of the breakdown of resistance? Will the companies agree to compensate them for the loss? This problem could become even more serious if companies incorporate genetic use restriction mechanisms known popularly as "terminator" in hybrids. Small farmers will then experience genetic enslavement since their agricultural destiny will be in the hands of a few companies.

A third issue relates to the potential impact of GM foods on biodiversity. This aspect has two dimensions – one dealing with the replacement of numerous local cultivars with one or two GM strains, thereby leading to genetic erosion, and the other relating to equity in benefit sharing. Modernisation of agriculture has resulted in a narrowing of the base of food security both in terms of the number of species constituting the food basket and the number of genetic strains cultivated (see Lost Crops of Incas 1989 : Little Known Plants of the Andes with promise for world wide cultivation. National Research Council, U S Board on Science and Technology for International Development, National Academy Press, Washington DC, pp. 415 and Lost Crops of Africa 1996 National Research Council, U S Board on Science and Technology for International Development, National Academy Press, Washington DC pp. 383). The local cultivars have often been the donors of many useful traits, including resistance to pests and diseases. Under small farm conditions, every farm is a genetic garden, comprising several crops, both annual and perennial, and several varieties of each crop. The need of the hour is to enlarge the food basket and not shrink it further.

The other aspect of GM foods and biodiversity relates to the equitable sharing of benefits between biotechnologists and the primary conservers of genetic resources and the holders of traditional knowledge. At present, the primary conservers remain poor, while those who use their knowledge (for example, the medicinal properties of plants) and material become rich. This has resulted in accusations of biopiracy. It is time that genetic engineers promote genuine biopartnerships with the holders of indigenous knowledge and conservers of genetic variability, based on principles of ethics and equity in benefit sharing.

Unless research and development efforts on GM foods are based on principles of bioethics, biosafety, biodiversity conservation and biopartnerships, there will be serious public concern in India as well as many other developing countries about the ultimate nutritional, social, ecological and economic consequences of replacing numerous local varieties with a few GMOs. Also, under conditions where the market is the dominant factor in determining research priorities, "orphans will remain orphans" in terms of investment of research funds, unless the public sector steps in. For example, it will be useful to apply the "terminator" mechanism to control alien invasive species like *Parthenium*. Such problems are unlikely to attract private investment. For deriving benefits from genetic engineering without undue risks, the Government of India should set up a multi-stakeholder **National Commission on Genetic Modification and Food and Health Security**.

B. The Ecotechnology Revolution

Knowledge is a continuum. There is much to learn from the past in terms of the ecological and social sustainability of technologies. At the same time, new developments have opened up new opportunities for developing technologies which can lead to high productivity without adverse impact on the natural resources base. Blending traditional and frontier technologies leads to the birth of ecotechnologies with combined strength in economics, ecology, equity, employment and energy.

For example, in the area of water harvesting and sustainable use, there are many lessons to be learnt from the past. In the desert area of Rajasthan, India, drinking water is available even in areas with 100 mm annual rainfall, largely because women are continuing to harvest water in simple structures called **kunds**. In contrast, drinking water is scarce during summer months in some parts of the North East India with an annual rainfall of 15000 mm. Thus, there is need to conserve traditional wisdom and practices, which are often tending to become extinct (Anil Agarwal and Sunita Narain, 1998, Dying Wisdom, Centre for Science and Environment, New Delhi). The decision of the World Intellectual Property Organization (WIPO) to explore the intellectual property needs, rights and expectations of holders of traditional knowledge, innovations and culture is hence an important step in widening the concept of intellectual property. Principles of ethics and equity demand that this invaluable component of IPR gets included when the TRIPS agreement (Trade-related Intellectual Property Rights) of the World Trade Organization comes up for review in 1999-2000. FAO has been a pioneer in the recognition of the contributions of farm families in genetic resources conservation and enhancement by promoting the concept of "Farmers Rights". Like WIPO, UPOV (Union for the Protection of New Varieties of Crops) should also undertake the task of preparing an integrated concept of breeders' and farmers' rights. UNESCO and FAO should develop a **Universal Declaration on the Plant Genome and Farmers' Rights** on the lines of the already approved declaration on the Human Genome and Human Rights.

New technologies supported by appropriate services and public policies have helped to prove doomsday predictions wrong and have led to the agricultural revolution (the green revolution) becoming one of the most significant of the scientific and socially meaningful revolutions of this century. Four thousand years of wheat cultivation led to Indian farmers producing 6 million metric tons of wheat in 1947. The green revolution in wheat helped to surpass in 4 years the production accomplishments of the preceding 4000 years, thus illustrating the power of technological change. There are uncommon opportunities now to harness the power of a new social contract among science, society and public policy to address contemporary development issues such as growing rich-poor divide, feminisation of poverty, famine of jobs, human numbers exceeding the population supporting capacity of ecosystems, climate change and loss of forests and biodiversity. Whether in economics or in ecology, experience has shown that a trickle down approach does not work. Fortunately, modern information technology provides opportunities for reaching the unreached. Computer aided and internet connected **Virtual Colleges** linking scientists and women and men living in poverty can be established at local, national and global levels for launching a knowledge and skill revolution. This will help to create better awareness of the benefits and risks associated with GMOs, so that both farmers and consumers will get better insights into the processes leading to the creation of novel genetic combinations.

The future of small farm families belongs to taking to precision agriculture, which involves the use of the right inputs at the right time and in the right way.

Biotechnology will play an important role in all the following six major components of integrated natural resources management and Precision farming:

- Integrated Gene Management
- Efficient Water Management
- Integrated Nutrient Supply
- Soil Health Care
- Integrated Pest Management
- Efficient Post-harvest Management

Ecotechnology based precision farming can help to cut costs, enhance marketable surplus and eliminate ecological risks. This is the pathway to an ever-green revolution in small farm agriculture.

C. The Yield Revolution

India's current position in the world in a few major crops is given in Table 1. In several crops and more particularly in wheat, our farmers have made striking progress. In 1947, we produced a little over 6 million tonnes of wheat; in 1999, our farmers harvested over 72 million tonnes, taking our country to the second position in the world in wheat production. The high position we occupy in the production of several crops is to a considerable extent due to the large area covered under those crops. Our position in productivity is however unenviable.

The position in pulses illumines the pathway for a new strategy in agriculture. We occupy the first position in the world in both area and production of pulses, but the 118th position in productivity (Table 1). A major reason for our low average yield is the cultivation of pulses mostly under conditions where soils are both thirsty and hungry. A Pulses Technology Mission now exists and it will be prudent to link it to the watershed development movement recently launched by Government. Our experience in organising **Pulses Villages** in the dry districts of Pudukkotai and Ramanathapuram in Tamil Nadu indicates that we can make rapid progress in improving the production and productivity of pulses, provided the farming families of the village/watershed cooperatively undertake the harvesting of every drop of rain water. There will be no cooperation in water harvesting, unless there is equity in water sharing. This is where high value but low water requiring crops play an important role in ensuring that the resource poor farm men and women get maximum income from the available water.

Productivity improvement will be possible only if we pay greater attention to improving the efficiency of input use, particularly nutrients and water. To cite just one example, our cotton yields are less than 20 per cent of the yields achieved in several other countries like Egypt and USA. However, we are using 25 times as much water to raise a ton of cotton as compared to California. Normally to produce 1 ton of grain, about 1000 tons of water may be needed but in most cases we are using much more.

To bridge the gap between actual and potential yields prevailing at the currently available levels of technology, we have to undertake a multi-disciplinary analysis in different regions and farming systems. For example, in rice which occupies the largest area among food crops, the opportunities for rapid progress are great. The States where speedy progress can be made are indicated in Table 2. The growing availability of rice hybrids increases the prospects for rapid progress in productivity improvement.

Wheat

We now occupy the second position in the world in wheat production. We grow bread, *durum* (macaroni) and dicoccum wheats. In order to meet the needs of our growing population it will be necessary to produce about 110 million tonnes of wheat by 2020. This is entirely feasible both because of the untapped yield potential in several wheat growing areas and because of the possibility of introducing hybrid wheat. Our aim should be to become the number one country in wheat production by 2010. We should also produce in the Himalayan region hard wheats to satisfy the needs of automatised bakeries.

If we enable farmers with appropriate technologies, services, prices and markets, our average yield can go up to 42 quintals per ha. from the present 27 quintals per ha. Yield improvement should be our goal in every crop and farm animal. This will be possible if our farm families can be helped to improve the efficiency of use of water, nutrients, plant protection chemicals, and post harvest technology.

IV. INTEGRATED NATURAL RESOURCES MANAGEMENT

Integrated natural resources management holds the key to sustainable food and livelihood security. There is need for new management systems, involving partnerships based on principles of equity and ethics, to conserve and improve natural resources. Policies are urgently needed to conserve prime farm land for agriculture and to ensure the sustainable use of the ground water.

I wish to take biodiversity, one of the key components of our basic life support systems, as an example to illustrate what changes are needed in the management of our biological resources. It is now widely realized that the genes, species, ecosystems and traditional knowledge and wisdom that are being lost at an increasingly accelerated pace limit our options for adapting to local and global change, including potential changes in climate and sea level. The Global Biodiversity Assessment published in 1995 by the United Nations Environment Programme (Cambridge University Press) estimates that about 13 to 14 million species may exist on our planet. Of this, less than 2 million species have so far been scientifically described. Invertebrates and micro-organisms are yet to be studied in detail. In particular, our knowledge of soil microorganisms is still poor. Also, biosystematics as a scientific discipline is tending to attract very few scholars among the younger generation.

Another important paradigm shift witnessed in recent decades in the area of management of natural resources is a change in the concept of "common heritage". In the past, atmosphere, oceans and biodiversity used to be referred to as the common heritage of humankind. However, recent global conventions have led to an alteration in this concept in legal terms. Biodiversity is recognised under the CBD as the sovereign property of the nation in whose political frontiers it occurs. Under the UN Convention on the Law of the Sea, nations with coastal areas have access to a 200 mile Exclusive Economic Zone (EEZ). For example, the ocean surface available to India under the EEZ provision is equal to two thirds of the land surface available to the country. The Climate Convention and the Kyoto protocol provide for both common and differentiated responsibilities to countries. Thus, the global commons can be managed in a sustainable and equitable manner only through committed individual and collective action among nations.

While we have some knowledge of variability at the ecosystem and species levels, our knowledge of intra-specific variability is poor, except in the case of plants of importance to human food and health security. The

Global Biodiversity Assessment warns, "Unless actions are taken to protect biodiversity, we will lose forever the opportunity of reaping its full potential benefit to humankind". What kind of action will help us to ensure not only the conservation of biodiversity, but also its sustainable and equitable use? In my view, we must foster an **Integrated Gene Management System** in every State of the country.

The Integrated Gene Management system includes *in situ, ex situ* and community conservation methods (Fig. 1). The traditional *in situ* conservation measures comprising a national grid of National Parks and protected areas are generally under the control of government environment, forest and wildlife departments. The exclusive control of such areas by government departments has often led to conflicts between forest dwellers and forest dependent communities, and forest officials. The non-involvement of local communities in the past in the sustainable management of forests has resulted in a severe depletion of the forest resources in India. It has become clear that sole government control alone will not be able to protect prime forests or regenerate degraded forests.

Participatory Forest Management (PFM)

The Participatory Forest Management (PFM) procedure initiated first at Arabari in the State of West Bengal in India in 1972, largely at the initiative of a young Forest Officer, Mr Ajit Kumar Banerjee, and later extended to the other States in the country became a significant turning point in the history of forest management in India as well as other Asian countries (see Samar Singh *et al*, 1997 and Mark Poffenberger *et al*, 1997). The essential feature of this system is that the State and community become partners in the management of the forest resource. The State continues to own the resource but the benefits are shared. Access to non-timber forest products becomes an important avenue of sustainable livelihoods to the forest-dependent communities. Thus, the community develops an economic stake in the preservation of forests, leading to conservation and sustainable use becoming mutually reinforcing components of a Forest Management Policy. The experience gained in India during the last 25 years shows that the process of natural forest degradation can be reversed through PFM and that forests can provide to the local community non-wood forest products on a continuous or seasonal basis, if there is a more widespread understanding of their regenerating capacity. Since forests are the home for a large proportion of naturally occurring biodiversity, saving forests results in saving genes.

Community Gene Management

Both *in situ* on-farm conservation of intra-specific variability, particularly in plants of food and medicinal value, and *ex situ* on-farm conservation through sacred groves have been part of the cultural traditions of rural and tribal families in India. In the Old Testament also, there are several references to sacred groves. Among the important trees usually preserved in Indian Sacred Groves are *Ficus religiosa, Saraca asoca, Shorea robusta, Alstonia scholaris* and many other species of ecological, economic and spiritual value. Unfortunately, several of these traditions are now tending to wither away. It is only by giving explicit recognition to the pivotal role of community conservation in strengthening ecological, food and health security systems that we can succeed in the revitalization of these traditions. In national integrated gene management systems, *in situ*, *ex situ* and community conservation methods should receive adequate and concurrent attention. A recognition and reward system, based on FAO's concept of Farmers' Rights and CBD's provisions for ethics and equity in benefit sharing, should become an integral part of national legislation in the fields of biodiversity conservation and plant variety protection.

Structure of Community Gene Management System (CGMS)

CGMS comprises the following three steps.

- Field Gene Bank (FGB) at the village level
  In situ on-farm Conservation by rural and tribal families
  Participatory Breeding (Selection)

- Area Seed Bank (ASB) at the level of a cluster of villages
  Conservation of representative samples of seeds in seed stores
  Insurance against total loss of seeds during drought years

- Community Gene Bank and Herbarium (CGB)
  Cryogenic preservation
  Evidence for getting tribal families reward and recognition under the proposed Plant Variety Protection and Farmers' Right Act

Field Gene Banks are basically *in situ on-farm* centres of conservation. Landraces and location-specific plant genetic resources (PGR), usually identified on a participatory basis with the local families, are conserved in FGBs. The local landraces are periodically grown in their native habitats for seed renewal. They serve as effectively decentralized and highly cost-effective arms of a community gene management system. Those FGBs which are situated at higher elevations are likely to have cooler and conducive climates for storing seeds without loss of viability. No expensive infrastructure is needed to set up FGBs. The FGBs would ensure *in situ on-farm* maintenance of landraces and preservation of the cultural and cropping practices under which PGR acquired their distinctive traits. The communities in natural resource areas would be stake-holders in the FGBs. They can then provide active support not only to prevent genetic erosion in those areas, but can add new genetic stocks as and when they discover them. FGBs serve the cause of both conservation and continuous evolution through selection, recombination and mutation.

Several FGBs can be linked to an Area Seed Bank (ASB) taking into account factors like distance, communicability, conservation space and the like. There could be 2 or 3 seed banks in a district. The Area Seed Bank will help to strengthen coping mechanisms for facing the problem of seed scarcity caused by drought-induced crop failures. The ASB would be managed by a committee of the stake-holders. Logically this committee can monitor IPR issues such as prior informed consent and benefit sharing. One or more ASBs will be linked to CGB. The CGB would hold *ex-situ* the seed stocks of landraces etc, along with herbarium sheets and other information needed to get the primary conservers reward and recognition under the proposed Indian Act for Plant Variety Protection and Farmers' Rights. The CGB could deposit duplicate sets of accessions in the National Gene Bank (NGB) of the National Bureau of Plant Genetic Resources at New Delhi.

The operational efficiency, low transaction costs and conservation potential of such a network will encourage their replication and implementation at State and National level. Thus, an effective integrated national gene management system can be built up.

The Community Gene Management System is being introduced in three predominantly tribal districts of Orissa, namely Koraput, Khandamal and Kalahandi. The Koraput district is a primary center of origin (particularly of the *aus* ecotype) of rice. Though genetic variability in rice has been well chronicled, there is a need to purify the seeds of landraces, multiply and evaluate them as they are direct and indirect sources of valuable genes.

Biosphere Management

Methods of fostering the sustainable management of Biosphere Reserves were discussed at a conference convened by UNESCO in 1995 at Seville in Spain. The Seville vision states:

> "Rather than forming islands in a world increasingly affected by severe human impacts, biosphere reserves can become theaters for reconciling people and nature. They can bring the knowledge of the past to the needs of the future"

To convert the above vision into field level reality, we have been working for a shift in the nature of management of Biosphere Reserves in India. India has currently 85 National Parks covering 3.6 million sq.kms and 448 wildlife sanctuaries covering about 120,000 sq kms in the major biogeographic zones. The total extent of protected areas includes 5 designated as World Heritage sites, 9 biosphere reserves and 6 Ramsar sites. With support from the Rajiv Gandhi Foundation the M S Swaminathan Research Foundation (MSSRF) conducted in 1994 a detailed study of the threats to the Gulf of Mannar Biosphere Reserve in Tamil Nadu. The study showed that unless the livelihood security of the families without assets living in that area are strengthened, unsustainable exploitation of biological resources will continue.

During 1996-98, MSSRF initiated a proposal for bringing the Gulf of Mannar Biosphere Reserve under a participatory management mode. This proposal was submitted by the Governments of Tamil Nadu and India to GEF through UNDP for support. The project proposal was considered by the Council of the Global Environment Facility (GEF) at its meeting held in May 1999. The GEF Council approved the proposal and commended it as a model project which deserves to be widely emulated by those preparing similar projects. The Scientific and Technical Advisory Panel (STAP) of GEF endorsed the project for approval with the following remarks:.

> "The project addresses a major challenge, namely the conservation of coastal biodiversity of the highest ecological value in a large area subject to considerable pressure from poor populations upon the sole resources that appear to be at their disposal. To meet this challenge, the project follows the only framework which can succeed, namely to combine the necessary protection of the threatened ecosystem and ecological processes with economic and social benefits which will meet the essential needs of local people, through providing appropriate institutional, financial and managerial arrangements".

The management structure through which people and Nature will get united in the Gulf of Mannar area is through a **Gulf of Mannar Biosphere Trust**. Representatives of fishermen and rural communities as well as all the principal civil society stake-holders will, together with government representatives, form the members of the Trust. The members of the Trust will hold this unique biological treasure in trust for present and future generations. Considerations of intra- and inter-generational equity will guide the work of the Trust, which will provide a horizontal dimension to the numerous vertically structured activities undertaken in this area both by Central and State governments and private sector organizations. It is hoped that the Gulf of Mannar Biosphere Trust, whose long term sustainability will be ensured with the help of a Trust Fund, will show the way for promoting a **management by partnership** system of governance in all the other Biosphere Reserves in the country.

Project Tiger launched in 1972 with support from WWF-International was the first step in institutionalized efforts in the conservation of habitats of endangered animals. An all-India tiger census conducted in 1972 revealed only 1827 tigers in the country as against an estimated 40,000 at the beginning of the 20th century. The 1993 census placed the tiger population at 3750.

One of the tiger reserves is in the Similipal Biosphere Reserve in Orissa. This Reserve was set up in June 1994 and covers an area of 2750 sq. kms. Of late, threats to this Reserve have increased due to both population pressure and commercial greed. Experience with participatory or joint forest management programmes in India has shown that wherever the local communities are fully involved as partners in saving forests, the results have been very encouraging.

It is proposed to develop a similar plan for the Similipal Biosphere Reserve in collaboration with the Government of Orissa. The output of this programme will be the establishment of a **Similipal Biosphere Trust** managed jointly by the major stake-holders.

A major need in such programmes is the strengthening of the livelihoods of the poor families living in the vicinity of the Biosphere Reserve. For this purpose, the Biovillage model of livelihood security will be introduced in the villages around the Reserve. The Biovillage concept of human-centred rural development aims to address concurrently the challenges of natural resources conservation and poverty eradication (Swaminathan M S 1999, A Century of Hope. East West Books(Madras) Pvt. Ltd.). Market-driven livelihood opportunities will be identified and local families will be assisted in taking to them with the help of institutional credit.

V. ATTRACTING AND RETAINING YOUTH IN FARMING

According to the standard UN definition, youth comprises the age group between 15 and 24. According to the 1991 Census of India, youth in this age group numbered 153.6 million. 71.6 per cent of youth lived in rural areas and 51.9 per cent of them were males. Our population now is 1000 million and the youth population in the age group 15 to 24 is likely to be over 200 million. It is a common observation that there is an increasing disinterest among youth in taking to farming as a profession. A vast majority of those who join Agricultural Universities do so to get salaried jobs and not to take to a career of self-employment in agriculture. How can this situation be reversed? I would like to suggest the following steps, as an integral component of the National Agricultural Policy for the early part of the 21st century:

A. Launching an Ever-green Revolution Movement

The ecological and equity issues associated with the green revolution technologies of the seventies and eighties resulting at present in a yield plateau or even decline in the major cereals are now well known. It is equally clear that in the decades ahead, higher production through higher productivity is a necessity. This will call for an agricultural production strategy which is rooted in the principles of conservation and enhancement of natural resources and social and gender equity, resulting in what I have been describing as an **ever-green revolution**. The transition from the green revolution to an ever-green revolution involves a shift from a commodity-centred to a farming systems-centred approach. It also involves the adoption of precision farming technologies which help to maximise yield and income per every unit of land, water and nutrients.

Experience shows that educated youth will be attracted to farming as a profession only if agriculture becomes both intellectually satisfying and economically rewarding. This is why the new eco-farming procedures based on eco- and precision farming methodologies will foster the more extensive participation of young women and men in launching the ever-green revolution movement. We cannot afford to relax on the food production front. An agricultural production strategy based on natural resources conservation spearheaded by youth will help to bridge the prevailing gap between potential and actual yields in most farming systems, and at the same time ensure the long term sustainability of high yields.

B. Implementation at the local level of the provisions of global conventions

Several provisions of the Global Conventions on Biodiversity (CBD), Climate and Desertification will have to be converted into local level action plans. For example, the ethics and equity provisions of CBD and the Farmers' Rights Concept of FAO involve recognising and rewarding the invaluable contributions of tribal and rural families to genetic resources conservation and enhancement. Educated youth can help rural families to revitalise their *in situ on-farm* conservation traditions and promote the improvement of local crops through participatory breeding. They can help to restore the biological potential of degraded land. They can help to enhance carbon sequestration through a programme for the greening of vacant lands. They can help tribal and rural communities in developing and adopting eco-technologies based on a blend of traditional wisdom and frontier science. In particular, they can help farming families to understand the risks and benefits associated with modern biotechnology. They can help them to take anticipatory steps to avoid the adverse impact of global climate change. Youth can through such action help in the practical realisation at the local level of the principles and goals contained in global agreements.

VI. LOCAL BODIES AND A HUNGER-FREE INDIA

The 73rd and 74th constitutional amendments provide a unique opportunity to remedy the ills of our past centralised and top-down approach to national development. The eleventh schedule of the Constitution has indicated 29 items for the economic and ecological regeneration of India. **If Panchayats and Nagarpalikas can be empowered legally, financially and technologically to undertake these 29 tasks, we can achieve speedily seemingly impossible tasks**. It is my conviction that we can achieve Gandhiji's and Nehruji's goal of a hunger-free India, if Panchayats and Nagarpalikas can be enabled to implement the following 7-point action plan. Decentralised implementation by local bodies will help to eliminate corruption, reduce transaction costs and foster transparency. It will help educated local youth to participate actively in improving human well being in their respective areas.

- **Identifying the hungry**

Gram Sabhas / Panchayats / Nagarpalikas should identify the ultra-poor families suffering from chronic under- and mal-nutrition in the areas. This will help in setting priorities in direct intervention measures. The ultra poor usually are illiterate, have no assets and are unskilled. The maximum they can expect to earn is the minimum wage.

- **Reaching the unreached: Household Entitlements Database**

In all the States / Union Territories, there are numerous programmes sponsored both by government and non-governmental organisations as well as multilateral and bilateral donors which are designed to assist families living in poverty to overcome endemic hunger. The hungry who are often illiterate tend to remain ignorant of such programmes. Information empowerment can be achieved by introducing a *household entitlements card* and by establishing computer-aided village knowledge centre. Reaching the unreached and voicing the voiceless should be the goals of such a programme for the information empowerment of the poor.

- **Fighting protein-calorie under-nutrition / malnutrition**

All available avenues of helping the hungry to overcome under- and mal-nutrition should be explored. Targeted public distribution, *Annapoorna*, Food for Work and other similar programmes should be harnessed to overcome calorie deprivation. In the design of local level farming systems, the nutrition security of the local population should be an important criterion.

- **Eliminating hidden hunger caused by the deficiency of micro-nutrients in the diet**

Deficiencies of iron, iodine, vitamin A and other micro-nutrients and vitamins cause anaemia, blindness and other nutrition-related health problems. Through a two pronged strategy consisting of raising horticultural home gardens and of direct interventions like the supply of iodised and iron-fortified salt and oral administration of vitamin A, local bodies can be involved in a movement for the total elimination of hidden hunger. Our goal should be the total elimination of hidden hunger in the country by the year 2007.

- **Ensuring the availability of safe drinking water and the promotion of environmental hygiene**

Clean drinking water and environmental hygiene are essential for ensuring the biological absorption of the food consumed. Children in particular lose the benefit of the food consumed due to diarrhoeal diseases and worm infection. Local bodies should launch a social mobilisation movement for keeping the environment clean through better sanitation and through efficient treatment and recycling of garbage and sewage water. Priority attention should be given to the provision of clean drinking water.

Improving economic access to food

Based on a study of opportunities for remunerative marketing, micro-enterprises supported by micro-credit can be promoted on the biovillage model in operation in Pondicherry. The use of Geographical Information System (GIS) methodologies can help to initiate scientific land and water use based farming systems. Integrated attention should be given to on-farm and non-farm employment and to production and post-harvest technologies. The entire biomass of plants and animals should be used for producing value-added products from every part of the plant or animal. Increasing urbanisation leads to the diversification of food habits and to a greater demand for animal products. Therefore, new opportunities are becoming available for establishing urban green belts and for linking the rural producers and urban consumers in a symbiotic manner. Small scale food processing and agribusiness enterprises based on the methodology of decentralised production supported by key centralised services, will however need for their long term economic sustainability, understanding and support from Members of the World Trade Organization. **WTO should include a livelihood impact assessment among the criteria to be used for determining unrestricted access to the markets of developing countries**. Similarly the Government of India should attach a livelihood impact statement while announcing its import and export policies, with particular reference to agricultural commodities.

- **Special attention to women and children**

Local bodies consisting of 33% women members should pay special attention to the effective implementation of programmes designed to assist children and pregnant and nursing women. Every third child born in the country is characterised by low birth weight (LBW). Such children suffer many handicaps in later life including impaired brain development. The 21st century will be a knowledge century with intellectual property becoming the most valuable property. Inequity at birth in opportunities for brain development is the cruelest form of inequity. **We should therefore lose no further time in eliminating the disadvantage of low birth weight caused by maternal and foetal under-nutrition**. Another area which needs attention is the reproductive health of women. Computer-aided and internet connected Village Knowledge Centres operated by young women can help to achieve rapid progress in reducing infant and maternal mortality rates.

To sum up, elected local bodies should accord priority attention to developing action plans to address the following issues:

- Foetal and maternal under-nutrition induced low birth weight in children is avoided and gender equity is promoted

- Environmental hygiene is improved and clean drinking water ensured.

- Hidden hunger caused by micro-nutrient deficiencies is totally eliminated

- Knowledge systems are established for ensuring sustainable food security at the level of each individual and for fostering an ever-green revolution on the farms.

VII. REVOLUTION IN HUMAN DEVELOPMENT: GOALS FOR THE 60TH ANNIVERSARY OF INDIA'S INDEPENDENCE

Poverty persists under conditions where the human resource is undervalued, and land and other material resources are over-valued. The late Dr Mahbul Huq and the UNDP have rendered a valuable service by initiating the Human Development Reports. **Focus on human development is the best way of getting national priorities right**. In the 1999 Human Development Report of UNDP, India ranked 132 among 174 nations in Human Development Index (HDI). Life expectancy at birth, adult literacy rate, school enrollment and per capita income are the criteria used for estimating HDI. In 1999, India occupied the 112th position in gender development index (GDI). GDI uses the same parameters as the HDI, but adjusts for gender inequality. In the area of gender empowerment measure, which is based on seats in Parliament held by women, female administrators, managers, professionals and technical workers, and women's real GDP per capita, India's position is 95 among the nations included in the 1999 Report.

Without added and urgent attention to human development, it will be difficult to make progress during the coming **knowledge century** in improving food, livelihood and ecological security. **In my view, we should aim to reach the 50th rank in HDI by the year 2007**. Given the necessary political will and action, this is entirely feasible. I shall explain why.

During the last 50 years, the country has made considerable progress in areas relating to human well being. Life expectancy now is nearly 60 years, as compared to 44 years in 1960. Adult literacy is 50% compared to 34% in 1970. Infant mortality is 89 per 1000, down from 165 in 1960. The population of poor people fell from over 50% in 1950 to just over 30% now. However, the actual numbers of persons living in poverty virtually doubled, due to population increase. We have today as many illiterate and poor persons as the entire population of India in 1947.

Our record in the field of education is poor. Only two thirds of children aged 6-14 attended school in 1992-93 and only just over half the population completes more than 4 years of education. Our population is now one billion and the descried goal of TFR 2.1 is yet to be achieved in most States. The United Nations medium projection indicates that our population may reach 1.5 billion in 2050 and could reach 1.6 billion before it gets stabilised.

To quote Jawaharlal Nehru, "We have to run twice as fast to stay where we are". Approximately 350 million adults in India cannot read or write. Infant mortality rates can and need to be reduced by another 80% or more. We will have another 500 million young people to educate and nurture by the middle of the 21st century. So while in percentage terms we may be better off, in absolute numbers the work left to be done is even greater than what we have accomplished till now.

I believe that looking back 100 years from now, India will have achieved rates of life expectancy, literacy and infant mortality comparable to that now prevalent in the most economically advanced countries of the world. But I am not heartened by this expectation. I find it totally unacceptable to say that our descendants a century from now will achieve that which we are perfectly capable of providing to our children and grandchildren today. I believe it is our solemn duty and obligation to make life better for all Indian citizens as fast and as much as possible.

This is why I propose that we set ourselves the inspiring goal of raising our position on the Human

Development Index from 134th place to 50th place by 2007, the 60th year of our independence. This will require that we raise life expectancy by another 10 years, reduce infant mortality by two-thirds and raise adult literacy to 90%. We should also strive to reach the 40th position in GDI and 10th position in Gender Empowerment Measure.

Some of you may think that such goals are unrealistic and unachievable. But I beg to differ. India has already demonstrated in at least one field that our nation can accomplish what many people both inside and outside the country considered impossible. In 1965-66 we lived a "ship to mouth" existence. Two successive years of drought threatened the lives of upto 10 million people with famine. Few people believed that India was capable of feeding itself. Very few could conceive that millions of uneducated traditional Indian farmers could take to modern agricultural methods speedily and enthusiastically.

These were the conditions under which India launched the Green Revolution. A fully committed government launched a massive programme to promote modern methods of cultivation. Hundreds of demonstrations were laid in small farmers fields. New organizations were established to produce hybrid seeds and fertilizers, build warehouses, purchase and distribute foodgrains. Agricultural education and research were revamped to focus them on meeting the nation's urgent needs. Above all, assured and remunerative prices were offered to farmers. Within five years foodgrain production increased by 50% and the country had achieved foodgrain self-sufficiency at the prevailing level of purchasing power. Now the production of foodgrains has reached the level of 200 million tonnes.

This remarkable feat was accomplished at a time when the country was far less equipped in terms of power generation, roads, ports, science and technology, educational levels, industrial facilities, management capabilities and financial resources than it is today. If so much could be achieved by our people at that time, surely the nation is capable of far greater feats of accomplishment today, particularly since we can mobilise "people power" through Gram Sabhas and elected local bodies.

We see a glimpse of that untapped capability in the rise of India's software export industry, which grew 400-fold from a mere $10 million in 1983 to $ 4 billion this year and is projected to reach $ 12 billion by 2002. How do we extend such accomplishments to the basic area of human development?

First it is important to recognize that the country already possesses the needed technological, organizational and financial resources to raise our position on the HDI to 50th place within seven years. We do not need to import technology. We do not need to borrow ideas or money from other countries. There is nothing wrong with borrowing strategies and techniques that have been successful elsewhere, but the idea that we should rely on foreign aid is totally counterproductive. If the **Education Guarantee Scheme** is implemented in such a way that it becomes community owned and managed, we can realise soon the goal of education for all.

Second, we need to redirect our national energies away from preoccupation with political competition to focusing on national development. Today, the bulk of the country's energy is being sapped and wasted in political battles which enable parties and candidates to gain temporary victories, but which contribute nothing to the advancement of the poor people or the country. Let there be a common minimum political agenda agreed to by all political parties for raising literacy, employment, income, gender equity and life expectancy, for stabilising the population and for safeguarding the environment.

Third, we should recognize that we already have formulated and put in place many plans and programmes that are capable of achieving the goal I propose. We have excellent examples in education, as for example the schooling revolution in Himachal Pradesh and Mizoram and the Education Guarantee Programme in Madhya Pradesh (Public Report on Basic Education in India, The PROBE Team, Oxford University Press, 1999). Similarly the success of Kerala, Tamil Nadu, Andhra Pradesh, Goa and Mizoram in stabilising population has many lessons for the rest of the country. We however need more and better implementation and execution. The country should declare a national emergency for achieving a corruption free administration characterised by a high degree of accountability and efficiency.

Four, we should enable local bodies to:

- extend programmes to provide safe drinking water and vaccinations in every village and town and to develop a health security system based on the principle "prevention is better than cure".

- double the investment in education, especially pre-school, primary and adult educational programmes in rural areas.

- provide incentives to encourage the adoption of innovative teaching methods, including the establishment of Virtual Schools and Colleges involving the use of computers, and active involvement of the private sector in rural pre-schools and primary schools.

The challenges facing India today are not for the weak or timid. Nor is it a time to be discouraged or lose heart. This is the time for bold innovative action. Revolutionary efforts that destroy the status quo are necessary to remove barriers to progress, but they are not sufficient to realise our potential. Revolution needs to be supplemented by evolutionary efforts to consciously create the organizational foundations for a better common present and future. It is time to abandon half-hearted, half-way measures and show ourselves and the world what this nation can accomplish. In the emerging Knowledge and Innovations century, human development will hold the key to sustainable food and livelihood security as well as to national well being. It is time to inspire our youth, which represent the best part of the population to fire them with enthusiasm to build a better India. This will be the most appropriate way of fulfilling Jawaharlal Nehru's vision of a new India when he referred to our tryst with destiny.

Table-1
INDIA'S CURRENT POSITION

Crop	Area (1000 HA)			Production (Million Tonnes)		
	India (rank)	Highest		India (rank)	Highest	
Wheat	25122 (3)	China	29001	72.4 (2)	China	109.005
Rice	42700 (1)	India		82.2 (2)	China	190.100
Maize	6150 (5)	USA	29602	8.66* (9)	USA	236.604
Sorghum	11700 (1)	India		10.50* (2)	USA	20.39
Potato	1089 (3)	China	3502	17.94* (6)	China	46.05
Pulses	25604 (1)	India		14.8 (1)	Australia	
Cotton	8300 (1)	India		14.0 (3)	China	18.75
Sugarcane	3870 (2)	Brazil	4826	289.7 (2)	Brazil	324.435

Footnotes:

Production figures for India are 1998-99 estimates taken from the Economic Survey. For the rest of the world, production figures correspond to the year 1996.

Figures pertaining to Productivity and Area correspond to the year 1996.

*1996 production figures used for these items.

Table-2

DISTRIBUTION OF DISTRICTS BASED ON RICE PRODUCTIVITY (1994-95)

	Yield Range (t/ha)						
	>3.5	3.0–3.5	2.5–3.0	2.0–2.5	1.5–2.0	<1.5	Total
Number	17	18	35	60	77	206	414
Percentage	4.3	4.3	8.5	14.5	18.6	49.6	100
<1.5	–	M.P. (43); Bihar (38); Orissa (26); Maharashtra (22); Rajasthan (20); U.P. (18); Assam (13)					
1.5–2.0	–	U.P. (24); Assam (10)					
> 3.0	–	Tamil Nadu (16); Punjab (13); Haryana (3)					

Figure-1

INTEGRATED GENE MANAGEMENT

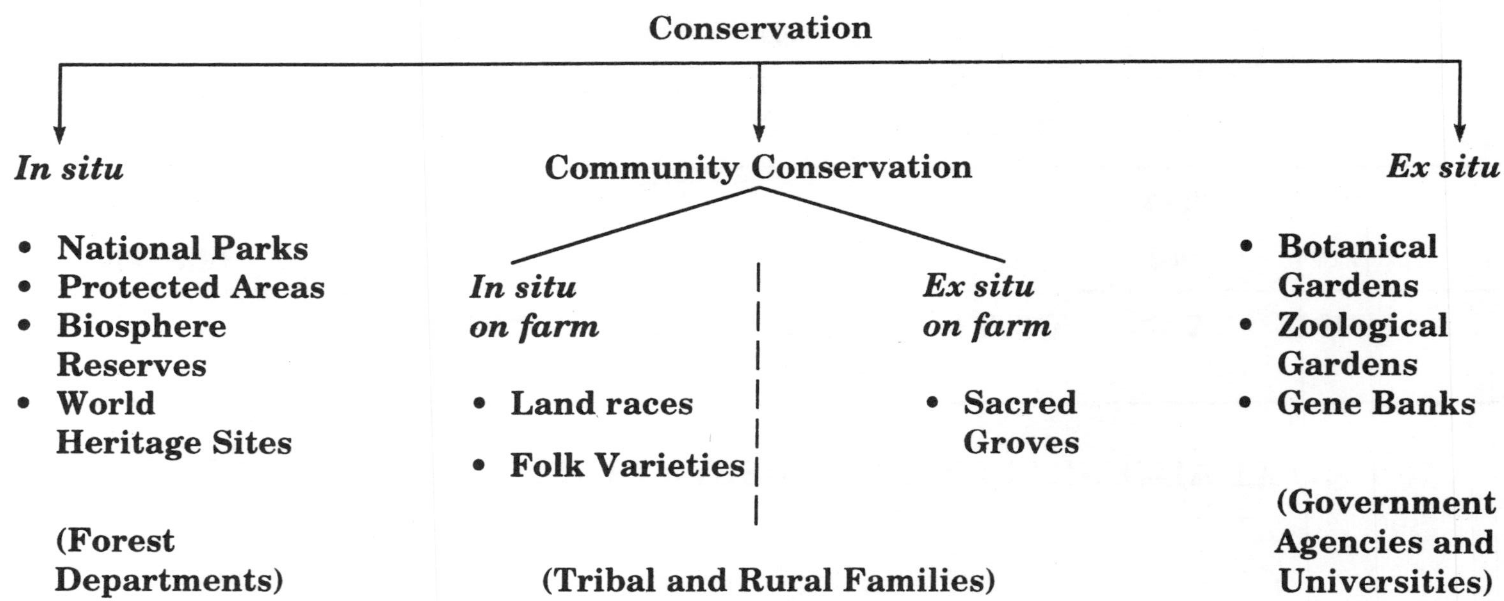

INDIAN AGRICULTURE AT THE CROSSROADS*

M. S. SWAMINATHAN†

Planning Commission, Yojana Bhavan, New Delhi 110 001, India

INTRODUCTION

I AM grateful for the privilege of being asked to deliver the Tenth J.N. Tata lecture. J.N. Tata was known both for vision and action. Where vision is limited, action is equally circumscribed. In a lecture instituted in the memory of a great visionary, it will be useful to discuss the future of our agriculture, since ours is primarily an agricultural country with land and water-based occupations providing most of the jobs in addition to food and income. The most challenging task ahead for a country of our population size and structure is to win the battle against the famine of work.

In the developed countries, the new wave of technological change spearheaded by automation and computerization tends to displace labour from its traditional position as the dominant factor of production. Adoption of this pathway will make our rich human resource a liability than an invaluable asset. The challenge of unemployment can be met only by developing agriculture and rural occupations in a scientific manner. If farming does not lead to expanded internal and external trade, the desired degree of diversification of employment opportunities will not occur in rural and urban areas. Without diversified income earning opportunities, the plight of marginal farmers and the landless poor will become worse as population increases. Most calculations show that over 70% of our work force will have to depend upon agricultural and rural professions for their living until the end of this century.

The importance of land and water-based occupations involving agriculture, horticulture, animal husbandry, fisheries, forestry and agro-industries in providing not only food but also jobs and income cannot, therefore, be over-emphasised.

What is the crossroads we have arrived at in agriculture? On several earlier occasions I had stated that our country can either become a major bread basket of the world or take to the begging bowl to meet our food requirements depending on the nature of public policies adopted and community endeavour generated. In a recent book on "International Trade in Grain and the World Food Economy" by Bastin and Ellis[1], the choice before us in wheat has been stated as follows:

"The USDA model highlights the position of India as a wheat producer of global significance. According to one set of assumptions, the country could be importing 8.6 million tonnes of wheat in 1985, making it the world's largest single wheat importer. According to another alternative set of assumptions, India could be exporting almost 5 million tonnes of wheat in 1985, making it one of the largest wheat exporters."

Should we take to the 'import' or 'enhanced home production' pathway for meeting our requirements of foodgrains, sugar, edible oils and cotton? In my view, the "import pathway" in agriculture, even if we can afford it, will take us to a serious "famine of work" leading to an accelerated migration of rural poor to urban areas seeking jobs. By import pathway. I do not mean occasional imports necessitated by natural factors but making imports a habit to make good for inaction at home. What action points will mark the difference between these two pathways of development? To answer this question, it would be useful to trace briefly the nature of our agricultural evolution during this century.

OUR AGRICULTURAL EVOLUTION DURING THIS CENTURY

For the sake of convenience, I would like to recognise three major evolutionary phases. The first phase, extending from 1900 to 1950, was a period of stagnation in our agricultural and rural economy. The average growth rate in agriculture during this period was about 0.3%. Agricultural stagnation meant rural stagnation and this in turn led to the drain of both brain and resources from the village to the town. The neglect of rural areas was so great that at the time of Independence, most villages did not have even one source of safe drinking water. Even now we have nearly 2,30,000 villages which are yet to be provided with potable drinking water. Stagnation also meant that no attempts were made to provide any form of commercial energy in support of agricultural occupations. Even today, the farm sector which provides nearly 40% of the export earnings and sustains over 70% of the population, consumes only about 10% of the commercial forms of energy (Table I). Thus, the contribution of this sector to our economic well-being is immense, considering the insignificant investment of resources and energy.

† Member in charge of science and technology

* Tenth J. N. Tata Lecture delivered at the Indian Institute of Science, Bangalore on 25 September 1981.

TABLE I*

Relative Shares of different forms of Commercial Energy for various sectors in 1978-79

(in percentage)

Sector	Energy Form			Percentage of total commercial energy consumption
	Coal	Oil	Electricity	
Household	10·0	71·2	18·8	13·7
Agriculture	0·0	61·8	38·2	10·6
Industry	44·5	7·9	47·6	38·5
Transport	13·3	83·9	2·8	31·7
Others	11·9	36·2	51·9	5·5

*Source: Report of the Working Group on Energy Policy, Planning Commission, 1979.

Whenever agriculture failed due to unfavourable weather, famine conditions prevailed. This is why, considerable thought was given in the pre-independence days to the development of Famine Codes and Scarcity Manuals. Above all, colonial rule tended to freeze social evolution. There was total social failure in the field of redistribution of agricultural assets like land and livestock and in the prevention of exploitation of primary producers at both the production and post-harvest phases of farm operations.

The second phase of our evolution covering the period 1950 to 1980 was one of growth in the development of the infrastructure necessary for the modernisation of agriculture. The outmoded land tenure system at last received some attention, resulting in a gradual elimination of the zamindari and other intermediary tenures. Impressive progress took place during this period in developing powerful instruments for technology development, technology transfer and technology sustenance. The Indian Council of Agricultural Research, the network of Agricultural Universities and several associated institutions in the country provided an effective mechanism for carrying out location-specific research and for developing appropriate technologies to suit different agro-ecological, socio-economic and socio-cultural conditions. Technology transfer mechanisms included the development of extension networks supported by input delivery systems. For sustaining agricultural advance, policy and operational instruments such as the Agricultural Prices Commission and Food Corporation of India were developed.

We can see three distinct trends during the period 1950-80, in terms of programme development. The first major step in promoting accelerated agricultural advance was taken in 1961 when the Intensive Agricultural District Programme (IADP) was initiated. This programme was designed to maximise the production in areas endowed with irrigation facilities. Later, the introduction of high yielding varieties of seeds was superimposed over the area approach, since the earlier strains did not respond to higher levels of nutrition and irrigation. Through this step, the missing link in the package programme (IADP), namely genetic strains which can respond well to the rest of the package, was provided. The area approach was also structured to meet specific ecological conditions like drought-prone, hill and desert areas. The area and crop-centred approaches, however, were found to be inadequate for the purpose of enabling the economically handicapped sections of farmers to take to new technology. It is this lacuna which was attempted to be remedied through the introduction of special programmes for small and marginal farmers, agricultural labour and tribal people. Thus, gradually, an individual farmer-cum-a specific area model of programme development was born.

Infrastructure development during the period 1950-80 has been particularly impressive in the fields of agricultural research and education, irrigation, fertilizer production and in the storage and distribution of grains. Credit systems for the rural poor were also improved. As a result of the various steps taken during this period, the growth rate in agriculture rose to about 2.8% per annum, thus keeping slightly above the growth rate in population. Also, land use patterns which were predominantly based on the home needs of the farming family tended to shift towards market requirements.

We have now entered the third and the more challenging phase of agricultural evolution. Thanks to the infrastructure for modernisation developed so far, a higher proportion of farmers have something to offer to the market. Consequently public policies in the field of post-harvest operations have assumed great importance. Also, growing disparities and unevenness in agricultural progress between regions, crops and farmer categories are creating various kinds of tensions and distortions. The rapid growth in population resulting in increasing fragmentation of holdings is making the size of farms smaller and smaller. The size of an operational holding has now reached about one hectare in States like Uttar Pradesh and Bihar (Table II). Therefore, the number of families which have to make a contribution for achieving a specific production target is increasing. The size of the landless labour population is also growing. While in a state like Punjab, foodgrain production increased between 1960-61 and 1978-79 by 8·01% per year, this growth was only 1·19% in Orissa. The annual growth rate for total agricultural

TABLE II

State-wise average size of operational holdings for 1970-71 & 1976-77 as per agriculture Census

(Hectares)

Sl. No.	State	1970-71	1976-77
	Group 1: Land Records States		
1.	Andhra Pradesh	2·51	2·34
2.	Assam	1·47	1·37
3.	Bihar	1·5	1·1
4.	Gujarat	4·11	3·71
5.	Haryana	3·77	3·58
6.	Karnataka	3·20	2·98
7.	Madhya Pradesh	4·0	3·60
8.	Maharashtra	4·28	3·60
9.	Rajasthan	5·46	4·65
10.	Tamil Nadu	1·45	1·25
11.	Uttar Pradesh	1·16	1·05
	Total Group I	2·49	2·18
	Group II: Non Land Records States		
12.	Kerala	0·57	0·49
13.	Orissa	1·89	1·60
14.	Meghalaya	1·7	1·5
15.	West Bengal	1·20	0·99
	Total Group II	1·26	1·04
	Group I & II (11+4) States	2·30	2·00
16.	Himachal Pradesh	1·53	1·65
17.	Tripura	1·02	1·20
18.	Manipur	1·15	1·2
19.	Nagaland	5·40	7·61
20.	All UTs except Mizoram	3·14	2·78
	Total 16–20	1·99	2·16
21.	Punjab	2·89	2·74
22.	Jammu & Kashmir	0·94	0·94
	Total 21–22	2·08	2·03
	Grand Total 1 + 22	2·28	2·00
23.	Sikkim	..	2·56
24.	Mizoram	..	1·49
	Total 23–24	..	2·28
	All India Total	2·28	2·00

production for Punjab during the period 1952-53 to 1969-70 was 6.6% as against 0.7% for Bihar. Minimising regional disparities and extending the benefits of new technology to all classes of farmers, therefore, become areas of priority concern.

Another major issue facing us now is the whole area of land reform and agrarian structure. Land reform has so far been mainly looked at from the point of land ownership, land ceiling and security of tenure. For enabling small and marginal farmers to produce more, land reform will have to be given an ownership-cum-production interpretation covering steps which will facilitate the more efficient use of land. For example, without land consolidation and levelling, it becomes difficult to manage water properly and thereby derive full benefit from our investment on irrigation. Human and animal demographic pressures on land are now threatening basic life support systems like soil, flora and fauna. The Himalayan ecosystem, for example, which sustains over 150 million people, is threatened with ecological disasters. While nature takes anywhere between 100 to 400 years or more to generate 10 mm of top soil, it has been calculated that nearly 1.4 million km^2 of our country are now subject to various forms of soil erosion. Sedimentation as a result of careless use of watershed forests is endangering the economic life of reservoirs, hydroelectric facilities and irrigation systems. Diminished quantities of irrigation water finally lead to smaller command areas and loss of food production. Recent surveys have shown that the storage capacity of Nizamsagar lake in Andhra Pradesh was reduced by 63% in 44 years; of Mayurakshi and Maithon reservoirs by 10% in 20 years; of Panchet reservoir by 12% in 19 years and of Bhakra reservoir by 6% in 16 years. In almost every case the originally estimated life of the reservoir will need re-assessment. Floods ravage the Indo-Gangetic plains almost every year taking their toll in terms of human life and property. Because of poor soil structure and alkalinity, the infiltration of water in the soil profile is poor. Consequently, several flood-affected areas tend to become drought affected soon after the recession of floods and withdrawal of monsoon. Hence land reform in our country should include steps not only relating to the equitable distribution of available land resources among all sections of the rural community but also measures such as consolidation of holdings, land levelling and soil health care which are essential for the sustained agricultural productivity. Agriculture being primarily a solar energy harvesting enterprise is the most important source of renewable wealth. There will, however, be no difference between agriculture and sources of non-renewable wealth, if the renewable base of farming is destroyed, as is happening in several parts of our country.

Another important question facing us now is the need for looking at problems relating to the cost, risk and return structure of the agrarian economy in an integrated manner. Thanks to the availability of new technology and additional sources of irrigation, more and more farmers are using their land, not only for satisfying their home needs but also market requirements. Thus today more farmers have something to offer to the market. When farmers invest more on purchased inputs, the best fertilizer to them is assured together with remunerative marketing. Market reform hence becomes an urgent necessity, so that the gap between what the producer receives and the consumer pays gets minimised to the utmost possible extent. Market reform is often not possible without credit reform since the merchant and the money-lender frequently is the same person. Crops thus get automatically bonded when farmers have to take loans from money-lenders for purchasing production inputs. The whole area of post-harvest technology relating to transport, marketing, pricing, storage and distribution thus assumes great importance. Where there is a mismatch between production and post-harvest technologies, neither the producer nor the consumer gets benefit from increased production. This is why we now witness various forms of both farmers' unrest and consumers' agitations.

Under conditions of small farms, the net "take home" income of the farmer is very small. The risks become correspondingly greater when input costs are high. For example, in our country the price of nutrients has been going up (Table III). Fertilizer prices have increased by about 62% during the past 15 months. What should be our input-output pricing policies? Should input cost be kept high and should farmers be compensated by a higher procurement price? In my view, it is more important that inputs are subsidised and output price kept low, so that consumption levels can be improved, Since there is no crop insurance worth the name and since farming is the riskiest profession in the world, the poor farmer will not be able to derive benefit from the new technology based on purchased inputs, if the price of these inputs is kept high. A majority of our farmers also live in dry farming areas. The soils in such areas have been described as even more 'hungry' than 'thirsty'. Soils in dry farming areas hence respond well to nutrient application. When fertilizer prices are high, balanced fertilization becomes a casualty. If the average productivity is to be improved, more farmers will have to have the capacity to purchase good seeds, fertilizer and appropriate forms of commercial energy. It is not with reference to crops only that we need a rational input-output pricing policy. Animal husbandry provides opportunities for providing supplementary income to the rural poor. Here again farmers are facing difficulties due to a doubling in the cost of feed during the last seven years. In contrast, the cost of egg has not risen by the same margin, although 80% of the cost of production goes to the cost of feed. Since we import edible oils and other farm products

TABLE III

Physical Returns on Wheat and Paddy 1971-72 and 1981-82 (11-7-1981)

(Data compiled by the Fertilizer Association of India)

Particulars	1971-72	1981-82 11-7-81	+% 1981 82 over 1971-72
I. Paddy			
1. Kg. of paddy required to buy 1 kg. of N (through Urea)	3·79	4·87	28·5
2. Kg. of paddy required to buy 1 kg. of P_2O_5 (through DAP)	3·51	5·55	58·1
3. Kg. of paddy required to buy 1 kg. of K_2O (through MOP)	1·68	2·07	23·2
II. Wheat			
1. Kg. of wheat required to buy 1 kg. of N (through Urea)	2·64	3·93	48·9
2. Kg. of wheat required to buy 1 kg. of P_2O_5 (through DAP)	2·45	4·48	82·9
3. Kg. of wheat required to buy 1 kg. of K_2O (through MOP)	1·17	1·67	42·7
Nutrient Prices (Rs./Kg.)			
1. Urea	2·01	5·11	154·2
2. DAP	1·86	5·83	213·4
3. MOP	0·89	2·17	143·8
Output Prices (Rs./Kg.)			
1. Paddy	0·53	1·05	98·1
2. Wheat	0·76	1·30	71·1

which we can and should grow in our villages, there is a tendency to export feed grains and proteins for animals to earn foreign exchange, although we should really be using them for enabling landless labour families to take to poultry rearing and other animal based occupations.

The current phase of our agriculture is hence highly dependent on public policies for its success or failure. Community co-operation and endeavour will be the other major requirement. Under conditions of small and fragmented holdings, it becomes essential that farmers carry out some aspects of farm operations such as water management, plant protection and post-harvest technology on a watershed basis. Fertilizer use

Estimated average fertilizer consumption per hectare and yield of wheat in some selected States (1971-72 to 1974-75)

(Data of ICAR)

State	Area	Irrigated area (million hectares)	Fertilizer consumption (kg/ha)				Yield kg/ha	Response in kg/kg of nutrient
			N	P_2O_5	K_2O	Total		
Uttar Pradesh	6·05	4·14	21·8	5·3	3·6	30·7	1,148	5·6
Punjab	2·35	2·07	58·3	18·7	4·7	81·7	2,307	15·4
Haryana	1·19	1·00	38·8	4·2	1·8	44·8	1,770	18·4

efficiency, for example, varies a great deal depending upon the level of management. In the Punjab and Haryana, application of one kg of nutrient to wheat on the average gives a respose of 12–15 kg of grain. However, in Uttar Pradesh the increase in yield was only 4–5 kg per kg of nutrient until a few years ago (Table IV). Due to lack of appropriate community action, we find that the efficiency of water use is not as high as it could be and also the intensity of cropping is much below the desired levels. Plant protection is poor due to the absence of an area approach in land use, planning and pest control. In black soils, water logging and other soil health problems arise. Above all, poor soil and water management and inadequate soil health care are tending to make the dividing lines between flood and drought narrow. Thus we have entered a stage of agricultural evolution where more detailed attention will have to be given to the co-operative organization of farmers and farming, to the management of both the production and to post-harvest phases of agricultural operations and above all to the public policy package essential for .increasing production as well as consumption. While dealing with these issues, I would like to refer briefly to the emerging trends in world agriculture.

Pathways Of Agricultural Development: Super Farms And Small Farms

The world today is witnessing two contrasting trends of evolution of farm ownership and management systems. In terms of ownership, the land may be either socially or individually owned. In both, free enterprise countries like the United States, Canada and Australia as well as in socialist countries like the USSR, Eastern Europe and China, the size of an operational holding is getting larger and larger. In contrast, in many of the developing countries of Asia and Africa, the size of a farm holding is tending to get smaller and smaller. We can thus recognise, on the one hand, large super farms exceeding 1,000 hectares in size based on capital and energy intensive and labour displacing technology, and on the other, small farms of 2 hectares and below operated largely with family labour and with very little capital or energy investment.

Super Farms

Among super farms, three broad categories can be recognised:

(a) U.S. Model: The United States today serves as a major bread basket of the world. A good example of the trend in U.S. farming methodology is provided by the rice culture operations in California[2]:

"The production of rice in California has evolved into one of the most highly mechanized agricultural operations in the world, including the levelling of fields by laser beam, the sowing of seed by airplane and the harvesting of the crop by special combines that do not bog down in the mud of the rice paddy. The 1979 California yield of 6,450 pounds of rice per acre was 50% higher than the average in the other rice-producing States in the U. S. and nearly three times higher than the world average of 2,360 pounds per acre. Although the acreage devoted to rice in the U.S. is only 0.9% of the world total, the U.S. produces 1.7% of the world crop and in many years is the world's largest rice exporter."

Super farm management technology of this kind is based on a heavy consumption of non-renewable forms of energy. In this system, energy consumption is high not only in the production phase but also in the post-harvest phase of the production-consumption chain. Serious thought is hence being given to the methods of reducing the high dependence on fossil fuels in achieving productivity improvement. For example, a study conducted on agricultural production efficiency by a Committee of the U.S.

National Academy of Sciences showed that modern cropping systems in the United States yielded less than 5 calories of digestible energy per total calorie of agricultural energy. Out of various systems studied, the most energy efficient agricultural systems were labour-intensive vegetable growing in New Guinea and rice culture in the Philippines[3]. However, these comparatively efficient cropping systems were also the least productive in terms of digestible energy per heactare. In other words, the pathway of productivity improvement chosen in the super farm management system involves an exponential rise in energy consumption.

(b) Soviet Model: The Soviet model of super farm operations largely consists of big State-collective farms. Since land is socially owned, a collective farm is run like any other State enterprise. The farms are highly mechanised and because of weather conditions in the USSR, energy consumption in these farms also tends to be very high. Many developing countries have also experimented with such large mechanised farming methods. We in India have a few large State Farms started with the help of USSR. These are mostly used now for seed production.

(c) Cooperative agro-industrial complexes: The third kind of super farm model is the one found in several Socialist countries in Eastern Europe like Bulgaria. This involves the organisation of a large contiguous area into an agro-industrial complex. Each agro-industrial complex has a land use model based on considerations of economics and ecology. Production, processing and marketing are linked together into an integrated system. The Complex provides the needed facilities to the farming families who are members of the cooperative which manages the complex. Land is not individually owned but the members of the complex have a sense of belonging because of the institutional devices introduced for promoting cooperative management of farm operations. Other Communist countries like China have also introduced cooperative systems of management of farm operations within a framework of social ownership of land.

Thus the super farm organisation of agriculture is assuming different manifestations based upon the land ownership pattern, *i.e.*, private *versus* State ownership. Whatever the administrative or ownership structure may be, the super farm operation provides opportunities for introducing scientific management systems such as the following:

1. Crop planning in accordance with agro-meteorological and soil-moisture-retention data on the one hand and marketing opportunities, on the other;
2. Minimum or appropriate tillage;
3. High plant density leading to dense crop canopies;
4. Weed-free environment;
5. Controlled release of fertilizer, use of nitrification inhibitors, foliar feeding and use of low cost, anhydrous ammonia;
6. Use of bacterial, algal and other microbial fertilizers;
7. Integrated nutrient supply involving an appropriate blend of organic, inorganic and biological sources of fertilizer;
8. Better on-farm management of water, including drip irrigation in arid land;
9. CO_2 fertilization for maximising production in glass houses;
10. Integrated pest and disease management involving crop sanitation and agronomic, genetic, biological and chemical methods of control.
11. Use of hormones and growth regulators in fruit trees and ripeners in sugarcane;
12. Organic recycling leading to crop-livestock, crop-fish, and crop-livestock-fish integration.

The super farm technology can also generate the necessary resources to overcome problems arising from risk and return and can help in efficient marketing. However, this kind of farming also faces a few major problems:

(a) high consumption of non-renewable forms of energy;
(b) environmental pollution arising from the application of large doses of chemical fertilizers and pesticides; and
(c) problems of motivation and soil productivity maintenance under conditions where the individual or firm has no long term stake in the land.

Confidence has been expressed by several experts that the energy-intensive super-farm technology could be reoriented gradually towards a zero-energy farming system. For example, Leach[4] describes what can be done in North America and Western Europe as given below:

"The energy-intensive Western farmer could, with little trouble, reduce his overall energy consumption dramatically—and without any loss of yield or return to labour-intensive methods. Recycling animal wastes partially to replace artificial fertilizers, or using them to generate methane, are two frequently urged solutions since they solve the sometimes critical waste disposal problems too. But by far the largest impact would be made by using sensibly a natural fuel that many farmers now burn wastefully and destructively in their fields—cereal straws.

Most farms grow cereals. An average acre of cereals in the U.K. yields roughly one ton of straw with a gross energy value of 3 million kcal (2,000 kcal/lb, 30% moisture content in the field). This is roughly equal to the *total* energy input per acre for most farms, and is more than twice the energy used per acre as fuel and electricity for all farms except pig and poultry enterprises. Preliminary work at Nottingham University shows how this straw might be harvested and turned into a fuel for a very low energy cost. The entire cereal plant is cut with a forage harvester (instead of a combine for the grain plus a tractor and cutter-baler for the straw) and dried. After drying, the grain and straw are separated, and the straw could be burned either in a furnace to provide heat for drying, or in an external combustion engine (e.g. Stirling cycle) to provide mechanical power or even electricity. The straw drying would consume about 12 to 15 gallons of fuel (600,000 to 700,000 kcal) under average conditions, giving a net energy yield of 2·3 to 2·4 million kcal. Alternatively, the straw could be fermented (without drying) to a liquid fuel such as alcohol with a net energy yield of around 1·5 million kcal/acre plus by-product material for feeding ruminants. This last figure equivalent to 36 gallons of diesel fuel, is higher than the liquid fuel consumed/acre for all farms except pig and poultry, and is over 50% higher than the consumption for dairy and cereal farms. With a judicious mixture of fermentation. and direct straw-burning, including burning to generate power, and with animal wastes providing methane or nitrogen, phosphorus and potassium, it seems entirely probable that many farms might become self-reliant for all fuel and power, and would need to import energy only in the form of some fertilizers, machinery and other technical equipment, and sprays."

The super-farm free-enterprise technology is not without its own socio-economic problems. A recent study on the causes which led to farmers' protest in the State of Kansas in the United States indicates that uncertain farm prices in a time of inflating farm costs is tending to get farming resources concentrated into fewer hands and thereby lead to a decline of the rural economy and society. This study also showed that while farmers of Kansas State were ideologically committed to "the free market concept", they favoured such non-market solutions as target prices, favourable loan rates, production controls and collective bargaining[5]. The primary need of the system was identified as an yearly orderly marketing programme.

Under the large mechanised farming operations, the profit per farmer will obviously be high. For example, according to USDA, one American farmer feeds 56 people and by this yardstick the American farmer can be stated to be most productive. But when we measure productivity in terms of land or capital or energy, this superiority is no longer true. This is where small farms have both an advantage and an opportunity.

Small farm development is also taking place in different directions. More than 90% of all tropical farms are less than 5 hectares in size. National averages in Asia are often less than 3 hectares. Land is mostly individually owned. Because of erosion and top soil loss, the nutrient status of many soils in the tropics and subtropics is not favourable for high productivity. The highly skewed rainfall pattern makes the land not only hungry but also thirsty during large parts of the year. Thus low soil fertility, poor soil structure, low energy input, poor seed, uncertain water availability, extreme temperatures, lack of access to inputs and markets, all reduce the productivity of small farms in the tropics and sub-tropics. However, under conditions where land is limiting, very highly productive and energy efficient systems of farming have been developed during the last 20–30 years. Japan, Taiwan and North and South Korea have all reached high levels of productivity in crops like rice under small farm conditions. China has also reached an average yield level in rice of over 3 tonnes per hectare in an area exceeding 30 million hectares. Although land is socially owned in China, the density of farming population is high. Systems of cooperative management, particularly of irrigation and plant protection, have helped in ensuring a high average yield. High yields of cotton have become possible in Egypt under small farm conditions through the introduction of collective management of farm operations, like plant protection within the confines of individual ownership of land. Data collected from selected villages in Asia show that there is no significant difference in the average yield of rice per hectare between large and small farmers adopting modern varieties except in Indonesia. These data are consistent with the hypothesis of scale neutrality of new technology. Ruttan[6] after a review of the available micro-level data in Asia concluded "neither farm size nor tenure has been a serious constraint to the adoption of new high yielding varieties" and "neither farm size nor tenure has been an important source of differential growth of productivity."

Harwood[7] has, in a recent book on "Small Farm Development", dealt with the various factors which are involved in improving the productivity of small holdings. For achieving productivity improvement under small holders' conditions, there is need for developing institutional structures which can promote collective or group management of certain farm operations like water management, plant protection and post-harvest technology, without affecting the individuality of ownership. If this does not happen, there will always be co-existence of certain unique farmers who produce very high yields while at the same time the average productivity of the area is very low. In other words, the challenge lies in making the unique into universal.

HELPING SMALL FARMERS

Even at the currently available levels of technology, the gap between potential and actual farm yields is high in most small holdings. The reasons responsible for this gap will have to be identified in every area through an interdisciplinary constraints analysis. While the small farm offers potential for intensive agriculture the small farmer faces numerous problems arising from the cost, risk and return structure of farming. Only by helping the small farmer to overcome the problems facing him, can the potential productivity of small farms be realised. Several steps will have to be taken to assist small farmers to improve their economic conditions by raising the profitability of farm operations as well as security of income. A few such steps are described below:

Organisation of Land Use Boards

In order to help in re-structuring land use patterns on scientific lines, it would be desirable to organise Land Use Boards with high-level inter-disciplinary expertise. Each Board can cover a specific agro-ecological area. Such Land Use Boards should assist farmers in optimising the economic benefits from cubic volumes of soil, water and air through attention to the following major ingredients of scientific land and water use:

(a) Ecology: Land use based on ecological considerations will help to maximise the economic benefits from a given environment and minimise damage through man-made as well as natural processes of damage to the biological potential of land. Agro-meteorological research data will have to be integrated in crop planning models, so that contingency plans suited to different weather probabilities can be prepared. Also, an action-reaction analysis will have to be done in irrigation projects not only to avoid salinization and waterlogging but also the spread of new vector-borne diseases which impair human productivity and efficiency.

(b) Economics: For reorienting land and water use on the basis of sound principles of economics, it is essential that production, storage, processing and marketing are viewed as a total system. The prevailing mismatch in many farming systems between these two areas of the production-consumption chain is harming both producers and consumers. High risks and uncertain returns prevent poor farmers from investment on cash inputs and thus the asset base of the farm remains sub-optimal from the point of view of productivity improvement. The scope for introducing insurance and other forms of protection will have to be investigated by the Land Use Board in order to stimulate the widespread adoption of scientific agriculture. The various possibilities in the field of crop insurance have recently been summarised by Ray.[8] When post-harvest technology is neglected, opportunities for the preparation of value-added products are lost. For example, food production statistics simply state that during 1978-79 India produced about 131 million tonnes of foodgrains. This ignores the fact that the plants represented in this statistics produced over 400 million tonnes of dry matter, out of which grains constituted about 131 million tonnes. If the entire biomass is viewed as an asset and is utilised effectively, new avenues of income generation can be opened up. By looking at the dry matter yield part by part and by introducing techniques of preparing value-added material, rural incomes can be enhanced.

(c) Energy: The energy needs of agriculture will have to be carefully worked out and an integrated energy supply system involving a suitable blend of renewable and non-renewable forms of energy will have to be introduced in each agro-ecological area. So far the pathway of productivity improvement has tended to rely heavily on a growing consumption of non-renewable forms of energy. We will have to reverse this process through the promotion of organic re-cycling techniques and through the wide-spread use of biological sources of fertilizers like *azolla*, blue green algae and symbiotic and non-symbiotic forms of nitrogen fixing organisms. Also the current tendency to cultivate energy-rich crops like grain legumes and oilseeds under conditions of energy deprivation has to be corrected. "Nitrogen farming" should become an integral part of farm technology. Phosphorus conservation and re-cycling will be particularly important since phosphorus is a non-renewable resource. These steps will be facilitated if the farmers in each area organize "Renewable Energy Associations" and adopt a scientific energy conservation, generation and utilization methodolcgy.

(d) Employment: Famines in India are often famines of work than of food, since when work can be had and paid for, food is always forthcoming. The situation today in the field of nutrition is one of providing the wherewithal to purchase food rather than just the availability of food in the market. All estimates of employment potential show that a majority of our people will have to depend upon agriculture, agro-industries and small-scale industries as the major source of income until the end of this century. Hence the employment impact of land and water use policies needs to be studied. The impact on the employment opportunities for rural women should receive specific attention in all development programmes.

(e) Nutrition: Each Land Use Board should have on its staff a nutrition scientist who could help to meet

nutritional criteria in land use patterns. This would help local communities to improve the level of nutrition through steps such as:

(i) Preparation of home-made weaning foods;

(ii) Providing some critical missing nutrients in the diet, like Vitamin A, Vitamin C, iron, etc.;

(iii) Developing a cereal-grain-legume combination so that all the essential amino-acids can be provided in the diet;

(iv) Introducing appropriate fodder legumes and shrubs which could provide the needed calories and proteins to farm animals, thereby enabling the introduction of larger quantities of animal food, like milk and milk products, eggs, etc., in the diet;

(v) Developing suitable agriculture-cum-aquaculture techniques which could help in promoting dietary combinations, like rice-fish, potato-fish, etc.; and

(vi) Promoting agro-forestry systems of land management where appropriate botanical remedies to the specific nutritional maladies of the region get incorporated.

A majority of our farmers have less than one hectare of land to cultivate. They have to meet their own food requirements and, in addition, should have some surplus produce for sale. An important method of obtaining supplementary income in such cases is the integration of animals in the farming system. It is necessary that suitable technologies are developed for preparing fortified feed material from all cellulosic wastes and from agricultural raw material. Fortification of straw with molasses and urea as well as microbiological enrichment of starchy material, like cassava will have to receive much wider adoption.

(*f*) *Crop-Weather Watch*: State and Central Land Use Boards will have to possess the capacity to advise farmers on the scientific handling of different weather situations. An inter-disciplinary crop-weather watch Group should be an integral part of the Board. With the help of the India Meteorological Department and Agricultural Universitites, this Group should give continuous advice to farmers on:

(i) Risk aversion and crop life saving techniques;

(ii) Contingency plans to suit different weather conditions;

(iii) Compensatory programmes in irrigated areas.

Techniques for mitigating the effects of floods and drought are now available. For example, in the drought affected areas of the U.S. in 1980, it was found that plots subjected to chemical tillage retained more water and gave nearly normal yields.

Rural Development

Rural development in the ultimate analysis involves the provision of opportunities for the optimum utilisation of the human resource in rural areas. Human resource development in its turn can take place only on the foundation of adequate nutrition and work opportunities. It is, therefore, necessary to base rural development programmes on the primary aim of providing opportunities for the human population to achieve optimum expression of their physical and mental potential. Such programmes should have the following four major components:

(a) Economic emancipation of the family with particular attention to provision of adequate employment opportunities to women who are largely engaged at present in unpaid and underpaid jobs, often characterised by physical drudgery;

(b) Education of children and adults;

(c) Provision of minimum needs, such as safe drinking water supply, health care, rural communication, etc.; and

(d) Promotion of a small family norm through population and contraceptic education.

Experience with programmes such as the "Food for Work" and the "National Rural Employment Project" of the Government of India and "Employment Guarantee Scheme" of the State Government of Maharashtra has provided useful insight into methodologies for nutrition improvement and rural infrastructure development. For example, the Employment Guarantee Scheme of Maharashtra has revealed that there is predominance of females in 52 out of 87 selected works. The female participation was on an average 57% in many rural works programmes. A similar observation has been made in several areas where the "Food for Work" programme has been in progress. The fact that women will have additional income and, in the case of "Food for Work" programme, will also receive directly foodgrains has an important implication from the point of view of nutrition of children. Thus, carefully designed rural development programmes which can generate diversified opportunities for employment and help to provide adequate nutrition will hasten agricultural advance.

Systems Approach to Research and Development

Mixed farming involving crop-livestock integration and mixed cropping involving the simultaneous cultivation of two or more crops in a field and multiple cropping leading to taking 2 to 4 crops per year in the same plot of land are possible in the tropics and sub-tropics. In the past, both research and developmental efforts were organised more on the basis of individual crops or farm animals. We now know that what the farmer needs is advice and assistance relevant to the entire farming system around which the lives of the farmer and his family revolve. Fortunately an awareness of this need is now growing. Some of the major farming systems which require attention in this respect are the following:

(a) Multiple-cropping systems in irrigated areas: Various two-, three- and even four-crop sequences are now being followed. In promoting multiple-cropping systems, attention should be paid to ensuring that grain and fodder legumes find a place in the rotation. Also, crops having the same pests and diseases should not be grown in succession. Introduction of grain and fodder legumes in the rotation will improve human nutrition as well as soil fertility. A mung-bean-rice-wheat rotation is a good method of combining cereals and legumes in north-west India. Short-duration varieties of pigeonpea (*Cajanus cajan*) have made pigeonpea-wheat rotation possible. A jute-rice-wheat rotation is becoming popular in parts of Assam and West Bengal in India as well as in Bangladesh.

The introduction of relative insensitivity to photo-period and temperature through breeding has been responsible for the development of 'period fixed' rather than 'season bound' varieties. For purposes of breeding varieties for multiple-cropping, per-day yield has to be used as a selection criterion in segregating generations. Also, other factors such as seed dormancy will need attention, since if a crop ripens before the monsoon rains have ceased, the grain will sprout if there is rainfall at harvest time.

(b) Rain-fed farming: Production possibilities in high-rainfall areas are similar to those in irrigated areas. However, in the unirrigated semi-arid areas commonly referred to as dry-farming areas, considerable production risks exist. Grain legumes, sorghum, millets, cotton and oilseed crops are mostly grown in such areas. A wide variety of fruit trees can also be grown. Research thrusts in semiarid areas should lay stress on water and soil conservation and land use planning based on precipitation, evapotranspiration and the moisture holding capacity of the soil. At the organisational level, watershed management and the introduction of *Varabandhi* or rotational distribution of water saved in farm ponds will be necessary. Contingency plans should be developed and introduced so as to minimise the risk of total crop loss during aberrant weather. Seed and fertilizer reserves will have to be built so as to make the adoption of alternative cropping strategies and compensatory programmes in irrigated areas possible. It is also necessary to find more profitable crops for some of the semi-arid areas. There are many under-exploited plants with potential economic value.

(c) Mixed cropping and intercropping: Various crop combinations are used by farmers, particularly in unirrigated areas, but not all are scientifically sound. Therefore, intercropping systems based on complementarity between the companion crops have to be developed. Among the major components of complementarity are: (i) efficient interception of sunlight; (ii) ability to tap nutrients and moisture from different depths of the soil profile; (iii) non-overlapping susceptibility to pests and diseases; (iv) introduction of legumes to promote biological N-fixation and increase protein availability.

(d) Multi-level or three-dimensional cropping: In garden lands where a wide variety of plantation crops, fruit trees, coconut, oil palm and other tree crops are grown, it is possible to design a crop canopy in which the vertical space is utilised more efficiently. Plant architects will have to take into account the effective use of both horizontal and vertical spaces when breeding varieties for use in three-dimensional crop canopies. Efficiency in such a cropping system will again be based on the extent of the complementarity generated among crops in the system. For example, studies have shown that coconut, cocoa and pineapple form a good combination from the point of view of efficient interception of sunlight and extraction of nutrients and moisture from different soil depths. Studies of the root system of companion crops are of particular importance. The introduction of grain and fodder legumes in these three-dimensional crop canopies will provide opportunities for animal husbandry. A careful study of all the major garden land cropping systems based on the extent of symbiosis and synergy among the system components will be useful in developing specifications which plant breeders can use in developing ideotypes (i.e. conceptual plant types) for efficient performance in three-dimensional crop canopies.

(e) Kitchen gardening and home fish gardening: Kitchen gardening can be one of the most efficient systems of farming from the point of view of solar and cultural energy conversion. Vegetables rich in beta carotene and iron need to be developed and popularised. If planned intelligently and scientifically, backyard gardens, roof gardens and other methods of growing vegetables and fruits in whatever space is available around mud huts as well as brick houses can make a substantial contribution to improved nutrition. Where ponds are available in large numbers, home fish gardening can be an excellent method of supplementing income and source of food.

(f) Forestry and agro-forestry: The importance of improving the productivity of forest canopies cannot be over-emphasised. Agro-forestry has been defined as a sustainable management system for land which increases overall production, combines agricultural crops, tree crops, forest plants and/or animals simultaneously or sequentially. Sylvi-pastoral, sylvi-horticulture, sylvi-agriculture and other combined land-use systems can help to meet the food, feed, fuel and fertilizer needs of people in many hilly regions. Plant breeders have yet to give attention to breeding

varieties suitable for such systems of sylviculture. Shrubs and trees suitable for raising energy plantations in villages need to be identified and popularised.

(g) Mixed farming: Mixed farming systems may involve (i) crop-livestock; (ii) crop-fish, and (iii) crop-livestock-fish production programmes. In Kerala and Eastern India, fishing in rice fields is common. The minimal use of pesticides will be important in order to avoid problems of fish mortality and transfer of toxic residues through the food chain. This will involve maximum use of genetic resistance and the development of integrated pest-management systems in crops like rice and jute.

(h) Sea farming: There are considerable opportunities for the spread of scientific sea farming practices involving an appropriate blend of capture and culture fisheries. The rate of growth of oysters, mussels, prawns, lobsters, eels and a wide variety of marine plants and animals is high in tropical seas. If along with such integrated sea farming practices, the cultivation of suitable economic trees like casuarina, cashewnut and coconut can be popularised along the coast, thriving coastal agriculture-cum-mariculture systems can be developed. In addition to improving income and nutrition, such farming systems can help to arrest coastal erosion.

Constraints anaysis and Removal

Agriculture starts moving forward only when an appropriate package of technology is supported by suitable packages of services and public policies which can help all farmers irrespective of the size of their holding and their innate input mobilising and risk taking capacity to take to new technology. For identifying the precise constraints which are responsible for the prevailing gap between potential and actual farm yields, it is necessary to conduct an inter-disciplinary constraints analysis. To be meaningful, such a malady-remedy analysis will have to be done in compact areas with similar farming systems and socio-cultural factors.

As mentioned earlier, an important constraint under conditions of small holdings with individual land ownership is the difficulty of achieving a high level of farm management efficiency. For example, if two neighbouring farmers adopt totally divergent approaches in the field of pest control, the farmer who wishes to achieve high levels of production may have to resort to a larger number of sprays of pesticides than would otherwise have been necessary. Hence a challenge to development planners and administrators dealing with small farm conditions lies in introducing suitable packages of services which can help to introduce a community/area approach in management wherever this is necessary.

For optimum efficiency, a blend of cash and non-cash inputs will be necessary. Such services are best provided by farmers' own organisations supported by appropriate training, credit and marketing assistance from Government. The organisation of community nurseries in crops like rice where transplanting is done, the introduction of rotational distribution of water in the command areas of irrigation projects so that all farmers in the command area get equal quantities of water and the supply of credit and the needed inputs before the sowing season in properly organised credit-cum-input supply village fairs, are examples of the approaches which have been found useful.

Learning from successes is equally important in programme formulation. If Punjab made striking advances in crop production after the release of dwarf varieties of wheat and rice it is because the State already possessed the substrate requirements essential for new technology to find widespread adoption. Three of the major technology diffusion substrate needs which Punjab possessed in mid-sixties were: owner cultivation, rural communication and rural electrification. Roads and energy supply are exceedingly important inputs.

Basic Building Blocks of the Enhanced Home Production Pathway

The two key requirements in this respect are: first, *organisation* of small farmers both for improving farm management efficiency through community action in water and soil conservation and management, pest control and post-harvest technology and for ensuring that the producer gets a high proportion of the price paid by the consumer; secondly, a well-planned programme of *diversification* of employment and income generation opportunities in rural areas so that a part of the farm labour can get absorbed in the secondary and tertiary sectors. This will call for a more detailed planning of the scientific utilisation of local resources, so that value-added products can be prepared from all agricultural raw material and suitable small and village industries promoted. Institutional devices such as the organisation of Horticulture and Aquaculture Estates, on the lines of industrial estates, will have to be promoted. Farmers' Service Societies operated by rural youth will have to be fostered.

Given the above steps, I am confident that the small farm technology can, on the one hand, acquire the strengths of the super farm methodology and at the same time avoid the difficulties in terms of social, ecological and energy costs, inherent in the super farm procedures. Making small farmers acquire the strengths of super farm cultivations, while avoiding their weaknesses, is thus a major challenge to agricultural technologists and planners working in

India. If this task is successfully accomplished, it will be possible to raise the average productivity of all farming systems and thereby help the country to derive benefit from its vast untapped production reservoir. Our two major resources are sunlight and the human population. These two resources can be linked in a symbiotic manner through green plants. When this is done on an extensive scale, we can bid farewell to the import pathway of meeting our needs of basic agricultural commodities.

1. Bastin, Geoffrey and John Ellis, 1980. International Trade in Grain and the World Food Economy. The Economist Intelligence Unit Ltd., London, 174 pp.
2. Neil Rutger, J. and Marlin Brandon, D. 1981. California Rice culture, Scientific American (Feb. 1981), pp. 42.
3. Agricultural Production Efficiency, 1975, U. S. National Academy of Sciences, Washington, D. C. 199 pp.
4. Gerald Leach, 1975. The Energy Costs of Food Production. In "The Man/Food Equation", published by Academic Press, 139.
5. Michie, Aruna Nayyar and Jagger, C. 1980. Why Farmers Protest Kansas Farmers, the Farm Problem and the American Agriculture Movement, Kansas State University, 88 pp.
6. Ruttan, Vernon, W., 1977. The Green Revolution: Seven Generations. International Development Review, **19** 16.
7. Harwood, R. R. 1979. Small Farm Development, Westiview Press, Colorado, 160 pp.
8. Ray, P. K. 1981. Agricultural Insurance, Pergamon Press, Oxford, 419 pp.

Theor Appl Genet (1995) 90: 253–257

B. Ba Bong · M. S. Swaminathan

Magnitude of hybrid vigor retained in double haploid lines of some heterotic rice hybrids

Received: 1 February 1994 / Accepted: 29 April 1994

Abstract Sixty one double haploid (DH) lines were evaluated for yield components and yield and compared with the three corresponding hybrids from which they were derived through anther culture. Analysis of a 6×6 diallel experiment led to the selection of these hybrids based on their high heterosis and revealed the nature of gene action of the characters under evaluation. The DH lines along with the hybrids and parents were planted following a simple lattice design with two replications. The results show that in DH lines the values of the characters expressing predominantly additive genetic effects could reach the heterotic level of the hybrids. For characters including yield showing predominantly dominance effects, values in the DH lines were significantly lower than those of the corresponding hybrids. The promising DH lines, however, possess a higher yield potential than the better parents.

Key words *Oryza sativa* L. · Rice · Heterosis
Anther culture · Double haploid

Introduction

Heterosis in rice has been successfully exploited in China which has 14 million hectares planted in hybrid rice (Yuan and Mao, 1991). Hybrid rice technology, however, has not been adopted for commercial production by other rice-growing countries even though intensive research in this area has been carried out. One of the main reasons for this is the difficulty in producing hybrid seed due to cytoplasmic male sterility (He Gui-Ting et al. 1987; Toenniessen 1991). This difficulty would be overcome if apomixis could be successfully transferred to rice to provide hybrids with permanently fixed heterosis. While such an achievement is still remote, application of the anther culture technique to highly heterotic hybrids may help in exploiting the advantage of superior gametic genotypes. Theoretically, it is possible to obtain homozygotes with fixed heterosis provided that partial to complete dominance predominates (Sprague and Eberhart 1977). In practice, anther culture applied to hybrid rice has led to the release of several new varieties (Han Hu and Bin Huang 1987). There is, however, a lack of information on the magnitude of heterosis that is retained in the double haploid (DH) line relative to that expressed by the corresponding hybrid from which it is derived.

The study presented here compared the performance of DH lines with that of their corresponding hybrids from which they were developed by anther culture. The findings suggest the possibility of exploiting hybrid vigor in rice through the development of DH lines from anther culture of potential hybrids.

Communicated by G. S. Khush

B. Ba Bong (✉)[1]
Division of Genetics, Indian Agricultural Research Institute, New Delhi 110012, India

M. S. Swaminanthan
M. S. Swaminathan Research Foundation, 3rd Cross Street, Taramani Institutional Area, Madras 600113, India

Present address:
[1] Cuu Long Delta Rice Research Institute, Omon, Cantho, Vietnam

Materials and methods

Selection of heterotic hybrids and study of gene action

Six improved rice varieties, three developed at the Indian Agricultural Research Institute, 'Pusa 743-1-1', 'Pusa 33'and 'Pusa 205', and three from the International Rice Research Institute, 'IR66', 'IR72' and 'IR74', were crossed following diallel design to develop 15 hybrids. These hybrids along with the parents were planted in a randomized complete block design with three replications. Genetic variance components were evaluated following Jinks and Hayman (1953) and Hayman (1954). Three hybrids namely, 'Pusa 743-1-1'/'IR66', 'Pusa 743-1-1'/'IR72' and 'Pusa 743-1-1'/'IR74' were selected for anther culture based on their high heterosis in grain yield.

Table 1 Mid-parent heterosis (HM) and better parent heterosis (HB) for yield components and yield of the hybrids used for anther culture

Character	Hybrid 1[a]		Hybrid 2[a]		Hybrid 3[a]	
	HB	HM	HB	HM	HB	HM
Days to 50% flowering	−0.8	−2.4*	0.4	−1.0	3.2**	−3.0**
Plant height	−5.6**	−1.9	−4.6**	1.1	1.8	11.5**
Panicle no./plant	−13.4**	−3.6**	−1.6	17.1**	0.7	2.9**
Grain no./panicle	29.7**	33.1**	18.6*	26.8**	14.1	15.9**
1000-grain weight	−11.4**	6.5**	−13.8**	−1.6	−5.0**	10.0**
High density grain	8.5	81.2**	8.6	78.6**	131.2**	178.3**
Harvest index	1.9	9.2**	6.3**	12.8**	−7.4**	12.4**
Grain yield	34.1**	44.7**	51.9**	56.5**	19.0**	34.1**

*, **=Significant at 5% and 1% level, respectively
[a] Hybrid 1='Pusa 743-1-1'/'IR66', Hybrid 2='Pusa 743-1–1'/'IR72', Hybrid 3='Pusa 743-1–1'/'IR7

Development of DH lines

Anther culture of the 3 heterotic hybrids was done at the Biotechnology Center, Indian Agricultural Research Institute, following established techniques for india rice. Sixty one DH lines were obtained, 46 lines derived from 'Pusa 743-1-1'/'IR66', 8 from 'Pusa 743-1-1'/'IR72' and 7 from 'Pusa 743-1-1'/'IR74'. These lines represented the first progeny of distinct DH anther-derived plants.

Field evaluation

The experiment was conducted on the farm of the Indian Agricultural Research Institute, New Delhi, in the wet season of 1991. The test entries, 61 DH lines, 3 corresponding hybrids ('Pusa 743-1-1'/'IR66', 'Pusa 743-1-1'/'IR72' and 'Pusa 743-1-1'/'IR74'), and four parents ('Pusa 743-1-1'/'IR66' 'IR72' and 'IR74'), were evaluated in a 9×9 single lattice design with two replications (13 conventional varieties/lines were added to satisfy the required number of treatment being 81). Each plot contained five rows with 25 plants/row. Single seedling was transplanted per hill at 20×20 cm spacing.

Data collection

Observations were recorded on plants in the three middle rows (excluding border plants) for following characters:

- days to 50% flowering: number of days from seeding to the day 50% plants showed panicle exertion
- plant height: in cm, from the ground to the tip of the main panicle at maturity
- panicle number/plant: average number from 5 plants/plot at maturity
- number of filled and unfilled grains/panicle: average number from 5 panicles/plot
- 1000-grain weight: recorded from evenly oven-dried seeds of 5 plants/plot
- proportion of high density grain: the percentage of grains showing specific gravity above 1.2 following the method of Venkateswarlu et al. (1986)
- grain yield: weight of clean and dry grains harvested from the whole plot (excluding border plants) computed at 14% moisture content (in kg/ha)
- biological yield: weight of total biomass including grains after harvesting and drying
- harvest index: ratio of grain yield to biological yield

Data analysis

The analysis of the variance for simple lattice design was performed following Gomez and Gomez (1984). Mean values of the character of test entries were compared using the Least Significant Difference (LSD) test.

Table 2 Estimates of variance components due to additive effect (D) and to dominance effect (H_1), and degree of dominance $(H_1/D)^{1/2}$

Character	D	H_1	$(H_1/D)^{1/2}$
Days to 50% flowering	34.72**	22.88**	0.80
Plant height	61.42**	72.97**	1.09
Panicle no./plant	2.53**	2.12	0.95
Grain no./panicle	91.38	630.77**	2.63
1000-grain weight	12.64**	4.83**	0.62
High density grain	509.16**	916.37**	1.34
Harvest index	28.89**	75.01**	1.61
Grain yield	0.25	83.68**	18.26

*, **=Significant at 5% and 1% level, respectively

Results

Selection of hybrids for anther culture

Out of 15 hybrids, 3 namely, 'Pusa 743-1-1'/'IR66', 'Pusa 743-1-1'/'IR72' and 'Pusa 743-1–1'/'IR74' were selected for anther culture. The heterotic performance of these hybrids expressed as mid-parent heterosis and better parent heterosis is given in Table 1. Mid-parent heterosis for grain yield was highest in 'Pusa 743-1-1'/'IR72' (56.5%) followed by 'Pusa 743-1-1'/'IR66' (44.7%) and 'Pusa 743-1-1'/'IR74' (34.1%). Besides yield, these hybrids showed mid-parent heterosis for more than three yield components, of which grain number/panicle was highly heterotic (Table 1). Better parent heterosis for grain yield was 51.9% in 'Pusa 743-1-1'/'IR72', 34.1% in 'Pusa 743-1-1'/'IR66' and 19.0% in 'Pusa 743-1-1'/'IR74'. The best hybrid ('Pusa 743-1-1'/'IR72') possessed two yield components (grain no./panicle and harvest index) that showed positive better parent heterosis.

Data on estimates of genetic variance (Table 2) showed that both components due to additive and dominance effects were important for days to 50% flowering, plant height, grain weight, high density grain and harvest index. Meanwhile, additive effects predominated for panicle number/plant and dominance effects predominated for grain number/panicle and grain yield. The degree of dominance varied from partial dominance for days to 50% flow-

Table 3 Performance of parents, F_1 hybrid and DH lines derived from 'Pusa 743-1-1'/'IR66'

Character	Mean performance				LSD 5%
	P1[a]	P2[a]	F_1[a]	DH lines (Range)	
Days to 50% flowering	79	89	83	66–97	4
Plant height (cm)	114	102	105	86–127	4
Panicle no./plant	6.7	8.8	9.9	6.3–12.8	2.4
Grain no./panicle	96	94	125	46–102	21
Spikelet sterility (%)	18.8	15.2	16.5	13.6–23.7	3.6
1000-grain weight (g)	29.8	21.9	23.5	18.1–31.6	1.8
High density grain (%)	12.6	63.1	68.5	7.1–68.2	14.9
Harvest index (%)	52.2	51.4	51.8	28.4–54.6	8.5
Grain yield (kg/ha)	5506	5251	7047	2917–6624	702

[a] P1='Pusa 743-1-1', P2='IR66', F_1='Pusa 743-1-1'/'IR66'

Table 4 Performance of parents, F_1 hybrid and DH lines derived from 'Pusa 743-1-1'/'IR72'

Character	Mean performance				LSD 5%
	P1[a]	P2[a]	F_1[a]	DH lines (Range)	
Days to 50% flowering	79	85	81	77–92	4
Plant height (cm)	114	105	108	102–124	4
Panicle no./plant	6.7	9.3	12.4	7.4–12.8	2.4
Grain no./panicle	96	91	138	59–98	21
Spikelet sterility (%)	18.8	14.7	17.1	13.9–23.4	3.6
1000-grain weight (g)	29.8	22.4	26.5	20.4–27.2	1.8
High density grain (%)	12.6	58.4	63.4	15.6–54.7	14.9
Harvest index (%)	52.2	50.0	56.3	33.9–53.4	8.5
Grain yield (kg/ha)	5506	5450	7818	3866–6606	702

[a] P1='Pusa 743-1-1', P2='IR66', F_1='Pusa 743-1-1'/'IR72'

Table 5 Performance of parents, F_1 hybrid and DH lines derived from 'Pusa 743-1-1'/'IR74'

[a]Character	Mean performance				LSD 5%
	P1[a]	P2[a]	F_1[a]	DH lines (range)	
Days to 50% flowering	79	106	85	75–106	4
Plant height (cm)	114	107	112	90–114	4
Panicle no./plant	6.7	10.2	10.8	8.4–12.3	2.4
Grain no./panicle	96	77	121	64–82	21
Spikelet sterility (%)	18.8	18.3	15.4	13.6–22.3	3.6
1000-grain weight (g)	29.8	23.0	29.7	22.8–27.3	1.8
High density grain (%)	12.6	18.9	47.3	12.4–43.6	14.9
Harvest index (%)	52.2	41.4	49.4	36.5–51.5	8.5
Grain yield (kg/ha)	5506	4865	6332	4270–5157	702

[a] P1='743-1-1', P2='IR74', F_1='Pusa 743-1-1'/'IR74'

ering and grain weight, to complete dominance for plant height and panicle number/plant, to overdominance for grain number/panicle, high density grain, harvest index and yield (Table 2).

Evaluation of DH lines

The performance of DH lines and their corresponding hybrids and parents is presented in Tables 3, 4 and 5. It was recorded that there were DH lines showing a shorter duration of growth and shorter plant height than the hybrids and parents. The upper limit of the range of DH lines for panicle number/plant, grain weight, proportion of high density grain and harvest index more or less reached the heterotic level of the hybrids and often exceeded thevalue of the better parents. However, DH lines showed a poorer performance in grain number/panicle than the hybrid.

The DH lines derived from 'Pusa 743-1-1'/'IR66' showed a grain yield ranging from 2917–6624 kg/ha compared to 7047 kg/ha for the hybrid and 5506 kg/ha for the better parent (Table 3). Yield of the DH lines derived from 'Pusa 743-1-1'/'IR72' ranged from 3866–6606 kg/ha against the 7818 kg/ha of the hybrid (Table 4). Range in

Table 6 Percentage increase/decrease in values of promising DH lines compared to F_1 hybrids and better parents (BP)

Line	Panicle no./plant		Grain no./panicle		1000-grain weight		Harvest index		Grain yield	
	BP	F_1	BP	F_1	BP	F_1	BP	F_1	BP	F_1
DH lines derived from Pusa 743-1-1/IR66 (H1)										
A26-2	28.4*	17.7	6.2	–12.4*	1.0	11.1*	3.2	4.0	20.3*	–6.0
A27-2	39.8*	28.1*	2.1	–21.6*	–26.5**	–19.1*	–2.5	–1.7	5.1	–17.8*
A10-2	27.3*	16.7	–18.7	–37.6*	–15.0*	–6.6	0.0	0.0	14.7*	–10.4*
A14-2	7.9	–1.0	–4.2	–26.6*	–1.7	8.1*	–10.5	–9.8	11.1	–13.1*
A53-2	45.4*	33.3*	–8.3	–29.6*	–20.5*	–12.5*	4.6	5.4	12.8*	–11.9*
DH lines derived from Pusa 743-1-1/IR72 (H2)										
A4-2	21.5	–8.9	2.1	–29.9*	–19.1*	–9.0*	2.3	–5.1	19.9*	–15.5*
A1-3	37.6*	3.2	–11.4	–38.4**	–25.2**	–12.4*	5.0	–2.3	19.3*	–16.0*
A1-2	11.8	–16.1	7.3	–35.5**	–11.7*	0.0	–23.0*	–28.6*	12.9*	–20.5*

*,**=Significant at 5% and 1% level, respectively

yield of the DH lines from 'Pusa 743-1-1'/'IR74' was 4270–5157 kg/ha, while the yield of the hybrid was 6332 kg/ha (Table 5).

From 61 distinct DH lines, 5 promising lines from 'Pusa 743-1-1'/'IR66' and 3 from 'Pusa 743-1-1'/'IR72' were identified, while none of DH lines from 'Pusa 743-1-1'/'IR74' showed any improvement. The comparison of these promising lines with the hybrids and better parents is given in Table 6. The best DH line from 'Pusa 743-1-1'/'IR66' (A26-2) had a similar yield level as the hybrid thanks to its similar performance for panicle number/plant and harvest index. Its decrease in grain no./panicle was compensated for by the increased value of 1000-grain weight. Relative to the yield of better parent, the yield of this line significantly increased by 20.3%. Other promising DH lines showed yields lower than that of the hybrid by 10.4–17.8%. The best DH lines from 'Pusa 743-1-1'/'IR72' possessed a yield potential that was lower than that of the hybrid by 15.5–20.5% but higher than that of the better parent by 12.0–19.9% (Table 6).

Discussion

Anther culture of potential hybrids makes it possible to develop superior gametic genotypes that possess the vigor and fitness of the hybrid. The advantage of this method is that such heterotic attributes are fixed in the homozygous condition as a consequence of chromosome doubling. Previous studies have shown that it is possible to obtain as desirable genotypes from the anther culture of F_1 rice hybrids as from the F_2 or F_3 population (Li et al. 1978; Liu et al 1980; Yin et al. 1983). It is evident that new rice varieties have been successfully developed from the anther culture of F_1 hybrids as studies have been reported from China (Chen Ying 1986; Loo Shih-wei and Xu Zhi-Hong 1991; Liang Shouyi and Huang Shouyin 1991), Korea (Lee et al. 1989) and Bulgaria (Boyadzhiev 1990). Zapata et al. (1991) identified 6 out of 109 DH lines derived from three hybrids as being promising under saline and nonsaline conditions.

The approach used in this study was to obtain a direct comparison of the DH lines with the hybrids from which they are derived. The results show that the performance of some DH lines is almost comparable to that of the hybrids for the characters days to 50% flowering, plant height, panicle number/plant, 1000-grain weight, high density grain and harvest index (Tables 3, 4 and 5). For this group of characters, the variance component due to additive effects is highly significant (Table 2). All of these characters showed partial/complete dominance except for high density grain and harvest index, which express overdominance. In contrast for for grain number/panicle, none of the DH lines from three hybrids reach the heterotic level of the hybrids. The same trend was also found for grain yield in which all but 1 DH lines possess significantly lower values than the corresponding hybrids. The only DH line from 'Pusa 743-1-1'/'IR66' which has a yield level of similar to that the hybrid (but numerically lower) is also not comparable to the hybrid for grain number/panicle. The study on gene action has revealed that grain no./panicle and grain yield show predominantly dominance effects and possess a very high $(H_1/D)^{1/2}$ ratio (Table 2).

Heterosis can result from partial/complete dominance, overdominance and a combination of these (Comstock and Robinson 1953). Both types of heterosis are likely to involve fixable additive and additive×additive genetic effects, which are fixable in DH lines. In our study,the observation that the performance of DH lines for the first group of characters is comparable to that of the hybrid supports this genetic basis. However, the heterosis associated with overdominance may also involve dominance, additive× dominance and dominance×dominance genetic effects, which are not fixable. Perhaps this is the reason that almost all DH lines cannot reach the heterotic level of the hybrids for the characters grain number/panicle and grain yield, which showed predominantly non-additive genetic effects.

As compared to the better parents, the yield improvement of promising DH lines is encouraging with a yield advantage of up to 20% (Table 6). This improvement is brought about through the combination of desirable traits in DH lines. It is known to date that hybrid rice outyields conventional varieties by 15–20% (Yuan and Virmani 1988).

It is possible to isolate promising DH lines from a relatively small population, as has been demonstrated in this study. In China, var 'Zhonghua No. 1' was selected from 71 DH lines, 'Zhonghua No. 8' and 'No. 9' from 40 DH lines (Chen Ying 1986) and var 'Huayu 15' was selected from 19 pollen-derived plant clusters (Liang Shouyi and Huang Shouyin 1991). Liu et al. (1983) reported that a population of 50–75 pollen-derived lines from each cross was required for efficient selection.

This study suggests that it is possible obtain promising DH lines through the anther culture of highly heterotic hybrids. Although DH lines cannot reach the yield level of the hybrid due to the failure in fixing non-additive genetic effects, this approach would be of some help in improving the yield potential of rice.

Acknowledgements The authors wish to thank Drs. F. U. Zaman, V. P. Singh, A. R. Sadananda, Division of Genetics, Drs. S. K. Raina, N. Gupta, Biotechnology Center, Indian Agricultural Research Institute and Dr. S. Devadath, Central Rice Research Institute, Cuttack, India for technical advice and providing facilities for this study.

References

Boyadzhiev P (1990) Mariana, a new variety of rice obtained by anther culture (in Bulgarian, English summary). Rastenievud Nauki 27:111–113

Chen Ying (1986) The inheritance of rice pollen plant and its application in crop improvement. In: Han Hu and Hongyuan Y (eds) Haploids of higher plants *in vitro*. China Academic Publ Beijing, pp 118–137

Comstock RE; Robinson HF (1953) Estimation of average dominance of gene. In: Gowen JW (ed) Heterosis. Iowa State College Press, Iowa, pp 494–516

Gomez KA, Gomez AA (1984) Statistical procedures for agricultural research. John Wiley and Sons, New York

Han Hu, Bin Huang (1987) Application of pollen-derived plants to crop improvement. Int Rev Cytol 107:293–313

Hayman BI (1954) The theory and analysis of diallel cross. Genetics 39:251–271

He Gui-Ting, Zhu Xigang, Flin JC (1987) Hybrid seed production in Jiangsu province, China. Oryza 24:297–312

Jinks JL, Hayman BI (1953) The analysis of diallel crosses. Maize Genet Newsl 27:48–54

Lee YT, Lim MI, Kim HS, Shi HT, Kim CH, Bae SH, Cho CI (1989) An anther-derived new high quality rice variety with disease and insect resistance "Hwacheongbyeo" (in Korean, English summary). Res Rep of the Rural Dev 31:27–34

Li LC, Zhang L, Tian WZ (1978) Oberservation on progeny segregation from pollen plants in paddy. Proc Symp Anther Cult Science Press, Beijing, pp 184-188

Liang Shouyi, Huang Shouyin (1991) Huayu 15, a high-yielding rice variety bred by anther culture. In: Bajaj YPS (ed) Biotechnology in agriculture and forestry, Volume 14: rice. Springer, Berlin Heidelberg New York, pp 230–247

Liu DY, Wang SD, Ding SL, Zhang MG (1983) The utilization of anther culture method in breeding of rice. In: Shen JH, Zhang ZH, Shi SD (eds) Studies on anther culture breeding in rice (in Chinese, English abstract). Press, Beijing, pp 20–23

Liu J, Xue QZ, Shen ZT (1980) Genetic analysis of several characters of pollen plant in rice (*Oryza sativa* subsp. japonica) (n Chinese, English abstract). J Zhenjiang Agric Univ 6:11–17

Loo Shih-Wei, Xu Zhi-Hong (1991) Anther culture for rice improvement in China. In: Bajaj YPS (ed) Biotechnology in Agriculture and Forestry Volume 14: Rice. Springer-Verlag, Berlin Heidelberg New York pp 151–179

Sprague GF, Eberhart SA (1977) Corn breeding. In: Sprague GF (ed) Corn and corn improvement. Am Soc Agron, Wisconsin, pp 309–362

Toenniessen GH (1991) Potentially useful genes for rice genetic engineering. In: Khush GS, Toenniessen GH (eds) Rice biotechnology. Int Rice Res Inst, Philippines, pp 253–280

Venkateswarlu B, Vergara BS, Paras FT, Visperas RM (1986) Enhancing grain yield potentials in rice by increasing the number of high density grain. Philipp J Crop Sci 11:142–152

Yin GD, Zeng RY, Wang XY, Yen ZL, Liu HR (1983) Studies on utilization of anther culture breeding in rice. In: Shen JH, Zhang ZH, Shi SD (eds) Studies on anther culture breeding in rice (in Chinese, English abstract). Agri Press, Beijing, pp 55–60

Yuan LP, Mao CX (1991) Hybrid rice in China-techniques and production. In: Bajaj YPS (ed) Biotechnology in agriculture and forestry, vol 14: rice. Springer, Berlin Heidelberg New York, pp 128–148

Yuan LP, Virmani SS (1988) Status of hybrid rice research and development. Hybrid rice. Int Rice Res Inst, Philippines, pp 7–24

Zapata FJ, Alejar MS, Torrizo LB, Novero AU, Singh VP, Senadhira D (1991) Field performance of anther-culture-derived lines from F_1 crosses of Indica rice under saline and nonsaline conditions. Theor Appl Genet 83:6–11

Development of a Monosomic Series in an Indian Wheat and Isolation of Nullisomic Lines

M. S. Swaminathan, V. L. Chopra, B. C. Joshi and D. Singh
Division of Genetics, Indian Agricultural Research Institute, New Delhi, India

The series of twenty one monosomic lines in the wheat variety Chinese Spring, developed by Dr E.R. SEARS, has greatly helped in advancing our understanding the genetics of the hexaploid bread wheat. Utilizing the monosomics of Chinese Spring, the Canadian workers have transferred the monosomic condition to other wheat varieties (UNRAU and MCGINNIS 1958). This report deals with the production of a complete series of monosomic lines in an Indian wheat variety Pb. C591 and the isolation of several nullisomic lines in it.

Pb. C591 is a fully bearded vairiety susceptible to all the three rusts. The purpose of producing chromosome deficient lines in this variety is to utilise them for the study of quantitative traits (JOSHI and KUMAR 1967, JOSHI and SINGH 1968) and to locate genes for rust resistance on specific chromosomes. In such studies the susceptible genetic background of Pb. C591 will be helpful in revealing the critical chromosome of the donor variety by the resistant reaction caused by the gene or genes introduced by such a chromosome.

All the 21 monosomic lines of the variety Chiriese Spiring, obtained from Dr. E.R. SEARS, were crossed as female parent with Pb. C591. The monosomic F_1's (2n=41) in each of the 21 lines were identified cytologically and again crossed as female with Pb. C591. This backcrossing programme of the monosomic hybrids with Pb. C591 was repeated till the sixth backcross generation, ensuring that all chromosomes in the 21 monosomic series are from Pb. C591. All the chromosome deficient lines of Pb. C591 are morphologically similar to the parent variety.

It is interesting to report that in the sixth backcross generation of mono 5A x Pb. C591, all the 45 plants were monosomics and no disomes were obtained. These plants, in the absence of chromosome 5A on which the speltoid suppressor gene *Q* is located, were easily identified phenotypically as 'speltoids'.

From among the aneuploid series of Pb. C591, 14 different nullisomic lines (2n=40) have been isolated. The morphological characteristics of these nullisomics are given in Table 1.

All the 14 nullisomics of Pb. C591, with exception of nulli 1D, are dwarf, weak, thin stemmed and thin leaved and straggling plants. Nulli 1D possesses medium broad leaves and stem. Nulli 5A and 6B have suppressed awning whereas all other nullisomes are fully bearded like the disome. This suggests that factors for awning, in Pb. C591, are located on chromosomes 5A and 6B.

All the nullisomes isolated by us, except nulli 5B, seem to possess good female fertility because in all cases where adequate pollen was available the seed setting was fairly good.

Meiotic metaphase I in nulli 5B of Pb. C591 shows multivalent formation, characteristic of chromosome 5B deficient plants of Chinese Spring (Riley and Chapman 1958, Sears and Okamoto 1958).

Literature

Joshi, B.C. and S. Kumar 1967. *Indian J. Genet.* **27**: 86–89.

Joshi, B.C. and D. Singh 1968. *Indian J. Genet.* **28**(1), In press.

Riley, R. and V. Chapman 1958. *Nature*. London **182**: 713–715.

Sears, E.R. and M. Okamoto 1958. *Proc. Tenth Int. Congr. Genet.* **2**: 258–259.

Unrau, J. and R.C. Mcginnis 1958. *Proc. Genet. Soc. Canada* **3**: 14–24.

(Received June 30, 1968)

Ann. Hum. Genet., Lond. (1967), **31**, 141
Printed in Great Britain

Consanguineous marriages and the genetic load due to lethal genes in Kerala

BY SUSHIL KUMAR, R. A. PAI AND M. S. SWAMINATHAN
Division of Genetics, Indian Agricultural Research Institute, New Delhi (India)

The Hindu population throughout India is organized into castes and subcastes which are largely endogamous. The populations in the southern states of India (Andhra, Kerala, Madras and Maharashtra) are unique in the occurrence of a fair frequency of consanguineous marriages within the subcastes, and they thus provide an excellent opportunity for studying the detrimental effects of low levels of inbreeding and for estimating the genetic load. The socio-economic reasons for the high incidence of consanguineous marriages in India have been outlined by Dronamraju (1964) and Sanghvi (1966). Dronamraju & Meera Khan (1963) reported the morbidity pattern in the children of consanguineous and non-consanguineous marriages in a hospital population of Andhra Pradesh. During the present study, several hospital populations in Kerala were examined for the relative mortality and morbidity of offspring from the two types of marriages. Several authors have recently considered the effects of inbreeding on mortality and other traits determining fitness in human populations (Morton, Crow & Muller, 1956; Böök, 1957; Schull, 1958; Slatis, Reis & Hoene, 1958; Chung, Robinson & Morton, 1959; Morton & Chung, 1959; Morton, 1960; Salzano *et al.* 1962; Neel & Schull, 1962; Freire-Maia, Freire-Maia & Quelce-Salgado, 1963; Freire-Maia, Guaraciaba & Quelce-Salgado 1964; Freire-Maia & Krieger, 1963; Goldschmidt *et al.* 1963; Dewey *et al.* 1965). Morton *et al.* (1956) developed formulae for the estimation of genetic load in terms of A and B statistics which could be used for discrimination between the mutational and segregational components of loads. Their method has been used in the present paper for analysing the data on mortality.

MATERIAL AND METHODS

The consanguinity data were collected during May and June 1966, from several hospitals in Kerala at Ernakulum, Quilon and Trivandrum. The hospitals are: Government Hospital, Ernakulum, Government Hospital and Neendakara Indo-Norwegian Hopital, Quilon; and Medical College Hospital and Mental Hospital, Trivandrum. Most of the patients received in these hospitals are from within the limits of the respective districts. The Mental Hospital at Trivandrum, however, receives patients from all over the Kerala State. Almost all in-patients during the period of visiting these hospitals were included in this study. In-patients were questioned about the consanguinity and progeny size of their parents. Morbidity data were taken from the hospitals' records. In the Mental Hospital and in the paediatric wards of other hospitals, where a statement could not be obtained from the patient due to the nature of his illness, data were collected by interviewing parents or other relatives. Mortality data were recorded on the families of patients of paediatric wards at the Trivandrum and Quilon hospitals. Only those families were considered where histories of births and deaths could be obtained by direct questioning of the mother or grandmother of the patient. Socio-economic data on caste, profession and income available in the hospital records were utilized.

RESULTS AND DISCUSSION

Observations on consanguinity

Data on the relative proportions of various types of consanguineous marriages are given in Table 1. It will be noted that (1) the mean proportion of consanguinity based on 889 marriages is 0·2036; (2) the incidence of consanguineous marriages in Quilon and Trivandrum districts is significantly higher than in Ernakulum ($\chi^2_{(1)} = 13{\cdot}92$, $P > 0{\cdot}01$); (3) the most frequent types of consanguineous marriages are, first, a girl with her maternal uncle's son and, secondly, a girl with her paternal aunt's son; and (4) marriages between children of two brothers or two sisters which are socially forbidden among Hindus do occur in this area, although their relative proportions are low. No case of uncle-niece marriage which is the highest degree of consanguinity permitted in Hindu law and which is frequent in the other southern states was recorded in Kerala.

Table 1. *Relative frequencies and percentages of various types of consanguineous marriages recorded among inmates of some hospitals in Kerala*

Consanguineous marriages	F Value	Ernakulam		Quilon		Trivandrum		Mental Hospital Trivandrum		Mean	
		Frequency	Percentage	Frequency	Percentage	Frequency	Percentage	Frequency	Percentage	Frequency	Percentage
Maternal uncle's son	1/16	2	2	24	12·7	61	12·3	3	3	90	10·1
Maternal aunt's son	1/16	—	—	—	—	3	0.6	—	—	3	0·3
Paternal uncle's son	1/16	—	—	2	1·1	5	1·0	—	—	7	0·8
Paternal aunt's son	1/16	2	2	19	10·1	42	8·5	2	2	65	7·3
Second cousin	1/64	—	—	3	1·6	6	1·2	—	—	9	1·0
Third cousin	1/256	—	—	2	1·1	5	1·0	—	—	7	0·8
Unrelated marriages	0	98	96	139	73·4	375	75·4	96	95	708	79·7
Total		102		189		497		101		889	
Mean F and standard error		0·0024 ± 0·0012		0·0152 ± 0·0019		0·0142 ± 0·0013		0·0031 ± 0·0013			

The mean coefficient of inbreeding among the offspring of the 889 marriages studied, analysed without reference to the number of children produced by these marriages, is 0·01182. Using the fertility structure of marriages, the mean coefficient of inbreeding is 0·01056. The difference in the mean fertility of inbred and outbred marriages, indicated by the lower value of inbreeding coefficient when the analysis is based on the offspring of the marriages than when the coefficient is derived from a consideration of marriages alone, is not significant ($\chi^2_{(1)} = 0{\cdot}13$, $P = 0{\cdot}8$–$0{\cdot}7$). The actual coefficients of inbreeding for this population should be higher than those estimated, since the effects of inbreeding in the past generations have not been considered. It is known that the incidence of consanguineous marriages in the past was much higher than today, a decreasing trend being evident during the last seven decades (Dronamraju, 1964).

The coefficients of inbreeding in sympatric populations of Christians and Muslims have also been calculated for comparison and work out for 84 Christian and 53 Muslim marriages to 0·00351 and 0·01563 respectively. For assessing the possibility of correlating levels of inbreeding with specific illnesses, estimates of inbreeding coefficient values of the parents of patients have been calculated and the scores for F are given in Table 2. The coefficients of inbreeding for congenital anatomical malformations, diabetes mellitus, susceptibility to tubercular meningitis and pulmonary tuberculosis are higher than for other illnesses or the hospital group as a whole. Analysis of the incidence of tuberculosis in twenty families indicated that the more frequent occurrence of this disease may not be solely due to a greater possibility of exposure. The small number of patients on which these values are based, however, render detailed analysis and drawing of definite inferences difficult.

Table 2. *Estimates of coefficient of inbreeding F.*
Estimation of inbreeding coefficients

Groups	Number of cases encountered	F
Congenital malformations	18	0·022
Pulmonary T.B.	28	0·022
T.B. meningitis	30	0·019
Diabetes mellitus	48	0·019
Liver cirrhosis	23	0·015
Bronchial asthma	32	0·011
Anaemia	26	0·010

Table 3. *Mortality in proportion to the pregnancies in progenies of consanguineous and non-consanguineous marriages*

F	Families studied	Pregnancies	Still births + miscarriages	Infant + juvenile deaths
First cousin 1/16	83	391	59 (0·151)	72 (0·186)
Second cousin 1/64	8	34	4 (0·118)	3 (0·088)
Not related 0	185	770	231 (0·030)	67 (0·087)

Mortality in children of non-consanguineous and consanguineous marriages

The observed proportions of mortalities in terms of still births + miscarriages and infant + juvenile deaths (death prior to the age of 15) are given in Table 3. A significantly higher frequency of still births and miscarriages occur in the inbred progenies ($F = 1/16$ and $1/64$) as compared to outbred ($F = 0$). Infant + juvenile deaths are significantly higher in progenies of first cousin marriages ($F = 1/16$) than in second cousin ($F = 1/64$) or control ($F = 0$) marriages. The total mortality for first cousin, second cousin and unrelated marriages is 33·58, 20·58 and 11·69 % respectively. It is of interest that the still birth and miscarriage rate reported in Table 3 for the unrelated group is much below the WHO figures for India as a whole. This is probably because of the comparatively superior medical, public health and educational facilities available in Kerala, thanks in part to the work of various missionary organizations.

Estimates of the magnitude of genetic load

The data presented in Table 3 have been used to estimate the magnitude of genetic loads in the population studied in units of lethal equivalents, where a lethal equivalent is defined as a group of mutant genes which would cause, on the average, one death, if dispersed in different individuals and made homozygous (Morton *et al.* 1956). By the genetic load theory of Morton *et al.* (1956) the proportion of survivors, S, is expected to be $S = e^{-(12A+BF)}$ where F is the inbreeding coefficient. In the present study, estimates of A and B have been calculated from the simple relationship, $S = 1-A-BF$. Table 4 gives the estimates of A, B and B/A.

The total load, $A+B$, is between 3 and 4 lethal equivalents per gamete, or 6 to 8 lethal equivalents per zygote.

The value of A is 0·0565 for still births + miscarriages, 0·0762 for infant + juvenile deaths and 0·1328 for total mortality. These fall in the range of values obtained by Morton *et al.* (1956) on the basis of their analysis of the data of Sutter & Tabah (1952) on French populations and for the data of Arner (1908) on an American population. This correspondence is remarkable, considering that the calculation of A value (i) includes both the genetic and environmental components of mortalities, (ii) depends on the demographic system of the population, and (iii) is subjected to considerable sampling errors. A values given in Table 4 are smaller than those reported for Japanese (Neel & Schull, 1962) and Brazilian populations (Freire-Maia *et al.* 1963; Freire-Maia, 1963; Freire-Maia & Krieger, 1963). The B value of 3·3325 for total child mortality between the late foetal and early adult stages shows that the effect of inbreeding in the Kerala population described is significant. The ratio B/A is 25·1 which is comparable to the ratio of 18·12 for the French population of Sutter & Tabah (1952) obtained by Morton *et al.* (1956). Relatively much lower inbreeding responses for mortality were reported by Neel & Schull (1962) in two Japanese populations (B/A = 4·63 and 4·68), by Freire-Maia (1963) for Brazilian white (B/A = 1·2) and Brazilian Negro (B/A = 13·8) populations, by Morton *et al.* (1956) for the American populations (B/A 10·75 and 7·94) examined by Arner (1908) and Bemiss (1958).

Table 4. *Estimates of the A and B statistics and of the B/A ratios*

Deaths	A	B	B/A
Still births + miscarriages	0·0565	1·64388	29·0938
Infant + juvenile deaths	0·0762	1·6887	22·1614
Total mortality	0·1328	3·3325	25·0941

For F = 0 and 1/16, B = 3·5023, and for F = 1/64 and 1/16, B = 1·6209, χ^2 = 0·6056 (P = 0·5 – 0·3).

SUMMARY

1. The Hindu populations throughout India are organized into castes and subcastes which are largely endogamous. The populations in the southern states of India are unique in the occurrence of a high rate of consanguineous marriages within the endogamous groups. Several hospital populations in a southern state of India (Kerala) have been examined for the rate of consanguinity and the genetic lead due to lethal genes.

2. The incidence of consanguineous marriages is about 20 % and the estimate of mean coefficient of inbreeding 0·01056.

3. The most frequent types of consanguineous marriages are of a girl with her (*a*) maternal uncle's son, and (*b*) paternal aunt's son. The uncle–niece marriage is not preferred in Kerala.

4. The frequency of foetal and infant deaths is significantly higher in inbred progenies than in the outbred. The estimates of total mortality for first cousin, second cousin, and unrelated marriages are 33·58, 20·58 and 11·69 % respectively.

5. The total load is between 3 to 4 lethal equivalents per gamete. The estimates of B and B/A statistics are high.

Our gratitude is due to Drs K. V. Krishna Das and M. Thangavelu for the facilities provided to us at the Medical College, Trivandrum and to Mr S. Ramanujam for helpful criticism and discussions.

REFERENCES

Arner, G. B. L. (1908). Consanguineous marriages in the American population, vol. 1, 101 p. New York: Columbia University.

Bemiss, S. M. (1958). Report of influences of marriages of consanguinity upon offspring. *Trans. Am. Med. Assoc.* **2**, 319–425.

Böök, J. A. (1957). Genetical investigation in a north Swedish population. The offspring of first cousin marriages. *Ann. Hum. Genet.* **21**, 191–221.

Chung, C. S., Robinson, O. W. & Morton, N. E. (1959). A note on deaf mutism. *Ann. Hum. Genet.* **23**, 357–66.

Dewey, W. J., Barrai, I., Morton, N. E. & Mi, M. P. (1965). Recessive genes in severe mental defects. *Am. J. Hum. Genet.* **17**, 237–256.

Dronamraju, K. R. (1964). Mating systems of the Andhra Pradesh people. *Cold Spring Harb. Quant. Biol.* **29**, 81–84.

Dronamraju, K. R. & Meera Khan, P. (1963). The frequency and effects of consanguineous marriages in Andhra Pradesh. *J. Genet.* **58**, 387–401.

Freire-Maia, N. (1963). The load of lethal mutations in white and Negro Brazilian populations. II. Second survey. *Acta genet.* **13**, 199–225.

Freire-Maia, N. & Krieger, H. (1963). A Jewish isolate in southern Brazil. *Ann. Hum. Genet.* **27**, 31–9.

Freire-Maia, N. Freire-Maia, A. & Quelce-Salgado, A. (1963). The load of lethal mutations in white and Negro Brazilian populations. I. First survey. *Acta genet.* **13**, 185–98.

Freire-Maia, N. Guaraciaba, S. M. R., M. Z. S. H. & Quelce-Salgado, A. (1964). The genetic load in the Bauru Japanese isolate in Brazil. *Ann. Hum. Genet.* **27**, 329–39.

Goldschmidt, E., Cohen, T., Bloch, N. Keleti, L. & Wartski, S. (1963). Viability studies on Jews from Kurdistan. *The Genetics of Migrant and Isolate Populations*, pp. 183–95, ed. by E. Goldschmidt. Baltimore: Williams and Wilkins Co.

Morton, N. E. (1960). The mutational load due to detrimental genes in man. *Am. J. Hum. Genet.* **12**, 348–64.

Morton, N. E. & Chung, C. S. (1959). Formal genetics of muscular dystrophy. *Am. J. Hum. Genet.* **11**, 360–79.

Morton, N. E., Crow, J. F. & Muller, H. J. (1956). An estimate of the mutational damage in man from data on consanguineous marriages. *Proc. Natn. Acad. Sci. U.S.A.* **42**, 855–63.

Neel, J. V. & Schull, W. J. (1962). The effect of inbreeding on mortality in two Japanese cities. *Proc. Natn. Acad. Sci. U.S.A.* **48**, 575–82.

Salzano, F, M., Marcallo, F. A., Friere-Maia, N. & Krieger, N. (1962). Genetic load in Brazilian Indians, *Acta. Genet.* **12**, 212–18.

Sanghvi, L. D. (1966). Inbreeding in rural areas of Andhra Pradesh. *Ind. J. Genet.* **26**A, 351–65.

Schull, W. J. (1958). Empirical risks in consanguineous marriages: sex ratio, malformation and viability. *Am. J. Hum. Genet.* **10**, 294–343.

Slatis, H. M., Reis, R. H. & Hoene, R. E. (1958). Consanguineous marriages in the Chicago region. *Am. J. Hum. Genet.* **10**, 446–64.

Sutter, J. & Tabah, L. (1952). La mortalité, phénomène biometrique. *Population* **7**, 69–94.

NOTE ON PAPER BY S. KUMAR, R. A. PAI AND M. S. SWAMINATHAN

By CEDRIC SMITH

The estimation of the values of A, B and their ratio B/A can be improved by using a weighted regression. Let us suppose that the families are divided up into classes (i) according to their degrees of relationship (F_i); for example, class 1 might consist of all families with first cousin parents ($F_1 = 1/16$), class 2 of all families with unrelated parents ($F_2 = 0$), and so on. In class i suppose that there are altogether n_i (> 0) children, of which m_i are affected. The observed proportion of affected is $p_i = m_i/n_i$. The corresponding expected proportion we will denote by P_i. The hypothesis is that

$$P_i = 1-\exp(-A-BF_i). \tag{1}$$

Suppose that we find in all classes that $m_i < n_i$ (which will almost always be true), so that $p_i < 1$, and let us write

$$z_i = -\ln(1-p_i), \tag{2}$$

$$Z_i = -\ln(1-P_i). \tag{3}$$

Then the assumption is that the regression of Z_i on F_i is exactly linear

$$Z_i = A+BF_i. \tag{4}$$

We can reasonably estimate A and B by minimizing the weighted sum of squares

$$S = \Sigma_i w_i(z_i-Z_i)^2 \tag{5}$$

where the weight

$$w_i = n_i(1-P_i)/P_i, \tag{6}$$

although the assumptions for least squares are not absolutely exactly fulfilled (e.g. we have in reality a multinomial distribution instead of a Gaussian one). There is a slight complication here in that the weight depends on the value of P_i, which again depends on the values of A and B we are trying to estimate. The calculation must therefore be done iteratively; we start with provisional values of A and B, use these to find appropriate weights w_i, use the weights to find the new values of A and B, use these new values to find new weights w_i, and so repeatedly until the new values agree with the old to the order of accuracy required.

More exactly, we proceed as follows. Given the values of n_i, m_i, and F_i we first calculate $p_i = m_i/n_i$ and z_i from (2). We now assume provisional values of A and B, e.g. $A = \ln 2$ and $B = 0$; we also set $j = 1$, and ϵ = some arbitrary small quantity, such as 0·001. From these we calculate, for each class i, (*) $Z_i = A+BF_i$ (this will ordinarily be positive: if exceptionally it should be zero or negative difficulties would arise and the calculation would have to be modified);

$$P_i = 1-\exp(-Z_i)$$
$$= 1-\text{antilog}(-0{\cdot}4343Z_i)$$
$$w_i = n_i(1-P_i)/P_i;$$

and from these

$$W = \Sigma w_i,$$
$$\bar{Z} = (\Sigma w_i Z_i)/W,$$
$$\bar{F} = (\Sigma w_i F_i)/W,$$
$$\Delta_{FF} = \Sigma w_i (F_i-F)^2,$$
$$\Delta_{FZ} = \Sigma w_i (F_i-F)(Z_i-Z),$$
$$B' = \Delta_{FZ}/\Delta_{FF},$$
$$A' = \bar{Z}-BF,$$
$$j' = j+1.$$

We now see whether these new values A', B' differ appreciably from the provisional values A, B. More precisely if

$$[j < 2] \quad \text{or} \quad [10\,|A'-A| + |B'-B| > \epsilon]$$

we replace the provisional values A, B, j by the new values A', B', j', and repeat the calculation from (*) onwards. Otherwise we take the values A, B to be, nearly enough, the final estimates. From these we can find the ratio B/A. But it may be convenient to deal with log (B/A) (to base 10) since in small samples any inaccuracy in the denominator A may very considerably affect the ratio B/A, and one may reasonably expect the standard error of the estimated log (B/A) to be better than the standard error of B/A itself as a guide to the probable range of values of the true ratio B/A. The error variances are then found as follows:

$$\text{var}\,(B) = 1/\Delta_{FF},$$

$$\text{var}\,(A) = 1/W + B^2/\Delta_{FF},$$

$$\text{var}\,[\log\,(B/A)] = 0{\cdot}1886\,[1/WA^2 + (\bar{F}/A - 1/B)^2/\Delta_{FF}].$$

The standard errors are the square roots of the error variances. In addition we may note that

$$\chi^2 = \Sigma w_i\,(Z_i - A - BF_i)^2$$

is distributed on the null hypothesis approximately as χ^2 with degrees of freedom equal to (the number of different classes -2). This gives a simple and useful check on the correctness of the hypothesis expressed by equation (1).

The calculation can be done fairly conveniently manually, or it can be straightforwardly programmed on an electronic computer. The University College London computer was used to analyse the data of Table 3 of the paper of Kumar *et al.*, and the following values were found (Table 5).

Table 5.

Type of mortality	A	B	log. (B/A)	B/A	χ^2 (1 D.F.)
Still births and miscarriages	0·032 ± 0·0065	1·07 ± 0·18	1·52 ± 0·095	33	4·0
Infant and juvenile deaths	0·091 ± 0·011	0·90 ± 0·21	1·00 ± 0·21	10	0·1
Total mortality	0·13 ± 0·013	2·27 ± 0·31	1·25 ± 0·06	18	0·9

These differ appreciably from the values given by Kumar *et al.* and suggest an estimate of around 2·5 lethal equivalents per gamete, or 5 per zygote, rather than the 7 found by them.

THE EXPERIMENTAL MANIPULATION OF GENES*

M. S. SWAMINATHAN

Indian Agricultural Research Institute, New Delhi

IT will be 100 years in 1965 since Gregor Mendel presented his paper on "Experiments in Plant Hybridization", in which he propounded the laws of heredity. The science of genetics was, however, born only in this century largely because of the failure of earlier workers to separate continuous from discontinuous variation in their thought and work. Progress in our understanding of the structure, function and mode of inheritance of genes has been phenomenal in recent decades. This is because of the growing knowledge, that while (to quote Kihara) "the history of the earth is written in its layers, the history of living organisms is inscribed in the chromosomes". Chromosomes and the genetic factors they carry have been subjected to a variety of biophysical, biochemical and cytogenetical studies, as a result of which we now know the following. (1) The basic chemical substance of heredity is deoxyribose nucleic acid (DNA), replaced in certain viruses by ribose nucleic acid (RNA). (2) The information needed to specify all the species and individual characteristics of every organism is recorded in its DNA in terms of a non-overlapping triplet code of bases. (3) During gene replication this information is reproduced by the unidirectional assembly of new DNA strands, each complementary in base sequence to the single strand serving as template, the end result being the formation of two daughter molecules of DNA, identical with the parental molecule. (4) The translation of genetic information into protein structure is achieved by the transcription in the nucleus of a specific DNA code sequence into the complementary sequence of bases in single stranded RNA molecule. This messenger RNA leaves the nucleus and serves as the template for the assembly of amino-acids into a polypeptide chain, the ribosomes in the cytoplasm serving as the sites for this reaction. Each specific DNA base triplet would specify a particular amino-acid and the sequence of base triplets specifies the sequence of amino-acids in the final polypeptide product. The recent knowledge on genetic determinants (*see* Tatum[1] for a summary) has enabled the planning and execution of genetic experiments of great elegance and utility. While much of the most exciting work has been done in micro-organisms, in which the "resolving power" of genetic analysis is exceedingly high,[2] significant break-throughs are also being accomplished in higher organisms. I wish to refer in this short article to a few of the possibilities opened up in the experimental manipulation of gene recombination, mutation and function.

Recombination.—By employing techniques causing denaturation and renaturation of DNA the relationships in the base sequences in nucleic acids could be studied in bacteriophages.[3] It has been shown that any two micro-organisms that can exchange genetic markers by transformation or transduction possess DNA with similar base compositions and can form hybrids on denaturation and annealing. Since there is a close relationship between the ability to form *in vitro* molecular hybrids, and the taxonomy and genetic compatibility of micro-organisms, this criterion could be used as an elegant test of phylogenetic relationships. Such attempts at cross-breeding at the molecular level have been easy in organisms like *Escherichia coli* and bacteriophage T 2 because of the organisation of the genetic units on a single linkage group, the "chromosome" being a single continuous entity (in electron micrographs, the DNA appears filled with uniformly distributed filaments about 25 Å thick and is not combined with histone; Ris[4] has suggested the term "genophore" to describe the structural counterpart of a linkage group in viruses and bacteria). In higher organisms several distinct chromosomes occur in every genome and each homologous pair cannot be easily isolated. Hence, the denaturation-renaturation technique of forming hybrid DNA molecules does not lend itself for use in higher organisms.

Recombination of genetic factors at the intra-specific level can now be carried out in a purposeful way through the application of various plant breeding techniques (*see* Allard[5] for a recent survey). All the exciting progress that has taken place in plant and animal breeding is due to the purposeful scrambling of genes from diverse sources. Studies on the control of crossing-over have, however, been few, although several workers have shown that linkage can appreciably diminish the response to selection for quantitative characters. For example,

* Summary of a talk delivered at the 30th Annual Meeting of the Indian Academy of Sciences held at Poona.

Mather and Harrison[6] found that the selection response was low where there was linkage between genes affecting the selected character and those affecting fitness. Since the actual mechanism of crossing-over is still not understood fully, studies on the control of crossing-over have to be largely empirical. The use of physical and chemical mutagens to bring about crossing-over particularly in the proximal segments of the chromosomes and thereby help to release the hidden genetic variability, is a promising line of approach.[7] In cross-pollinated plants, linkage tends to be relatively more tight than in comparable self-pollinated plants and hence the possibility of releasing genetic variability through increased crossing-over can be exploited in both these groups (Gupta and Swaminathan, unpublished).

A unique genetic system capable of restricting crossing-over to wholly homologous chromosomes was discovered 6 years ago in bread wheat (*Triticum æstivum*) by Riley and Chapman[8] and Sears and Okamoto.[9] In the absence of a gene located on the long arm of chromosome 5 B, pairing occurred among the homologous chromosomes of the 3 genomes, while in the presence of this gene pairing was confined to the homologous pairs within each genome. This discovery has enabled various types of chromosome manipulations at the inter-specific as well as inter-generic level in Triticineæ. It is not yet clear whether this type of cytological diploidisation mechanism is of general occurrence among polyploid plants.

Developments in the field of applied embryology and organ culture have helped in extending the limits of genetic recombination beyond the frontiers of a species. A recent example of such work is the successful hybridization of the two jute-yielding species, *Corchorus olitorius* and *C. capsularis*.[10] Sears[11] opened up a new era in chromosomal engineering by using a radiation-induced translocation for transferring a gene resistance to leaf rust from *Aegilops umbellulata* to *Triticum æstivum*. Tazima[12] used a similar technique for marking the sex of silkworms at the pupal stage. Translocations serve as a source of gene duplications and Hagberg[13] has employed this tool to study the dosage effect of the gene controlling alpha-amylase activity in barley grains. Since it is known that a heterozygous condition at certain loci, like the *V* locus in barley, can cause heterosis, duplications will provide a means of creating true-breeding heterozygotes. Induced chromosome aberrations also render the breakage of tight linkages possible, as demonstrated by Konzak[14] in separating the gene conferring resistance to *Puccinia coronata* from that causing susceptibility to *Helminthosporium victoriæ* in oats.

Mutation.—The knowledge that the sequence and the number of base pairs in DNA determine genetic specificity has helped in planning mutation experiments in micro-organisms at the molecular level. Attempts can be made to attack the gene at 4 different levels—a base pair in DNA (the purines, adenine and guanine and the pyrimidines, cytosine and thymine), a nucleotide (nitrogenous base-sugar-phosphoric acid combination), a cistron (a section of a chromosome that has a unitary function and may consist of several thousand nucleotides) and a chromosome (the gene strings of one human chromosome have been estimated to contain some 4×10^9 nucleotide pairs). Mistakes leading to the alteration of the sequence of the base pairs, such as addition, deletion, inversion and substitution, may occur spontaneously or can be induced artificially. Freese[15] classified the substitution of base pairs into *transitions* (*i.e.*, the replacement of a purine by another purine or a pyrimidine by a pyrimidine) and *transversions* (replacements of a purine by a pyrimidine and *vice versa*). Some of the chemicals used for bringing about such changes artificially are listed in Table I. While the action indicated in Table I represents the most likely one, there could be other effects also. For example, Kreig[16] has suggested that

TABLE I

Types of changes in DNA base pair sequence induced by some chemicals

Chemical		Nature of action
5-bromo uracil	..	G–C ⟵⟶ A–T transition
2-amino purine	..	G–C ⟵⟶ A–T transition
Nitrous acid	..	(*a*) A–T ⟵⟶ G–C transversion
		(*b*) Cross-linking of DNA strands
Hydroxylamine	..	G–C ——→ A–T transition
Ethyl methane sulphonate		(*a*) G–C ——→ A–T transition
		(*b*) Deletion or addition of base pairs
Acridines	..	Deletion or addition of base pairs

EMS-induced mutations may arise from three causes—pairing errors of 7-ethylguanine with thymine, pairing of 3-ethyladenine with cytosine and replication errors at the site of gaps produced by the hydrolysis of 7-ethyl-guanine.

Chemical mutagens have helped to alter the ratios of the different types of chlorophyll mutations recovered in higher plants, although definite shifts in mutation spectrum are yet to be demonstrated. Swaminathan *et al.*[17] and Natarajan and Upadhya[18] found that the alkylating agent, ethyl methane sulphonate, breaks chromosomes preferentially at the centromeric and proximal regions. Correlating this observation with the incidence of a high frequency and wide spectrum of chlorophyll mutations in the M 2 progenies of EMS-treated barley, Swaminathan *et al.*[17] postulated that several of the genes relating to chlorophyll development in barley should be located near the centromere. This prediction has been proved to be correct by subsequent linkage analysis[19,20] (Fig. 1).

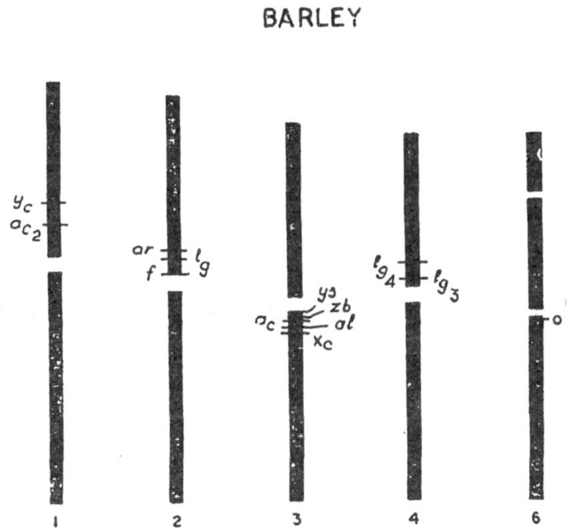

FIG. 1. Genes for chlorophyll development located near the centromere in 6 different chromosomes of barley (data from Robertson[19] and Nilan[20]).

A wide array of physical and chemical mutagens are now available and in most organisms a high frequency of mutations can be induced by devising suitable procedures of treatment and handling. Neutrons have proved to be extremely valuable in such work by virtue of their relative insensitivity to the action of modifying factors, absence of secondary physiological effects associated with their action and high RBE values for the induction of chromosome aberrations and mutations.[21] Through the use of different mutagens, all the known subspecies of *Triticum æstivum* have been recreated.[22]

Looking at the end-product of gene action, namely the protein chain, Ingram[23] first showed that in the abnormal hæmoglobin S, a glutamic acid residue is replaced by valine. In the A-protein of tryptophan synthetase of *E. coli*, a good correlation was found between induced mutations and amino-acid substitutions (*see* Yanofsky).[24] Thus, it is clear that the replacement of a single base in a code triplet may lead to the substitution of one amino-acid by another in the polypeptide product.

Function.—At the functional level, it has long been known that phenocopies simulating known phenotypes can be experimentally produced. The growing knowledge of the biochemical consequences of a mutation generates the hope that many of the hereditary infirmities in man can be artificially rectified. Such a hope has led Lederberg[25] to call attention, under the term "euphenics" to the enormous future possibilities in this field that can be opened up by molecular biologists through studies of gene expression and function.

The famous biologist Jacques Loeb in his book *The Dynamics of Living Matter* published in 1906 set "experimental abiogenesis" and "the transformation of one species into another" as the two major goals of 20th century biology (cited by Muller).[26] The research on the experimental manipulation of the gene carried out during the past decade has helped us in advancing rapidly towards the attainment of these goals. Those who work in the field of molecular genetics should, however, recognise that the living cell is a unit in which the constituent parts co-exist co-operatively. The knowledge of how genes beget proteins helps us only to understand how a random group of compounds are produced, and not how the many functions of the cell are so beautifully co-ordinated and integrated. Weiss[27] has, hence, rightly stressed the need for more research on how the molecular components of the cell are subordinated to ordered group co-existence in a system of "molecular ecology".

1. Tatum, E. L., *Proc. N.A.S.*, 1964, **51**, 908.
2. Pontecorvo, G., *Trends in Genetic Analysis*, Columbia University Press, New York, 1958, p. 145.
3. Marmur, J., Rownd, R. and Schildkraut, C. L., *Progress in Nucleic Acid Research*, 1963, **1**, 231.
4. Ris, H., In *The Interpretation of Ultrastructure*, Academic Press, 1962, **1**, 69.
5. Allard, R. W., *Principles of Plant Breeding*, New York, 1960, p. 485.
6. Mather, K. and Harrison, B. J., *Heredity*, 1949, **3**, 1, 131.
7. Lawrence, C. W., *Rad. Bot.*, 1961, **1**, 92.
8. Kiley, R. and Chapman, V., *Nature*, 1958, **182**, 713.
9. Sears, E. R. and Okamoto, M., *Proc. 10th Int. Genet. Congr.*, Montreal, 1958, **2**, 258.
10. Swaminathan, M. S., Iyer, R. D. and Sulbha, K., *Curr. Sci.*, 1961, **30**, 67.
11. Sears, E. R., *Brookhaven Symposia in Biology*, 1956, **9**, 1.

12. Tazima, Y., *Reports of the Silk Science Research Institute*, 1955, No. 5, 78.
13. Hagberg, A., Proc FAO/IAEA Symposium on "The Use of Induced Mutations in Plant Breeding," *Radiation Botany*. 1964 (In press).
14. Konzak, C. F., *Brookhaven Symposia in Biology*, 1956, **9**, 157.
15. Freese, E., *Molecular Genetics* Part I, Academic Press, 1963, p. 207.
16. Kreig, D. R., *Genetics*, 1963, **48**, 561.
17. Swaminathan, M. S., Chopra, V. L. and Bhaskaran, S., *Indian J. Genet.*, 1962, **22**, 192.
18. Natarajan, A. T. and Upadhya, M. D., *Chromosoma (Berl.)*, 1964, **15**, 156.
19. Robertson D. W., *Barley Genetics*, 1963, **1**, 159. .
20. Nilan, R. A., "The cytology and genetics of barley 1951-1962," *Monographie Supplement No.* 3, Washington State University, 1964, p. 278.
21. Gopal-Ayengar, A. R. and Swaminathan, M. S., In "*Biological Effects of Neutron and Proton Irradiations*," I.A.E.A., Vienna, 1964, **1**, 409.
22. Swaminathan, M. S., *L'energia nucleare in agricoltura*, C.N.E.N., Rome, 1963, p. 243.
23. Ingram, V. M., *Haemoglobins in Genetics and Evolution*, Columbia Univercity Press, New York, 1963.
24. Yanofsky, C., In *Informational Macromolecules*, Academic Press, New York, 1963, p. 195.
25. Lederberg, J., In *Ciba Symposium* "*Man and His Future*," 1963.
26. Muller, H. J., *Bulletin of the Atomic Scientists*, 1964 **20**, 3.
27. Weiss, P., *J. Theoret. Biol.*, 1963, **5**, 389.

NATURE OF POLYPLOIDY IN SOME 48-CHROMOSOME SPECIES OF THE GENUS SOLANUM, SECTION TUBERARIUM

M. S. SWAMINATHAN [1]
School of Agriculture, University of Cambridge, England

Received May 16, 1953

CHROMOSOME numbers have been reported for 116 species in the section Tuberarium of the genus Solanum. An euploid series with 24, 36, 48, 60 and 72 somatic chromosomes exists. PRAKKEN and SWAMINATHAN (1952) concluded from the available evidence that the basic chromosome number in the genus is 12 and not 6 as suggested by some authors. Species with 48 chromosomes will, on this basis, be tetraploids and tetraploid species are present in the taxonomic series Acaulia Juz., Conicibaccata Bitt., Longipedicellata Buk., and Tuberosa Rydb. Both native and cultivated species with 48 chromosomes occur in the series Tuberosa to which the commercial potato (*Solanum tuberosum*), a tetraploid, belongs. Nineteen tetraploid species are so far known in the section Tuberarium (SWAMINATHAN and HOWARD 1953).

STEBBINS (1947, 1950) has pointed out that it is by no means easy to classify a naturally existing polyploid species into one or the other of the two rigidly contrasting categories usually recognized, namely auto- and allo-polyploids. Such a task is further complicated in potato species since (1) any set of 12 chromosomes is usually capable of pairing with any other set in species hybrids (HOWARD and SWAMINATHAN 1952), and (2) genetic studies are difficult owing to various causes among which the following are some of the more important. First, all potato varieties are heterozygous for many of the characters studied and homozygous plants can only be obtained, if at all, by several years of selfing. Secondly, among the seedlings derived from selfing a variety or from crossing two varieties there is usually a varying percentage of degenerate individuals which may affect genetic segregations for the character being studied. Thirdly, there is the difficulty that among the offspring many may not flower, or, if they do, may have pollen and ovule sterility. Fourthly, segregations are often complex owing to the polyploid constitution of the potato; and fifthly, bud mutations and chimaeras occur frequently. Studies on this problem are nevertheless called for since any information obtained is of considerable interest to potato breeders.

MATERIAL AND METHODS

The following species and varieties possessing 48 chromosomes were used in the study.

[1] Present address: Department of Genetics, University of Wisconsin, Madison, Wisconsin.

Series Acaulia: *Solanum acaule* varieties Bukasov and Recoba.
Series Longipedicellata: *S. longipedicellatum* C.P.C. 28.
Series Tuberosa: *S. andigenum* C.P.C. 1384.
S. tuberosum varieties Duke of York, Gladstone and Cl.B. 72 (the last mentioned is a selection of MR. J. CLARK of Ireland which includes in its parentage the varieties Craigs Defiance, Ballydoon and Katahdin.)

The somatic chromosomes were studied following the use of the 8-hydroxyquinoline pretreatment method of TJIO and LEVAN (1950). The schedule adopted was somewhat modified (SWAMINATHAN 1950). Meiotic configurations were studied in microsporocytes which had been fixed in acetic alcohol (3 absolute alcohol : 1 acetic acid) to which a few drops of a saturated solution of ferric chloride had been added. Squashing in a drop of acetocarmine gave satisfactory preparations.

Chromosome doubling was induced in *S. acaule, S. longipedicellatum* and *S. tuberosum* by the colchicine-agar seed treatment method (SWAMINATHAN 1950).

Definition of terms

The following terminology as proposed by STEBBINS (1947, 1950) is considered to be the most satisfactory.

Autopolyploid: The constituent genomes are completely homologous, the progenitor being a fertile species.

Segmental allopolyploid: Contains two pairs of genomes which possess in common a considerable number of homologous chromosomal segments or even whole chromosomes, but differ from each other in respect to a sufficiently large number of chromosome segments, so that the different genomes produce sterility when present together at the diploid level.

Genomic allopolyploid: Contains two or more sets of very different genomes and the only type of pairing which normally occurs is that between similar chromosomes of the same genome.

Autoallopolyploid: Combines the characteristics of the preceding types. Plants belonging to this group will have the constitution AAAABB, etc.

These terms are used in this paper in the above senses.

RESULTS

a. *Somatic chromosomes.* The somatic chromosomes of *S. tuberosum* (var. Duke of York) and *S. longipedicellatum* (C.P.C. 28) were examined and classified using over-all length, position of centromeres, secondary constrictions and presence of satellites as diagnostic characters. The chromosomes fell into seven different groups (table 1). Each of these species had two chromosomes bearing satellites in their short arms and two chromosomes with secondary constrictions in their long arms (figs. 1 and 2). No conspicuous differences between the idiograms of the two species could be detected.

b. *Meiosis.* The chromosomes are usually associated as bivalents at diaki-

TABLE 1

Somatic chromosomes of S. tuberosum and S. longipedicellatum.

Description	Length in microns	No. of chromosomes in	
		S. tuberosum (Duke of York)	*S. longipedicellatum* (C.P.C. 28)
1. Long chromosomes with sub-median centromere and satellites in the short arm	3.0 to 3.5	2	2
2. Long chromosomes with sub-median centromere and secondary constriction in the long arm	3.0 to 3.5	2	2
3. Long chromosomes with sub-terminal centromere	3.0 to 3.5	8	6
4. Medium length chromosomes with sub-median centromere	2.0 to 3.0	8	8
5. Medium length chromosomes with median centromere	2.0 to 3.0	4	4
6. Short chromosomes with median centromere	1.5 to 2.0	4	0
7. Chromosomes of varying lengths with sub-median centromere	1.0 to 2.5	20	26

nesis and at metaphase I in *S. acaule* and *S. longipedicellatum*. Occasionally, 23_{II} and 2_{I} may be present (figs. 3 and 4). Multivalents occurred in *S. andigenum* and *S. tuberosum* and the mean pairing frequency at M I in these two species was $2.70_{I} + 19.74_{II} + 0.11_{III} + 1.37_{IV}$ (27 cells) and $2.227_{I} + 18.61_{II} + 0.166_{III} + 2.0_{IV}$ (18 cells) respectively. Two M I plates with univalents, bivalents, trivalents and quadrivalents are shown in figures 5 and 6. The chiasma frequency found at M I in the four species examined is given in table 2.

The quadrivalents observed in *S. andigenum* and *S. tuberosum* were of the following types:

Type	Number in	
	S. andigenum	*S. tuberosum*
	12	16
	4	5
	21	15

A dicentric bridge and an acentric fragment were twice observed at anaphase I in *S. longipedicellatum* (fig. 7). Metaphase II and subsequent stages were regular in *S. acaule* and *S. longipedicellatum* (fig. 8). In *S. andigenum* and *S. tuberosum,* however, restitution nuclei were formed in 2 to 3% of cells leading to the formation of dyads. Eleven percent of metaphase II plates in *S. andigenum* and 17% in *S. tuberosum* had unbalanced chromosome numbers.

c. *Studies in induced polyploids.* Chromosome doubling was induced by colchicine treatment of seeds in *S. tuberosum* (var. Gladstone), *S. acaule*

1

2

FIGURE 1.—Somatic chromosomes of *S. tuberosum*, variety Duke of York (× 1500). FIGURE 2.—Somatic chromosomes of *S. longipedicellatum*, C.P.C. 28 (× 1200).

(var. Recoba and Bukasov) and *S. longipedicellatum* (C.P.C. 28). Octoploid *S. acaule* and *S. longipedicellatum* plants had thicker leaves with a lower length/breadth index, with more prominent hairs and with larger stomata. They were late to commence flowering and had larger flowers with calyx more hairy, with the pedicels thickened and with the anthers thicker but shorter. These plants were vigorous, fertile and showed a good fruit setting both spontaneously and after crossing with *S. tuberosum* (SWAMINATHAN 1951). In contrast, octoploid *S. tuberosum* plants were dwarf, unhealthy in appearance and never flowered. The chromosome associations at M I of meiosis in octoploids of *S. acaule* and *S. longipedicellatum* were either quadrivalents, bivalents or univalents.

Seeds of F_1 *S. demissum* × *S. tuberosum* (2n = 60) were treated with colchicine to determine the nature of the plants with 120 chromosomes. They were, in general, more vigorous than octoploid *S. tuberosum* plants, formed flowers and set a few seeds (5 seeds per berry).

d. *Studies in hybrids between induced polyploids and S. tuberosum.* 1. Octoploid *S. longipedicellatum* × *S. tuberosum* (Katahdin). This cross succeeded easily whereas no seeds could be obtained following crosses between normal

TABLE 2

Chiasma frequency at M I in tetraploid species.

Species	No. of PMC's examined	No. of configurations with xta					Chiasmata		
		0	1	2	3	4	Total	Per cell	Per bivalent
S. acaule	20	4	173	305	0	0	783	39.15	1.63
S. andigenum	25	73	231	305	16	21	973	36.04	1.50
S. longipedicellatum	15	8	206	150	0	0	506	33.75	1.40
S. tuberosum	18	41	156	202	21	15	663	36.80	1.53

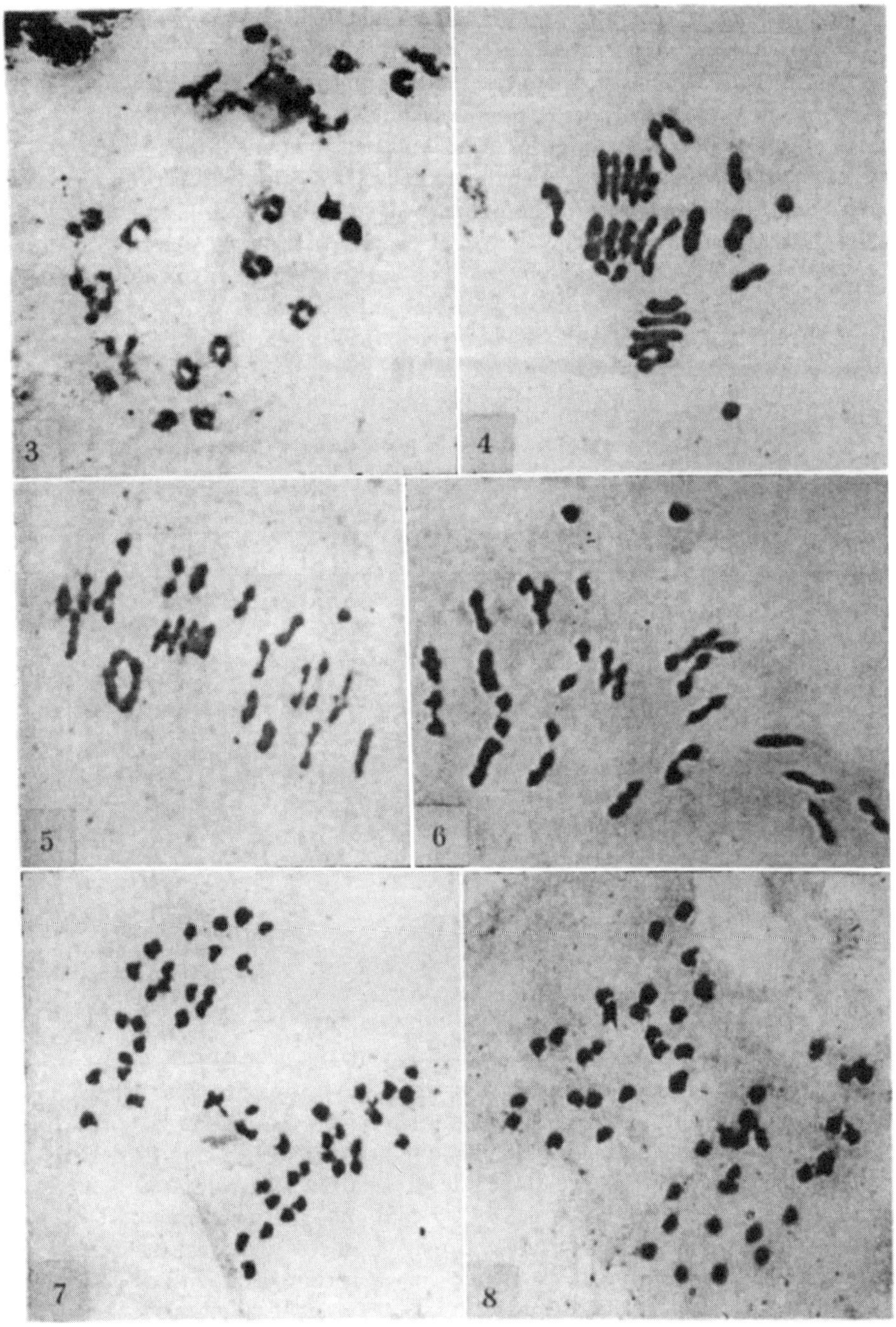

FIGURE 3.—Diakinesis in *S. longipedicellatum*. 24_{II}. FIGURE 4.—First metaphase in *S. longipedicellatum*. $2_{I}+23_{II}$. FIGURE 5.—First metaphase in *S. andigenum*. $1_{I}+18_{II}+1_{III}+2_{IV}$. FIGURE 6.—First metaphase in *S. tuberosum*. $2_{I}+19_{II}+2_{IV}$. FIGURE 7.—First anaphase in *S. longipedicellatum* showing a dicentric bridge and an acentric fragment. FIGURE 8.—Second metaphase in *S. longipedicellatum*. 24 + 24 segregation.

S. longipedicellatum (2n = 48) and *S. tuberosum*. The F_1 hybrids were healthy and flowered but did not produce seeds on selfing. Meiosis was studied in one hybrid plant which had the expected chromosome number 2n = 72. There was at M I a mean frequency per plate of $2.3_{I} + 34.65_{II} + 0.1_{IV}$ (20 cells). The average number of chiasmata per nucleus and per potential bivalent were 49.05 and 1.36 respectively. At M II, each of the 12 plates examined had 36 chromosomes. Seventy-five percent of the pollen was stainable.

The hybrid plants with 2n = 72 would have 48 chromosomes of *S. longipedicellatum* and 24 of *S. tuberosum*. The nearly regular bivalent formation in them may be due to the pairing of the chromosomes of the two species among themselves, since normal *S. longipedicellatum* and haploid *S. tuberosum* (IVANOVSKAJA 1939) form 24 and 12 bivalents respectively.

TABLE 3

Chromosome pairing at M I in (F_1 S. acaule 8x × S. tuberosum) × S. tuberosum, 2n = 60.

	No. of cells	Configuration				
		I	II	III	IV	V
	1	4	14	6	0	2
	1	3	15	6	1	1
	1	7	13	6	1	1
	1	4	17	6	1	0
	1	14	12	6	1	0
	1	7	17	5	1	0
	1	6	19	4	1	0
	1	8	17	6	0	0
	2	6	18	6	0	0
	1	9	18	5	0	0
	1	7	19	5	0	0
	3	5	20	5	0	0
	1	14	17	4	0	0
	3	8	20	4	0	0
	1	6	21	4	0	0
	1	5	23	3	0	0
	2	7	22	3	0	0
Total	23	159	422	111	6	4
Mean		6.913	18.347	4.826	0.261	0.1765

2. Octoploid *S. acaule* × *S. tuberosum*. Like the octoploid plants of *S. longipedicellatum,* those of *S. acaule* could be easily crossed to *S. tuberosum,* while crosses involving the undoubled plants did not succeed. Thirty-two berries were obtained from the 86 crosses made in 1950 betwen octoploid *S. acaule* and varieties of *S. tuberosum*. Each berry contained, on an average, 31 seeds. The F_1 hybrids, raised in 1951, had broader leaves with many interjected leaflets and a prominent pedicel articulation when compared with *S. acaule* in which the articulation is almost obsolete.

Meiosis was studied in one F_1 plant which had the chromosome number 2n = 72. There was at M I a mean frequency of $1.8_{I} + 34.7_{II} + 0.2_{IV}$ (20 cells). The mean chiasma frequencies per nucleus and per bivalent were 51.15 and

1.42 respectively. Sixty-four percent of the pollen was stainable. The hybrids could be backcrossed to *S. tuberosum* easily and four berries obtained in such crosses had 57, 62, 68 and 91 seeds respectively.

3. (F_1 *S. acaule* × *S. tuberosum*) × *S. tuberosum*. The plants of this cross resembled *S. tuberosum* in leaf shape, pedicel articulation and other morphological characters. One plant was examined cytologically and was found to have 60 somatic chromosomes. There was a mean frequency of $6.91_{I} + 18.347_{II} + 4.826_{III} + 0.26_{IV} + 0.176_{V}$ at M I. The detailed data are given in table 3. A striking feature of chromosome association at M I in this plant was the presence of several trivalents (range 3 to 6), which had not been observed in other pentaploid hybrids such as F_1 *S. demissum* × *S. tuberosum* (Howard and Swaminathan 1952). There were two pentavalents and six trivalents on one plate (fig. 9). All the trivalents observed were of the chain types

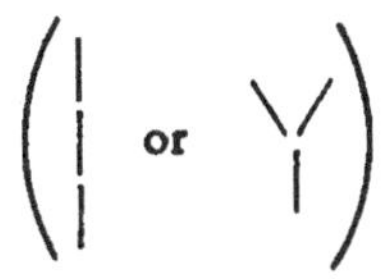

The pentavalents were also simple chains. The mean number of chiasmata per plate was 41.04.

Chromosome distribution was very irregular at M II. Among the 24 cells examined, two had 27 chromosomes; two, 28; four, 29; six, 30; four, 31; two, 32; three, 33; and one, 35. Two plants in which the pollen was examined had 52 and 64% good pollen. These plants could again be easily backcrossed to *S. tuberosum*. Each berry from such a cross contained, on an average, 41 seeds.

The 72-chromosome plants of the cross octoploid *S. acaule* × *S. tuberosum* would have 48 *acaule* and 24 *tuberosum* chromosomes. Their gametes will probably have 24 *acaule* + 12 *tuberosum* chromosomes in view of the fairly regular meiosis in the plants. A pentaploid hybrid resulting from the fusion of such a gamete with a normal *S. tuberosum* gamete will have 24 *acaule* and 36 *tuberosum* chromosomes. The simplest explanation of the occurrence of about 5 trivalents per plate at M I in the pentaploid hybrid would thus be that they are formed following multivalent association of some of the chromosomes of the 3 basic sets of *S. tuberosum* present.

e. *S. acaule* ($2n = 48$) × *S. simplicifolium* ($2n = 24$). The triploid F_1 plants from the cross *S. acaule* × *S. simplicifolium* were vigorous and had compound leaves (*S. acaule* has compound leaves and *S. simplicifolium*, simple leaves—fig. 12). There was a mean frequency of $8.85_{I} + 9.0_{II} + 3.05_{III}$ at M I (20 cells). The plants were both pollen and ovule sterile. Amphidiploid plants were produced by treating F_1 seeds with colchicine (Swaminathan 1950). There were $0.4_{I} + 35.8_{II}$ per plate (25 cells) at M I in the amphidiploid. At M II, 30 cells had 36 chromosomes, one had 37 and another 35. The subsequent stages were regular. Pollen fertility was about 61%. The plants were self-fertile and set berries spontaneously. The berries contained, on an average, 101 seeds.

The subsequent generations of this amphidiploid grown in 1951 and 1952 consisted of vigorous and fertile plants which were uniform in growth habits.

f. The backcross hybrid (amphidiploid *S. acaule* × *S. simplicifolium*) × *S. simplicifolium*. Crosses were made between amphidiploid *S. acaule* × *S. simplicifolium* and *S. simplicifolium* in 1950. Several of the berries obtained had nothing but shrivelled seeds whereas one berry had 3 good seeds which were sown in 1951. Two seeds germinated and gave rise to healthy plants. One

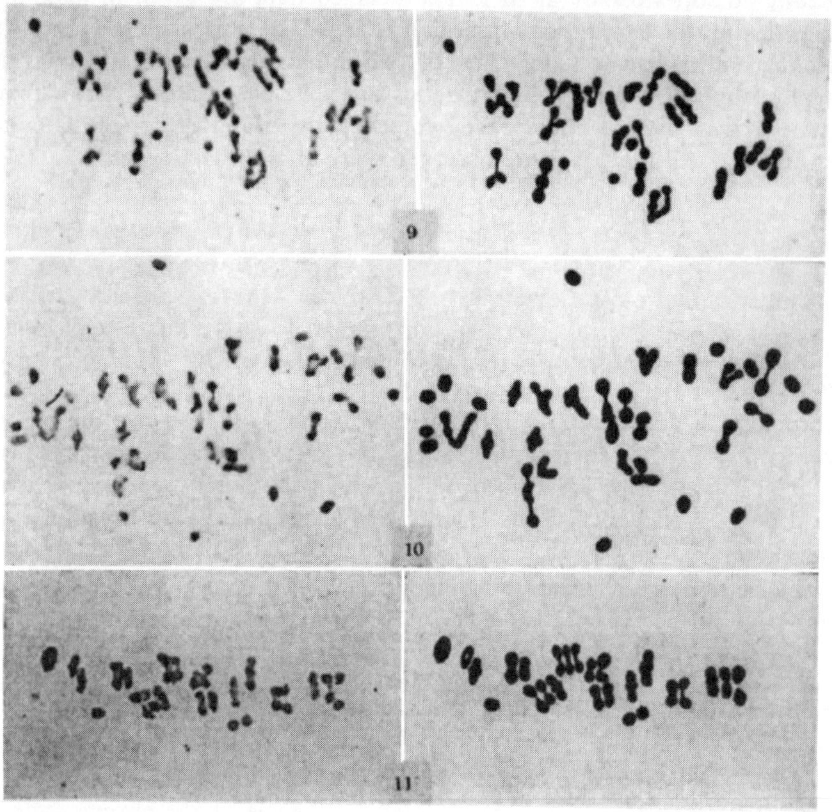

FIGURE 9.—First metaphase in (F₁ *S. acaule* × *S. tuberosum*) × *S. tuberosum*. $4_{I} + 14_{II} + 6_{III} + 2_{V}$. FIGURE 10.—First metaphase in (F₁ *S. acaule* × *S. tuberosum*) × *S. tuberosum*. $14_{I} + 12_{II} + 6_{III} + 1_{IV}$. FIGURE 11.—First metaphase in AD6.151. $3_{I} + 22_{II}$. (All photomicrographs × 1200.)

(plant no. A.D.6.151) had simple leaves and the other (A.D.6.152) had compound leaves (fig. 12). Chromosome counts in root tips and PMC squashes showed that A.D.6.151 had 2n = 47 and A.D.6.152, 2n = 48.

There was at M I in A.D.6.151 a mean frequency of $4.1_{I} + 20.1_{II} + 0.9_{III}$ (50 cells). A M I plate of this plant is shown in fig. 11. Fourteen of the 40 M II nuclei studied had 24 chromosomes whereas in the rest, the number ranged from 21 to 27. The plant had 8% good pollen and it produced no seed following either selfing or crosses with the sister plant and *S. tuberosum*. Plant A.D.6.152, on the other hand, was fertile and produced seeds in crosses

FIGURE 12.—Leaves of A. *S. acaule*; B. *S. simplicifolium*; C. Amphidiploid *S. acaule* × *S. simplicifolium*; D & E. Two plants of the cross (amphidiploid *S. acaule* × *S. simplicifolium*) × *S. simplicifolium*; D. Plant No. AD6.152; E. Plant No. AD6.151.

with *S. tuberosum*. There were $1.2_{I} + 23.08_{II} + 0.08_{III} + 0.1_{IV}$ at M I (50 cells). The chromosome number was regular at M II in 20 of the 24 plates examined. The plant had 61% good pollen.

It was surprising that one plant of this backcross should have 47 chromosomes in view of the rarity of unbalanced gametes in amphidiploid *S. acaule* × *S. simplicifolium*. The cross was repeated in 1951 using the second generation amphidiploid plants as mother parents. Three berries containing a total of 8 seeds were obtained and these seeds were sown in 1952. All of them germinated and the plants had compound leaves and a somatic chromosome number of 48. The plants had 60 to 70% good pollen and in one plant there were $3.6_{I} + 21.4_{II} + 0.4_{IV}$ (5 cells) at M I.

Origin of the simple leaf plant

It was found in studies on the inheritance of leaf type (compound or simple) in diploid species, that those with compound leaves have the genotype *LL* (*L* being a dominant factor for compound leaf) and *S. simplicifolium* has the genotype *ll* (SWAMINATHAN 1952). If *S. acaule*, a tetraploid, has the genotype *LLLL*, its gametes will have *LL* and the triploid hybrid from a cross between it and *S. simplicifolium* will be *LLl*. The amphidiploid will be *LLLLll*. A balanced gamete of the amphidiploid will have *LLl* on the asumption that the 36 bivalents usually formed in it result from pairing within the complements of *S. acaule* and *S. simplicifolium* respectively. The tetraploid plant resulting from backcrossing the amphidiploid to *S. simplicifolium* will have the genotype *LLll* and will have compound leaves. The 47-chromosome backcross plant will have the genotype *Lll* if the gamete of the amphidiploid parent had been deficient for one chromosome with the locus *L*, and should still possess compound leaves since it has been found from studies in amphidiploid *S. macolae* × *S. simplicifolium* that plants simplex for *L*, i.e., *Llll*, have compound leaves.

The formation of a simple leaf plant in the backcross can be explained if *S. acaule* is disomic for *L*. The F_1 hybrid in such a case will be *Ll* and the amphidiploid *LLll*. The balanced gametes of the amphidiploid will have *Ll*, while a deficient gamete with the chromosome carrying *L* absent will have only *l* and such a gamete on being fertilized by *S. simplicifolium* pollen will give rise to a plant with simple leaves. These results thus suggest that with reference to locus *L*, *S. acaule* behaves as a functional diploid.

DISCUSSION

STEBBINS (1950) has described in great detail the occurrence and characteristics of polyploids and has pointed out the difficulties in classifying them. He has recognized four types of polyploids: autopolyploids, segmental allopolyploids, true allopolyploids and autoallopolyploids. These are only modal types and may be connected by several intermediate categories.

Chromosome associations at the first division of meiosis are the usual source of information concerning the type of polyploidy in a given plant. Presence of multivalents is considered to indicate autopolyploidy and their absence, allopolyploidy. However, MÜNTZING and PRAKKEN (1940) and GILES and RANDOLPH (1951) have proposed for two different reasons that the absence of multivalents should not be treated as evidence of allopolyploidy. The former authors found a genotypically controlled tendency to form only bivalents in an autopolyploid Phleum. The latter authors made a study of the frequency of quadrivalents in a strain of autotetraploid maize at the beginning and end of a ten year period. They found that there were fewer quadrivalents and more bivalents at the end of the period than at the beginning.

The occurrence of tetrasomic ratios of inheritance provides evidence of autotetraploidy. Genetic ratios again may not provide absolute proof of the nature of origin of long established plants since the initially homologous genes lying in the different chromosome sets of a polyploid may mutate in different directions and gradually become so distinct as to be no longer allelic. Such a differentiation may eventually transform a polyploid into a species that has most genes represented only once in the gametes. Also, different loci may segregate differently in segmental allopolyploids so that any conclusions drawn from the study of one or two factors may be misleading.

These limitations will have to be borne in mind while discussing the mode of the origin of a polyploid plant from evidence obtained in present day studies. Such studies, however, show the relationship between the component genomes of a polyploid plant as it exists today, and in a polyploid which is also a crop plant, this knowledge facilitates the formulation of a rational approach to current breeding problems. The available data relating to four Solanum species are summarized below.

S. acaule

This species is highly fertile (pollen and seed) and regularly forms 24 bivalents and no quadrivalents at M I of meiosis. Octoploid *S. acaule* forms quadrivalents and as LAMM (1945) has pointed out, it is hence unlikely that the formation of bivalents in tetraploid *S. acaule* is determined by genotypic control. While no genetic study has been reported in this species, the data presented herein suggest that it behaves as a functional diploid with reference to locus *L*.

The triploid hybrids from crosses between *S. acaule* and diploid species have a high trivalent frequency (3 to 7), which led PROPACH (1937) to suggest that *S. acaule* is really an autotetraploid. However, the fact that amphidiploid *S. acaule* × *S. simplicifolium* forms only bivalents and the hybrid between octoploid *S. acaule* and *S. tuberosum* forms very few quadrivalents, suggest that the two genomes in a gamete of *S. acaule* are well differentiated. No diploid species has so far been found in the taxonomic group Acaulia. JUZEPCZUK (1937) considers Acaulia to be the most recently formed series, since the pedicel articulation is almost obsolete and the corolla lobes are very much fused together. It is probable that *S. acaule* arose as a segmental allotetraploid from diploid species belonging to other taxonomic groups in Solanum.

S. longipedicellatum

This species has two satellited chromosomes, usually forms 24 bivalents and is fertile. One clone was heterozygous for an inversion. Octoploid plants are vigorous, form quadrivalents and are seed fertile. The hybrid between octoploid *S. longipedicellatum* and *S. tuberosum* forms mostly bivalents. Also, as in Acaulia, no diploid species have so far been recorded in the series Longipedicellata. Hence this species seems to be a segmental allotetraploid or a genomic allotetraploid.

S. andigenum

The name *S. tuberosum* was given to the European potato by BAUHIN in 1596 and that name was retained by LINNAEUS. JUZEPCZUK and BUKASOV (1929) grouped the Andean tetraploid varieties into a new species, *S. andigenum,* since these varieties form tubers best under short day conditions and the authors considered that they had had an independent origin from that of *S. tuberosum.* SALAMAN (1937) and HAWKES (1944), however, suggested that the two species were not markedly distinct and that they had had a common origin. VAN DER PLANK (1946) adduced evidence to suggest that the first European potatoes were short day types and he concluded that the new name *S. andigenum* is not justifiable since the chief characteristic of the plants of this species is their short day reaction. HAWKES and DRIVER (1946) agreed with the views of VAN DER PLANK and recently CORRELL (1952) has also listed *S. andigenum* as a synonym of *S. tuberosum.*

Varieties of *S. andigenum* can be easily crossed reciprocally with *S. tuberosum* and the hybrids are fertile. One to two quadrivalents are formed at M I in *S. andigenum* (maximum—4 quadrivalents). The meiotic behavior is hence quite similar to that found in varieties of *S. tuberosum.* Since there could thus be no objection on taxonomic or cytogenetical grounds for *S. andigenum* being considered as a subspecies or variety of *S. tuberosum,* the following discussion as to the nature of polyploidy in *S. tuberosum* would likewise apply to *S. andigenum.*

S. tuberosum

Floral characters. FUKUDA (1927) suggested that *S. tuberosum* arose from a single species since the floral characters in the various varieties have remained constant during the past 300 years. LONGLEY and CLARK (1930) pointed out that there was no need to suppose that more than one type of calyx and corolla were involved in the early ancestry and that it would be possible for a dominant type of flower to be carried unchanged through several generations of inbreeding especially if it was linked with desirable economic characters. All the varieties investigated by LUNDEN (1937) segregated for flower color and this was considered by him to point to the autotetraploid nature of the cultivated potato.

Somatic chromosomes. ELLISON (1935) found that the somatic chromosomes of *S. tuberosum* could not be classified into groups of four by their size.

His results are very misleading since he reported differences between the chromosomes of the varieties Langworthy and Golden Wonder. These should have similar somatic chromosome complements since Golden Wonder has been shown to be a periclinal chimaera with an inner core of Langworthy (CRANE 1936). LAMM (1945) who found 4 chromosomes with satellites in autotetraploid *S. rybinii* suggested that the presence of only two such chromosomes in *S. tuberosum* might be due to amphiplasty (NAVASHIN 1928). It appears that the study of somatic chromosomes will not provide much evidence concerning the nature of polyploidy in Solanum species.

Meiosis. It would appear from the data of authors other than CADMAN (1943)—table 4—that in *S. tuberosum* there occurs a mean trivalent + quadrivalent frequency of between 2 and 3 per plate as compared with about 4 to 6 in artificially produced autotetraploids (SWAMINATHAN 1952) and 0 in *S. acaule* and *S. longipedicellatum.* THOMAS (1946) reported that in *S. tuberosum* only two true quadrivalents per plate occur and he suggested that several

TABLE 4

Frequency of multivalents at M I in S. andigenum and S. tuberosum.

Species and variety	No. of cells with a quadrivalent (+ trivalent) frequency of										Multivalents		Reference
	0	1	2	3	4	5	6	7	8	9	Mean per nucleus	Coefficient of realization	
S. tuberosum													
Flourball	1	0	4	5	13	15	15	11	4	2	5.24	0.436	CADMAN (1943)
Deodara	1	1	3	5	7	3	0	0	0	0	3.25	0.27	LAMM (1945)
36/200	1	8	8	2	1	0	0	0	0	0	1.70	0.14	" "
Cl. B72	1	5	6	3	2	1	0	0	0	0	2.16	0.18	Present study
S. andigenum													
C.P.C.1384	8	8	4	4	3	0	0	0	0	0	1.48	0.12	" "

workers had failed to distinguish between primary and secondary associations. It is difficult to judge whether the quadrivalents observed are formed in a random manner or whether particular chromosomes always form them, since it is not possible to identify the individual chromosomes from morphological characters.

LONGLEY and CLARK (1930), MEURMAN and RANCKEN (1932) and ELLISON (1936) have all suggested that the irregularities of meiosis frequently observed in *S. tuberosum* (like dyad formation, asynapsis, etc.) could be explained if it was assumed that the species had originated as a hybrid between two species whose chromosomes did not pair fully. LAMM (1945) pointed out that similar irregularities occur in synthetic autotetraploids.

Sterility. KOMAROV (1931) suggested that the sterility of many cultivated varieties as contrasted with the fertility of many of the wild species, indicates a hybrid origin of the former. ELLISON (1936) on the other hand, has shown that several cultivated varieties are highly fertile when grown under favorable conditions.

Study of a haploid. Ivanovskaja (1939) obtained a single haploid (technically " polyhaploid ") plant of *S. tuberosum* var. Aurora. Usually 12 bivalents were present at M I in this plant but occasionally 11 bivalents and 2 univalents occurred. Bridges and fragments were observed in 7% of the sporocytes at anaphase I, indicating some differentiation of the two genomes in the gametic set of *S. tuberosum.* The haploid produced 3 seeds in 12 berries. Ivanovskaja (1939) concluded from these observations that *S. tuberosum* is an allotetraploid. Lewis (1943) has, however, pointed out that it is to be expected that haploid *S. tuberosum* plants will be self-incompatible like the majority of diploid Solanum species.

Study of species hybrids. The pairing behavior of the chromosomes of the two genomes in the gametic set of *S. tuberosum* can be inferred from hybrids such as the tetraploids obtained from the cross *S. tuberosum* × diploid species (due either to the functioning of unreduced gametes in the diploid parent or the use of autotetraploids of diploids) and from the hexaploid hybrids, octoploid *S. acaule* × *S. tuberosum* and octoploid *S. longipedicellatum* × *S. tuberosum.* All the results agree in showing that the chromosomes of the two genomes may pair to form approximately 12 bivalents. The pentaploid hybrid, (octoploid *S. acaule* × *S. tuberosum*) × *S. tuberosum,* provides some evidence as to the nature of the chromosome association in a triploid complement of *S. tuberosum.* The maximum frequency in a cell of 6 trivalents and 2 pentavalents suggests that 8 chromosomes of each of the 3 sets of 12 can associate together.

Behavior of polyploids. Octoploid plants of *S. tuberosum* are dwarf and weak in contrast to such plants of *S. acaule* and *S. longipedicellatum,* which are large and vigorous. Decaploid (*S. demissum* × *S. tuberosum,* 2n = 120) plants have more vigor than octoploid *S. tuberosum* which may be considered as evidence in support of the general view that species hybrids suffer less from chromosome doubling than true species.

Effect of inbreeding. Krantz (1946) found that results from inbreeding commercial potato varieties are more explicable on an autotetraploid than on an allotetraploid basis.

Genetic results. Müller (1930) first pointed out that the potato, owing to its tetraploid nature, does not behave genetically in the same way as a diploid. Prior to that time, all results had been interpreted on a simple, disomic basis. Assejeva and Nicolaeva (1935) obtained evidence for the occurrence of several duplicate factors controlling plant and tuber color. It appeared probable that some of these duplicate factors were allelomorphic. The occurrence of tetrasomic inheritance in the commercial potato was clearly established by Lunden (1937). He found, besides typical tetrasomic ratios of inheritance for certain loci, double reductional segregation, i.e., random chromatid segregation, at the following loci. (a) In reciprocal crosses of the type *Pppp* × *pppp* and *Dddd* × *dddd* (*P* is a factor for purple pigmentation and *D,* a factor needed for the development of plant color), *P* segregated as 1 *P* : 1 *p* while *D* segregated as 13 *D* : 15 *d.* (b) Duplex *RRrr* individuals were found in backcross progenies of the type *Rrrr* × *rrrr* (*R* is a factor for red color). (c) Nulliplex *dddd* offspring occurred in *DDDd* × *dddd* crosses. CADMAN (1942) found double reductional division at the *Nx* locus (gene *Nx* controls the top necrotic reaction to virus X). Since the publication of LUNDEN'S (1937) paper, segregations for various characters in commercial varieties have been explained on a tetrasomic basis. Simplex genotypes have been assigned in many instances without any other evidence of tetrasomic inheritance and in such cases, it is equally likely that the inheritance is disomic.

To summarize the results of the cytogenetical studies in *S. tuberosum* outlined so far: (a) genetical results suggest that several presumably unlinked loci show tetrasomic inheritance, and (b) cytological results show that first, 12 bivalents are formed in a haploid plant; secondly, a triploid complement may form up to 8 trivalents; and thirdly, a maximum frequency of 5 quadrivalents (9 according to the data of CADMAN 1943) and a mean number of 2 to 3 quadrivalents per metaphase I plate occur in some varieties. The lower quadrivalent frequency at the tetraploid level in comparison with the bivalents in the haploid and trivalents in the triploid may be due to some degree of preferential pairing in the tetraploid caused by a slight differentiation of some of the chromosomes. The extent of differentiation may vary in different varieties probably depending upon the degree of hybridization involved in their evolution. This may lead to significant differences in the multivalent frequencies observed in different varieties. The low chiasma frequency (less than 2 per bivalent) found in Solanum species may also partly lead to the realization of potential quadrivalents in the form of bivalents or trivalents and univalents.

These results would suggest that *S. tuberosum* probably arose as an autotetraploid. However, at this distance from the period of its evolution and with much hybridization work having been done in the breeding of commercial varieties, it would be futile to expect representatives of this species to behave strictly in conformity with the classical definition of autopolyploids. For all practical purposes, we may consider the current commercial varieties as segmental allotetraploids. Whether the term " segmental allotetraploid " can be applied to *S. tuberosum* in the evolutionary sense, as STEBBINS (1950) has proposed, can only be decided after several haploid plants of this species have been obtained and data concerning their fertility are available.

The fact that *S. tuberosum* seems essentially to be an autotetraploid will have some major implications in potato breeding programs. The production of homozygous breeding lines will be a slow process and a good proportion of the inbred plants may be aberrant owing to the possibility for the accumulation of deleterious lethal factors in a vegetatively propagated crop like the potato. It may consequently be difficult to perform several generations of continuous inbreeding. Desirable genes like the ones giving top-necrotic reaction to virus X (COCKERHAM 1943) and resistance to wart (LUNDEN 1950) seem to be present in the varieties possessing them in the simplex state and the breeder will require both much space and patience to get these genes in the triplex or quadruplex conditions. On the other hand, the existence of the possibility to produce plants quadruplex for a particular gene is a definite advantage in instances where the genes have an additive effect.

SUMMARY

1. *Solanum longipedicellatum* and *S. tuberosum* have two chromosomes with satellites and two with secondary constrictions in their long arms.

2. *S. acaule* and *S. longipedicellatum* invariably form 24 bivalents at metaphase I of meiosis in microsporocytes. Some multivalents ($\pm$ 2) occur in *S. andigenum* and *S. tuberosum*.

3. Octoploid plants of *S. acaule* and *S. longipedicellatum* are vigorous and fertile, while those of *S. tuberosum* are dwarf and fail to flower. Decaploid plants of F_1 *S. demissum* × *S. tuberosum* with 120 chromosomes are vigorous but weakly fertile. Associations higher than quadrivalents have not been observed in octoploid *S. acaule* and *S. longipedicellatum*.

4. Thirty-six bivalents occur at M I in the hexaploid plants from crosses between octoploid *S. acaule* and *S. longipedicellatum* (pistillate parents) and normal *S. tuberosum* (pollen parent).

5. A maximum frequency of 2 pentavalents + 6 trivalents occurs in the pentaploid hybrid, (F_1 octoploid *S. acaule* × *S. tuberosum*) × *S tuberosum*. This suggests that 8 chromosomes of the 3 sets of *S. tuberosum* present in the hybrid may form trivalent associations.

6. The amphidiploid from the cross *S. acaule* × *S. simplicifolium* is fertile in contrast to the complete sterility of the triploid hybrid. Thirty-six bivalents are present at M I.

7. Two plants from the backcross (amphidiploid *S. acaule* × *S. simplicifolium*) × *S. simplicifolium* had the chromosome numbers 2n = 48 and 47 respectively. The 48-chromosome plant had compound leaves like *S. acaule* and the 47-chromosome plant had simple leaves like *S. simplicifolium*. This suggests that *S. acaule* has the factor *L*, needed for the development of the compound leaf, in only 2 of its 4 genomes.

8. The available data suggest that (a) *S. acaule* and *S. longipedicellatum* are segmental allotetraploids, (b) *S. andigenum* and *S. tuberosum* are autotetraploids though the current commercial varieties can be considered as segmental allotetraploids, and (c) both on taxonomic and cytogenetical grounds *S. andigenum* can be treated as a sub-species of *S. tuberosum*.

ACKNOWLEDGMENTS

The author wishes to express his sincere thanks to DR. G. D. H. BELL, Director, Plant Breeding Institute, Cambridge, for his interest in the problem and for the facilities provided for the investigation; to DR. H. W. HOWARD for many discussions; to PROF. R. PRAKKEN and DR. H. J. TOXOPEUS, Agricultural University, Wageningen, Holland, for their suggestions and for providing some of the material; and to DR. D. C. COOPER, Department of Genetics, University of Wisconsin, for his kind help in the preparation of the manuscript.

LITERATURE CITED

ASSEJEVA, T., and N. NICOLAEVA, 1935 Genetic nature of the coloring of potato tubers, sprouts and flowers (In Russian). Works Pot. Res. Inst. Moscow **9**: 107.

CADMAN, C. H., 1942 Autotetraploid inheritance in the potato; some new evidence. J. Genet. **44**: 33–52.

1943 Nature of tetraploidy in cultivated European potatoes. Nature **152**: 103–104.

COCKERHAM, G., 1943 The reaction of potato varieties to viruses X, A, B and C. Ann. Appl. Biol. **30**: 338–344.

CORRELL, D. S., 1952 Section *Tuberarium* of the genus *Solanum* of North America and Central America. U. S. Dept. Agric. Monograph No. 11.

CRANE, M. B, 1936 Note on a periclinal chimaera in the potato. J. Genet. **32**: 73–77.

ELLISON, W., 1935 A study of the chromosome numbers and morphology in certain British varieties of the common cultivated potato (*Solanum tuberosum* L.). Genetica **17**: 1–26.

1936 Meiosis and fertility in certain British varieties of the cultivated potato (*Solanum tuberosum* L.). Genetica **18**: 217–254.

FUKUDA, Y., 1927 Cytological studies on the development of pollen grains in different races of *Solanum tuberosum* with special reference to sterility. Bot. Mag. Tokyo **61**: 459–476.

GILES, A., and L. F. RANDOLPH, 1951 Reduction of quadrivalent frequency in autotetraploid maize during a period of ten years. Amer. J. Bot. **38**: 12–17.

HAWKES, J. G., 1944 Potato collecting expeditions in Mexico and South America. II. Systematic classification of the collections. Imp. Bur. Pl. Breed. Genet. Cambridge, pp. 142.

HAWKES, J. G., and C. M. DRIVER, 1946 Origin of the first European potatoes and their reactions to length of day. Nature **158**: 168.

HOWARD, H. W., and M. S. SWAMINATHAN, 1952 Species differentiation in the section *Tuberarium* of the genus *Solanum* with particular reference to the use of interspecific hybridization in breeding. Euphytica **1**: 20–28.

IVANOVSKAJA, E. V., 1939 A haploid plant of *Solanum tuberosum* L. C. R. (Doklady) Acad. Sci. U.R.S.S. **24**: 517–520.

JUZEPCZUK, S. W., 1937 New species of the genus *Solanum* L. in the group *Tuberarium* Dun. Bull. Acad. Sci. U.S.S.R. Ser. Biol. **2**: 295–331.

JUZEPCZUK, S. W., and S. M. BUKASOV, 1929 A contribution to the question of the origin of the potato. Proc. U.S.S.R. Congr. Genet. Pl. & Animal Breed. **3**: 593–611.

KOMAROV, V. C., 1931 Origin of cultivated plants. State Press of Agricultural & Cooperative Literature, Moscow & Leningrad, pp. 240.

KRANTZ, F. A., 1946 Potato breeding methods. III. A suggested procedure for potato breeding. Minn. Agr. Exp. Sta. Tech. Bul. 173, pp. 24.

LAMM, R., 1945 Cytogenetic studies in *Solanum*, sect. *Tuberarium*. Hereditas **31**: 1–128.

LEWIS, D., 1943 Physiology of incompatibility in plants. III. Autopolyploids. J. Genet. **45**: 171–185.

LONGLEY, A. E., and C. F. CLARK, 1930 Chromosome behaviour and pollen production in the potato. J. Agr. Res. **41**: 867–888.

LUNDEN, A. P., 1937 Arvelighetsundersokelser i potet. Meld. Norg. Landbr. Høisk. **17**: 1–156.

1950 Undersokelser over reaksjon mot kreft (*Synchytrium endobioticum*) hos potet. Meld. Norg. Landbr. Høisk. **137**: 1–48.

MEURMAN, O., and G. RANCKEN, 1932 Untersuchungen über die Chromosomenverhältnisse bei kultivierten Kartoffelsorten (*Solanum tuberosum* L.). Soc. Sci. Fennica Commentationes Biologicae **3**: 1–27.

MÜLLER, K. O., 1930 Ueber die Phytophthora-resistenz der Kartoffel und ihre Vererbung (Zugleich ein Beitrag zur Frage der Polyploidie bei der Kartoffel). Angew. Bot. **12**: 299–324.

MÜNTZING, A., and R. PRAKKEN, 1940 The mode of chromosome pairing in *Phleum* twins with 63 chromosomes and its cytogenetic consequences. Hereditas **26**: 463–501.

NAVASHIN, M., 1928 "Amphiplastie"—eine neue karyologische Erscheinung. Z. Ind. Abst. Ver. Suppl. **2**: 1148–1152.

PRAKKEN, R., and M. S. SWAMINATHAN, 1952 Cytological behaviour of some interspecific hybrids in the genus *Solanum*, sect. *Tuberarium*. Genetica **26**: 77–101.

PROPACH, H., 1937 Cytogenetische Untersuchungen in der Gattung *Solanum*, sect. *Tuberarium*. II. Triploide und tetraploide Artbastarde. Z. Ind. Abst. Ver. **73**: 143–154.

SALAMAN, R. N., 1937 The potato in its early home and its introduction into Europe (Masters Lectures, 1936). J. Roy. Hort. Soc. **62**: 61–266.

STEBBINS, G. L., 1947 Types of polyploids: their classification and significance. Adv. Genetics **1**: 403–429.

1950 Variation and evolution in plants Oxford University Press, pp. 643.

SWAMINATHAN, M. S., 1950 Einige Verfahren für die Verwendung wilder *Solanum*-Arten zu Zuchtzwecken. Züchter **20**: 358–360.

1951 Notes on induced polyploids in the tuber-bearing *Solanum* species and their crossability with *S. tuberosum*. Amer. Potato J. **28**: 472–489.

1952 Species differentiation and the nature of polyploidy in certain species of the genus *Solanum*, sect. *Tuberarium*. Ph.D. dissertation, University of Cambridge, England.

SWAMINATHAN, M. S., and H. W. HOWARD, 1953 The cytology and genetics of the potato (*Solanum tuberosum* L.) and related species. Bibliographica Genetica **16**: 1–192.

THOMAS, P. T., 1946 36th Ann. Report John Innes Hort. Institution, pp. 11.

TJIO, J. H., and A. LEVAN, 1950 The use of oxyquinoline in chromosome analysis. Anales Estacion Experimental de Aula Dei **2**: 21–64.

VAN DER PLANK, J. E., 1946 Origin of the first European potatoes and their reaction to length of day. Nature **157**: 503–505.

888 NATURE November 5, 1955 VOL. 176

Overcoming Cross-Incompatibility among some Mexican Diploid Species of *Solanum*

SPECIES of *Solanum* belonging to the section *Tuberarium* and occurring in Mexico have long been of interest to potato breeders, as several of them possess resistance to the late-blight caused by *Phytophthora infestans* de Bary. Like the tuber-bearing species of *Solanum* found in South America, the Mexican species also form a polyploid series comprising diploids ($2n = 24$), triploids ($2n = 36$), tetraploids ($2n = 48$), pentaploids ($2n = 60$) and hexaploids ($2n = 72$). These species have been classified in eight distinct taxonomic series, and our information concerning the cytogenetic affinities among them is still very fragmentary. As a preliminary towards gathering such data, a systematic survey of ability to cross among the diploid species occurring in Mexico was undertaken by me, during the summer of 1953, at the Inter-regional Potato Introduction Station, Sturgeon Bay, Wisconsin.

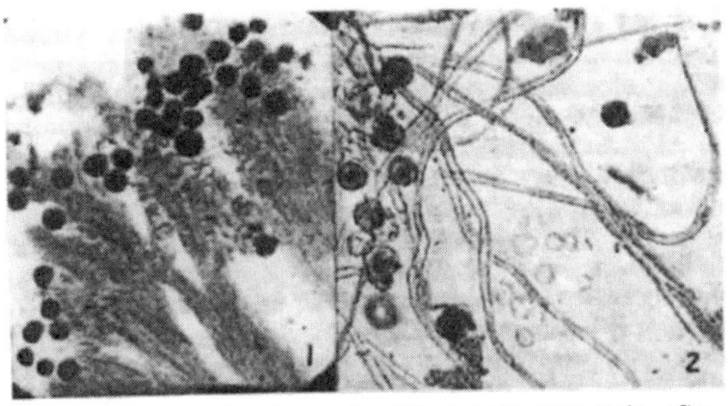

Fig. 1. Stigma and part of style of *S. pinnatisectum*, 72 hr. after pollination with *S. lanciforme* pollen. There is practically no germination of pollen

Fig. 2. Pollen of *S. bulbocastanum* germinated on an agar–sucrose–gelatin medium

The following diploid species were used in the study: *S. pinnatisectum* (series *Pinnatisecta*), *S. bulbocastanum* (series *Bulbocastana*), *S. lanciforme* (series *Cardiophylla*) and *S. polyadenium* (series *Polyadenia*). In the first set of crosses made in the usual way (emasculation followed by controlled pollination), only the cross *S. pinnatisectum* × *S. polyadenium* succeeded. A study was then made of the causes for the failure of seed-setting in the crosses *S. pinnatisectum* × *S. bulbocastanum* and *S. pinnatisectum* × *S. lanciforme*. For this purpose, the ovaries with style and stigma were fixed in Carnoy's solution (alcohol, acetic acid, chloroform, 6 : 1 : 3), 24, 36, 48, 72 and 120 hr. after pollination, sectioned and stained with acid fuchsin and light-green. Examination of the slides revealed that in both crosses there was little or no germination of the respective pollen on the stigma even after 72 hr. from the time of pollination (Fig. 1). This duration is usually found to be sufficient to allow fertilization in compatible crosses. In the few germinated pollen grains the length of the tubes was only about three times the diameter of the pollen.

To overcome the pollen-stigma incompatibility, two techniques were tried, namely, smearing the stigmatic exudation from the pollen parent on the stigma of the pistillate parent, before pollination, and applying a suitable artificial medium for pollen germination on the cut surface of the style after removal of the stigma. An intermediate method described below led to the formation of berries and seeds in both crosses. The stigma, with a small portion of the style, was removed with a pair of fine scissors from the flowers of *S. pinnatisectum*, and a drop of an agar–sucrose–gelatin medium was applied to the decapitated surface of the style. This medium was prepared by dissolving 0·5 gm. of agar and 2·5 gm. of sucrose in 25 c.c. of distilled water to which 0·5 gm. of gelatin was added. The pollen grains of many species of *Solanum* were found to grow well in this medium (Fig. 2). After applying pollen to this medium on the cut surface of the style, the cut style was covered with a piece of moist cotton wool. From the crosses *S. pinnatisectum* × *S. lanciforme* and *S. pinnatisectum* × *S. bulbocastanum* thus made, four and three berries containing on an average thirty-nine and eight seeds respectively were obtained. Seeds from these crosses were grown at the Inter-regional Potato Introduction Station, Sturgeon Bay, in the summer of 1954 by Dr. R. W. Hougas, who found that the plants were real hybrids.

Such crosses have been attempted previously by several workers in different countries but without success[1]. The hybrid between *S. pinnatisectum* and *S. bulbocastanum* may be of particular interest to potato breeders as a valuable breeding stock, since Niederhauser and Mills[2] have recently reported that *S. bulbocastanum* is not only resistant to late blight but is also probably the only species immune to it.

This investigation was supported by a grant from the Research Committee of the Graduate School with funds provided by the Wisconsin Alumni Research Foundation. I thank Drs. D. C. Cooper, G. H. Rieman and R. W. Hougas for advice and help during the course of the study.

M. S. SWAMINATHAN*

Department of Genetics,
University of Wisconsin,
Madison 6, Wisconsin. June 7.

* Now at the Division of Botany, Indian Agricultural Research Institute, New Delhi 12.

[1] Swaminathan, M. S., and Howard, H. W., *Bibliographia Genetica* 16, 1 (1953).

[2] Niederhauser, J. S., and Mills, W. R., *Phytopath.*, 43, 456 (1953).

480 NATURE March 2, 1957 VOL. 179

Polyploidy and Radiosensitivity

Conger and Johnston[1] have concluded, from studies of the effects of X-radiation in a flower bud of *Tradescantia paludosa* containing a mixed population of diploid and haploid microspores, that chromosomal radiosensitivity or sensitivity per unit length of chromosome is identical in haploid and diploid cells. The results of different workers in which radiation sensitivity of cells of polyploid plants has been found to be equal to, greater or less than comparable diploids have been attributed by them to intrinsic differences in species sensitivity rather than to the effect of polyploidy itself. In a comparative study of the frequency of chromosome aberrations induced by different types of radiations on diploid, tetraploid and hexaploid *Triticum* species, we observed that the results obtained varied with the type of radiation used.

Table 1. CHROMOSOME ABERRATIONS IN *Triticum* SPECIES IRRADIATED WITH X-RAYS, β-PARTICLES AND FAST NEUTRONS

Radiation and material	No. of cells studied	Aberrations per cell (exchange and deletions)	
		Range	Mean ± s.e.
X-rays			
Diploid	152	0–8	0·35 ± 0·087
Tetraploid	192	0–14	0·57 ± 0·137
Hexaploid	218	0–32	1·11 ± 0·212
β-particles from phosphorus-32			
Diploid	140	0–6	1·5 ± 0·279
Tetraploid	153	0–28	2·47 ± 0·388
Hexaploid	471	0–42	2·53 ± 0·318
Fast neutrons			
Diploid	568	0–20	0·498 ± 0·079
Tetraploid	540	0–28	1·460 ± 0·169
Hexaploid	1,033	0–66	9·700 ± 0·517

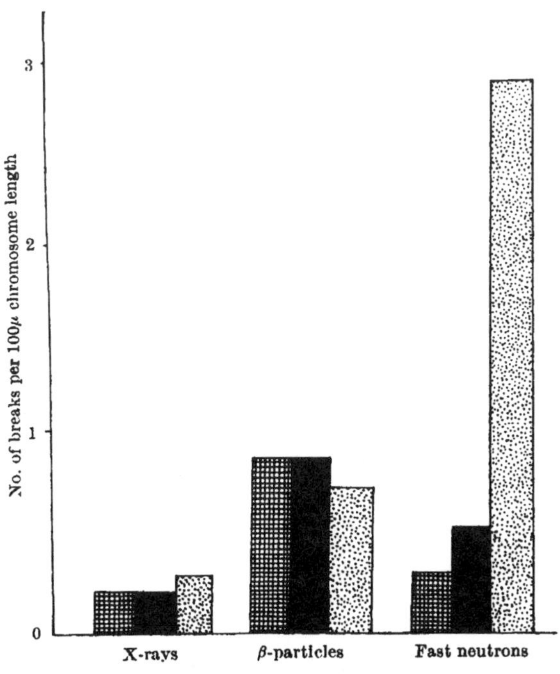

Fig. 1. Effect of the type of radiation on the number of aberrations per unit chromosome length in diploid, tetraploid and hexaploid *Triticum* species. Lined, diploid; black, tetraploid; stippled, hexaploid

Dry seeds of *T. monococcum*, *T. dicoccum* and *T. aestivum* were exposed to 11,000 r. of X-rays, and to fast neutrons of source strength 10^9 n_f/cm.^{2}/sec. over a 2π solid angle for three hours. For β-radiation, the seeds of the three species were soaked in phosphorus-32 solution for 24 hr., the dose used being 5 μc. per seed. Cytological studies were carried out in root tip smears prepared from material fixed 24, 48 and 72 hr. after germination (Table 1). The total lengths of the chromosome complements of diploid, tetraploid and hexaploid wheats determined in comparable preparations were 172·2, 282·0 and 353·2μ, respectively, and the number of aberrations per 100μ of chromosome length was calculated for all the treatments. The data were subjected to a test of significance by Fisher's Behrens test. This showed that sensitivity per unit length of chromosome was: (*a*) identical in all the three species in the case of X-ray treatment; (*b*) identical in the diploid and tetraploid but significantly different in the hexaploid with phosphorus-32 treatment; (*c*) significantly different in all the three species irradiated with fast neutrons. Differences in the rate of absorption of phosphorus-32 may be one of the factors influencing the results obtained with this isotope. In the case of fast neutron radiation, however, the dosage given was exactly similar and it is clear that the hexaploid and tetraploid species had very many more breaks than can be accounted for by their polyploid chromosome length (Fig. 1).

The results obtained by Konzak and Singleton[2] in diploid, tetraploid and hexaploid *Avena* and *Triticum* species irradiated with X-rays and thermal neutrons are of interest in this connexion. They observed that while the hexaploids showed the greatest resistance to a given dose of X-rays, the tetraploids were more resistant than the hexaploids to the thermal neutron radiation. While no cytological data are reported in their study, it is likely that the occurrence of extensive chromosome aberrations, as found in our material exposed to fast neutrons, offsets the advantage of ploidy in the hexaploids irradiated with thermal neutrons. It thus seems possible that the simple relationship between ploidy and chromosomal radiosensitivity outlined by Conger and Johnston[1] in X-ray treated material may not hold good in experiments involving the use of more densely ionizing radiations. It will no doubt be interesting if a comparative study of the relationship between polyploidy and chromosome sensitivity to different types of radiations could be undertaken in such ideal material as used by them.

M. S. SWAMINATHAN
A. T. NATARAJAN

Indian Agricultural Research Institute,
New Delhi 12.
Dec. 13.

[1] Conger, A. D., and Johnston, A. H., *Nature*, 178, 271 (1956).
[2] Konzak, C. F., and Singleton, W. R., *Genetics*, 37, 596 (1952).

Disomic and Tetrasomic Inheritance in a *Solanum* Hybrid

In the course of a cytogenetic study of the nature of differentiation of diploid species ($2n = 24$) of the section Tuberarium, genus *Solanum*, the mode of inheritance of three loci could be followed in a cross between *S. macolae* and *S. simplicifolium*, two species of the series Tuberosa occurring in Bolivia and Argentina. The *simplicifolium* parent used in the cross had simple leaves, a prominently winged stem and white corolla, in contrast to the compound leaves, rudimentary stem wing and purple corolla possessed by the *macolae* parent. Except for corolla colour in *macolae*, the two species bred true for their characters. The *macolae* parent was found to be heterozygous for P, a dominant factor necessary for the formation of purple colour in the corolla.

Meiosis was regular in the F_1 plants, and the plants set berries with viable seeds. These plants had compound leaves and prominently winged stems. There was segregation for corolla colour; six plants had purple and four plants had white corollas. In F_2 progenies from F_1 plants with compound leaves, prominent stem wing and purple corolla, there was monogenic segregation for all the characters. In backcrosses between the F_1 and recessive and dominant parents, 1 : 1 and ∞ : 0 segregations respectively were obtained. Thus, the genotypes of the *macolae* and *simplicifolium* parents were determined as *LLwwPp* and *llWWpp* (L, for compound leaf; W, for prominent stem wing and P, for purple corolla) respectively. There was no evidence of linkage among factors L, P and W in the F_2 and backcross data.

Seeds from the initial cross were treated with colchicine, and tetraploid hybrid plants were thus obtained. At first metaphase in the doubled hybrid, there was a mean frequency of 1·19 quadrivalents 0·095 trivalents, 21·025 bivalents and 0·905 univalents per cell. The number of quadrivalents per cell ranged from 0 to 4. The tetraploid plants were fertile. When the C_2, C_3 and C_4 plants (corresponding to the F_2, F_3 and F_4 plants of the diploid hybrid) were grown in following years, it was found that all the plants in the different generations had broad stem wings like the C_1 and *simplicifolium* parents. Since the C_1 plants would have had the genotype *WWww* for stem wing character, the lack of segregation in subsequent generations would mean that the *simplicifolium* chromosomes carrying factor W were pairing preferentially at the tetraploid level. The segregation ratios for leaf type, however, were typical of tetrasomic inheritance; simplex and duplex genotypes were identified by the 3 : 1 and 35 : 1 ratios obtained on selfing C_2 plants (see Table 1) and a 5 : 1 ratio was observed on backcrossing a duplex to a nulliplex type. The chromosome numbers of the simple leaf plants in C_2, C_3 and C_4 were determined to make sure that the simple leaf character in them is not due to the loss of chromosomes carrying the L factor. The number in all the plants was $2n = 48$, and it is thus clear that the locus L in tetraploid *macolae* × *simplicifolium* segregates tetrasomically.

For colchicine treatment, seeds from the initial cross had been used, and consequently there was segregation for corolla colour in the C_1 plants (three coloured and two white). In the C_2 progeny grown from a sib-cross between two C_1 plants, one with coloured and another with white flowers, the forty-three plants which flowered all had coloured corollas. Two C_3 progenies were raised: one from seeds of a C_2 plant selfed and another from a C_2 sib-cross. Both segregated for corolla colour in the ratio 3 coloured : 1 white (observed: 29 : 11 and 12 : 5 respectively). The genotypes of the C_1 plants with purple and white corollas would be *PPpp* and *pppp* respectively, and the ratios suggest that the chromosomes in which the P locus is situated pair preferentially at the tetraploid level. From the data concerning the inheritance of factors L, W and P in the diploid and tetraploid populations of *macolae* × *simplicifolium*, it can be concluded that the quadrivalents and bivalents observed in the tetraploid are not formed in a random manner but result from pairing among specific chromosomes. These observations thus lend support to the inference drawn earlier from cytological studies[1] that cryptic structural differentiation of chromosomes has played an important part in the origin of species differences in the section Tuberarium.

Table 1. Inheritance of Leaf-type in *S. macolae* × *S. simplicifolium*

Material	Genotype	Segregation				Deviation	
		Observed		Expected			
		L	l	L	l	X^2	P
F_2 progeny	*Ll* × *Ll*	45	16	5	1	0·0492	0·9–0·8
F_1 × *simplicifolium*	*Ll* × *ll*	52	44	1	1	0·6666	0·5–0·3
C_2 progeny	*LLll* × *LLll*	87	2	35	1	0·0920	0·8–0·7
C_3 progenies i. C_2 206 self	*Llll*	18	4	3	1	0·5454	0·5–0·3
ii. C_2 203 × 205	several possibities	132	0	X	0		
iii. C_2 211 (compound) × C_2 198 (simple)	*LLll* × *llll*	73	19	5	1	1·0464	0·4–0·3
iv. C_2 198 self	*llll*	0	28	0	X		
C_4 progeny	*LLll*	39	1	35	1	0·0112	0·95–0·9

This work was begun at the Laboratory of Genetics, Wageningen, Holland, in 1950, with material kindly supplied by Dr. H. J. Toxopeus, and was continued at the Plant Breeding Institute, Cambridge, and the Department of Genetics, University of Wisconsin. My thanks are due to Prof. R. Prakken, Dr. H. J. Toxopeus, Dr. H. W. Howard and Dr. R. W. Hougas for facilities and advice.

M. S. Swaminathan

Division of Botany,
Indian Agricultural Research Institute,
New Delhi.

March 23.

[1] Swaminathan, M. S., and Howard, H. W., *Bibliographia Genetica*, 16, 1 (1953).

The Green Revolution in Indian Agriculture from an Environmentally Sound Technology Point of View

Monkombu S. Swaminathan

Swaminathan, M.S. (1992). The green revolution in indian agriculture from an environmentally sound technology point of view. Environmentally sound technology for sustainable development. Advanced technology assesment system Bulletin (7). Dirk Pilari and David Philips Eade (eds). United Nations Publication, New York, pp. 143–148

Introduction

In India, the green revolution is synonymous with land-saving agriculture, which measures yield per unit area per unit of time. The application of land-saving technology in India has had dramatic results over the past two decades. In 1964, for example, India produced 12 million tons of wheat from 14 million hectares (ha) of land, and in 1990 it produced 55 million tons of wheat from 23 million ha. Without land-saving agriculture, an additional 30 ha of land would have been needed to produce those 55 million tons at 1964 yield levels, and such large areas could only have been made available by clearing forest land. On a global scale, the diversion of forestland for crop husbandry is a major source of its depletion.

While saving forests is the positive side of the green revolution, concerns, are often expressed about the excessive use of pesticides and mineral fertilizers and the loss of genetic heterogeneity due to the cultivation of the same genetic strain over vast areas. Therefore, agricultural strategy in India has been reoriented to meet the challenge of integrating considerations of ecological sustainability with those of unity and economic efficiency.

Meeting the Challenges

The following list contains some of the principles that should form the basic guidelines for India's agricultural strategy in the 1990s:

- *Water, energy and cost-saving crop husbandry*: that is, productivity advance and water and energy-use efficiency become the major pathways for enhanced production;
- *Grain-saving animal husbandry*: that is, the avoidance of animal husbandry practices that require large quantities of cereal grains, and the initiation of a programme for the establishment of rural biomass refineries;
- *To the factory, but not to the town*: that is, the generation of adequate numbers of skilled jobs in the secondary agro processing and rural industries sector and the tertiary services sector, so that those without assets or work need not migrate to towns and cities;

- *Sustainable management of basic agricultural assets*: namely, land, water, flora, fauna, and the atmosphere; for this purpose, current patterns of productivity measurement should be changed, as will be pointed out later;
- *Income and employment orientation to agricultural research and development*: to enable agriculture to become not only an instrument for the production of more grains and other commodities, but also a catalyst for the production of more jobs and income in rural areas.

Developing technologies by climatic zone

India has categorized its land mass into seven science and technology zones. Within each zone, technology should be precisely tailored to meet the needs of prevailing agro-climatic conditions.

Green revolution areas

These areas occur in most Indian States and are characterized by substantial improvements in the productivity of crops. A major challenge is to defend the progress already made in these areas and to raise the ceiling to further yields. Such actions call for the introduction of frontier technologies, including the development of crop varieties that can give higher yields per cubic metre of water. Post-harvest technology will also need additional attention so that all links in the producer-consumer chain are integrated. Through cereal-legume crop rotations, animal husbandry programmes can be supported at a higher level of efficiency for stall-fed animal rearing.

Green, but not green revolution areas

In several parts of India, the availability of water is not the primary constraint, and the land is green most of the year. In those areas, underground water sources have not been extensively tapped and the scientific management of water and nutrients holds the key to improving productivity. New methods of feeding crops are necessary.

Dry-land farming

The areas of saving and sharing water, new approaches to breeding and feeding of crops, agro forestry systems with particular focus on leguminous shrubs and trees, animal husbandry and of T-farm employment will have to receive much greater attention. Steps to tap ground water on a sustainable basis, and the development of alternative cropping strategies based on different rainfall and moisture availability patterns are needed. Computer simulation models for these processes should be supported by appropriate seed reserves of alternative crops. Just as grain reserves are essential for food security, seed reserves are essential for crop security.

Arid ecosystems

The desert areas of the State of Rajasthan provide unique opportunities for high-quality seed production, arid horticulture and animal husbandry

Coastal area planning

The coastal areas afford opportunities for the integration of agriculture and aquaculture production systems. Also, coastal agriculture can be given an export orientation, since transport costs can be minimized. For the purpose of planning coastal agricultural systems, 20 kilometres (km) of sea surface and 20 km of land surface can be considered an integrated entity.

Wetlands

India has a rich endowment of wetlands. Detailed planning is necessary for the sustainable management of these wetlands in order to produce a variety of aquatic products and avoid biological impoverishment.

Wastelands

The rehabilitation of degraded lands must be undertaken using technologies most appropriate to each situation. Every effort should be made to restore the full biological potential of wetlands by the year 2000. A simple method of measuring the biological potential of land must be developed and popularized.

Sustainable development systems for small farmers

The concept of a small farm varies from country to country, and in India, a small farm is below 2 ha in size. The farmers who cultivate farms below 1 ha are classified as marginal farmers. The cultivatable land per capita declined in India from 0.48 ha in 1951 to 0.26 ha in 1981. By the year 2000 this area is expected to decline to 0.15 ha. In China, the per capita land availability is 0.11 ha, and in the semi-arid, rain-fed areas of Africa, small farmers ay often be cultivating 10 to 20 ha of land, but yields are low and uncertain.

A study conducted by the Food and Agricultural Organization of the United Nations shows that in most developing countries the average size of an operational holding will continue to decline by 2000. In addition to the growth in population, the factors contributing to this trend are:

- Withdrawal of farmland for industrial, communication, housing and other uses;
- Inadequate opportunities for employment in the non-farm sector.

Between 60 and 90 per cent of most developing country populations live in rural areas, and for them, land and water-based occupations-consisting of crop and animal husbandry, fisheries and forestry-constitute both a way of life and a secure livelihood. If current global trends continue, two contrasting systems of agriculture will evolve by the end of the century. In one system, farmers will disappear and be replaced by large agribusiness. In the other, millions of families that operate small holdings will struggle to improve the productivity, profitability, stability and sustainability of the farming systems that determine both the quality of their lives and the dependability of the food security system of their countries.

To help small farmers, specific methods must be introduced into the scenario, but first, it will be useful to define the concept of sustainability.

Measurement of sustainability

The core of the sustainability concept lies in equal and concurrent attention to intra-generational and inter-generational equity issues. Intergenerational equity, which aims to give a fair deal to socially and economically disadvantaged groups, involves attention to issues such as adding a dimension of resource neutrality to scale neutrality in technology development and transfer. By using a more democratic method to disperse technology, all farmers can be helped, irrespective of the size of their holdings or their innate input mobilizing and risk taking capacity to derive economic benefit from new technologies.

Intergenerational equity involves protection of the agricultural productive base, and implies that today's progress-using the basic agricultural assets of land; water, flora, fauna and atmosphere-does not ruin tomorrow's prospects. Thus, sustainability has both economic and ecological dimensions.

Operational implications

At a meeting organized at the International Rice Research Institute in December 1987, the following definition was agreed on for the purpose of integrating economic and ecological sustainability in one measurable equation:

Productivity:

$$\frac{\text{Output value}}{\text{Input value}} + \text{Changes in environmental capital stocks}$$

Sustainability:

Stability of productivity over time

The value of output and input can be calculated with precision; however, there is no readily usable method for measuring the effect of environmental capital stocks with reasonable accuracy. At least three areas need measurement tools:

Soil health: what is the effect of a given level of productivity on the biological potential of land'! Soil is the sum of its physical, chemical, microbiological and erodibility characteristics. Can a reliable, speedy and low-cost method of measuring the biological potential of land be developed?

Water quality and availability: suitable criteria are needed for measuring the effects of irrigation and crop intensification on the water resources of an area, such as the rate of aquifer exhaustion or the increase in salinity;

Biological diversity: what is the effect of new technologies on genetic erosion, both in crops and farm animals? Can methods be developed that give expression to the effect of the spread of new technologies on biological diversity?

The standardization of reliable, low-cost methods of measuring and expressing productivity is urgently needed to measure the effect of productivity on environmental assets, on which sustained agricultural advances depend. The development of this methodology merits technical and financial support; however, the emphasis should be on the simple tools of measurement that farmers can readily understand and adopt.

Measuring sustainability by estimating the stability of productivity over time is best done by using an entire village or a watershed, the command area of an irrigation project or the catchment area of a river system as the unit of measurement.

Support Services

Package of technology

To promote agricultural production on an economically and ecologically sustainable basis under conditions of small-fann holdings, there is a need for reorientation of agricultural research towards the following areas.

Yield enhancement: Yield is the result of interaction between the genetic efficiency of the crop plant or farm animal and the management efficiency of the farm family.

Land/water management: Scientific land and water management is the key to sustainable agriculture. As mentioned earlier, simple methods of measuring the biological potential of land are needed. Agronomic and agricultural engineering research are necessary to help the watershed community to upgrade degraded lands through appropriate land management systems, including agro-forestry systems. There is inadequate attention now to the social engineering aspects of soil health management. Social engineering research is necessary in devising public policy packages that can fostering cooperation among members of a watershed/command area community in !operations such as soil conservation, soil health care, and water harvesting land sharing.

Energy management: Energy management is needed in the areas of: *(a)* irrigation, land preparation and crop management, including sowing and harvesting; *(b)* pest management; and *(c)* nutrient supply. In irrigated areas, energy management, particularly for lumping ground water or distributing conserved water, holds the key to the success of farming in years of drought.

Crop management: Particular attention should be paid to breeding crop varieties that can optimize yield under given moisture regime, and also to feeding plants so that the genetic potential for yielding new strains can be fully realized. The breeding strategy, in particular, needs reorientation. A selection strategy should be initiated where F2, F3[1] and subsequent generations are grown alternately under conditions of moisture sufficiency and deficiency. Such selections, under divergent conditions of moisture stress, will help to breed varieties that have the capacity to give high yields when moisture is adequate, and reasonable yields in drought conditions. The crop feeding strategy should also involve the use of legumes in the crop rotation (including stem-nodulating green manure crops), organic recycling, and balanced fertilizer application in accordance with different soil types. In addition, cultural and cropping practices that include the date, density and manner of sowing and symbiotic crop growing, should help to distribute and reduce risks, while enhancing yield and income. Risk aversion and distribution should be important goals of agronomic research, particularly in dry-land areas, since small farmers often do not have access to protective irrigation.

Post-harvest management: While efficient post-harvest management includes the handling of crops that have direct human value, it also implies detailed attention to every component of the biomass, with particular stress on the preparation of value-added products from each of the components. Recent advances in bioprocessing, microbiological enrichment of cellulosic material, and chemical engineering have opened up new opportunities in biomass use.

Farming systems management: The packaging of the best available technologies into an economically and ecologically desirable farming system should be the end-point of all scientific efforts. For this purpose, it is necessary to promote the optimum use of the climatic, land, water, labour and credit resources available to the farm family. To achieve this goal, a two-pronged strategy is needed that includes: *(a)* appropriate blends of traditional and "frontier technologies"; the frontier technologies that merit particular attention are biotechnology (both molecular biology and bio-processing), satellite imagery, computer technology and microelectronics; and *(b)* a farming system design whether based exclusively on crops, crops and animals, or crops, animal husbandry, fisheries and forestry-that gives explicit consideration to promoting better linkages among the primary (agricultural), secondary (industrial) and tertiary (services) sectors of economic activity in rural areas. Fanning system research specifically designed to promote better growth linkages is imperative if the dimensions of income and employment generation are to be added in projects designed to increase food production.

Package of Services

Knowledge and ski]] transfer, credit supply, and the timely availability of the inputs needed for higher productivity are areas that have received attention in African countries. However, more research on delivery systems is needed, with particular reference to the following:

- Synchronization in space/time between knowledge transfer and the inputs needed for its application;
- Synchronization between credit supply and the inputs for which the credit is intended;
- Organization of the services sector in the villages-for example, seed production and distribution, integrated pest management (IPM), and improved post-harvest operations, particularly by rural women and youth. For this purpose, training women in seed technology, IPM and improved post-harvest operation and helping them to organize effective service organizations will be necessary;
- Organization of producer-oriented marketing arrangements and assisting farmers to avoid distress sales;
- Promotion of industries designed to produce value-added products from every part of the agricultural biomass available in a village or in a group of adjoining villages through training and appropriate infrastructure support. In general, research on delivery systems that enable small farmers to derive full economic benefit from new technologies has not received the attention it deserves. Such research should draw from the successes and failures of technology transfer, and help to design self-replicating systems of technology diffusion. Training procedures should also be largely based on learning by doing, particularly where illiteracy is still widespread. Adequate attention to the messages to be conveyed through the mass media is also essential if the investment on the hardware components of such media is to confer the anticipated benefit.

Package of public policies

A public policy package should include incentives for group cooperation in areas such as land and water management, IPM, integrated nutrient supply (INS), improved post-harvest technology, and organization of local level associations for sustainable development. In addition to reorienting extension services on a community basis, suitable incentives such as group insurance, group credit and group awards for community endeavor in the conservation of environmental assets would be helpful.

Public policy instruments for promoting community adoption of knowledge-intensive and ecologically desirable technologies are important, and an equally important, but often neglected, area of policy is the promotion of activities in areas where ecological linkages over widely separated areas determine a common agricultural future. This is particularly true in the Sahelian region of Africa with regard to measures for combating desertification. Public policy should also foster research efforts and, as stressed earlier, the integration of traditional and frontier technologies is relevant to sustainable agriculture. With the powerful tools of genetic engineering available today, agricultural technology can help to integrate scale and resource neutralities in technology development and transfer, and help to generate skilled jobs in the rural sector. These same tools also can help to generate technologies for women to help to reduce drudgery, and can introduce the twin concept of flexitime and flexilocation in relation to job opportunities for farmwomen. Emerging technologies that require particular attention from the point of view of integration with traditional technologies include the following:

- Molecular genetics and recombinant deoxyribonucleic acid (DNA) technology;
- Tissue and cell culture;
- Sero-diagnostic techniques and vaccines for animal diseases;
- Bio-processing and preparation of value-added products from agricultural biomass;
- Microbiological upgrading of cellulosic material;
- Satellite imagery, remote sensing;
- Computer technology;
- Microelectronics.

There is a distinction here between science and technology. Scientific research leading to new knowledge and material can be generated in research institutions and universities, whereas technology development is best done jointly by scientists and farmers in farmer's fields. The failure to make this distinction often leads to difficulties in the diffusion of new methods and approaches because of economic and/or ecological factors. Centres for technology blending and dissemination are urgently needed that are inspired by the goal of bringing the best in modern science to the service of the farm families. Non-governmental organizations, for example, can playa key role in this task.

The role of women

The concept of equity involves adequate attention to the needs of the economically, ecologically and socially disadvantaged sections of the population. Achieving growth with equity demands concurrent attention to economics, ecology, generations and gender. Emphasis must be placed on enhancing the productive role of women and reducing drudgery in the household and in the productive aspects of their lives.

Various studies have been under. Taken in Africa to identify and articulate the real issues confronting women' in their triple role. The agricultural field studies carried out so far indicate that in spite of the key role of women in areas including crop husbandry. Animal husbandry, fisheries, forestry, and post-harvest technology, both in the past and now the present-those in charge of formulating packages of technologies, services and public policies for rural areas have often tender to neglect their productive role. Consequently, training for women and the development of technologies tailoll to women-specific occupations and the involvement of women in technology development and transfer have received inadequate attention from both the scientific and administrative departments of Governments.

Several priority areas need urgent and integrated attention:

- The problems arising from unemployment, underemployment, certain employment and unpaid! Underpaid employment;
- The drudgery involved because of the predominantly unskilled nature of the work assigned;
- The lack of flexibility in the working hours, place of work, duration of work;
- The adverse effects of labour displacing technologies;
- The lack of adequate maternity, health care, crèche, and associated needs;
- Little or no access to credit and other production inputs;
- Inadequate opportunities for the acquisition of new skills and for extension assistance compounded by the handicaps of illiteracy;
- Discriminatory land or other asset ownership rights and social status;
- The scarcity of organizations that can help women with the needed management, expertise and collective strength.

From the scientific and educational viewpoints, the response to these challenging problems must be both short and long term. Many problems can be solved immediately with existing technologies and institutional infrastructures. Specific research and development programmes should be initiated within the next 5-10 years that can: *(a)* help to convert unskilled jobs to skilled ones; *(b)* diversify opportunities for employment; and *(c)* introduce flexibility in the place, time and duration of work.

Some immediate solutions

For the short term, we should build on the historic and current professional skills of rural women and enhance their beneficial effects on income and quality of life by removing the major handicaps and constraints. For example, since the dawn of agriculture, African women have played a major role in food crop production, seed production and management and conservation of biological diversity, renewable energy management, crop / livestock / fish management, and post-harvest management, including the use of agricultural by-products.

Research and development institutions should take the necessary steps to strengthen the role of women in the above areas.

Conservation of biological diversity

In seed production and management, for example, women have traditionally been seed selectors and preservers. Whether literate or illiterate, they have the ability to spot the healthy plants whose seeds they carefully preserve for sowing the following year. The skills of rural women in the production of high-purity hybrid seeds should be fully used.

Women have also traditionally been active conservationists. Their involvement in the organization and conservation of heritage sites, biosphere reserves, national parks and community gene banks will enhance the effectiveness of the conservation movement.

Renewable energy management

Women actively collect fuel for cooking and also find nutrition for both crops and farm animals. Organic recycling and collection of fodder and feed are their traditional occupations. Today there are many opportunities in the field of renewable energy management and use that women can handle effectively with assistance.

Crop, animal, fish and forest management

Experience in the Philippines and elsewhere has shown that women can very effectively organize IPM involving biological and cultural control, genetic resistance, and need-based application of chemical pesticides, thereby assisting in the prevention of pest resurgence and epidemics. Similarly, women can effectively handle bio-fertilizer production programmes and help to organize and implement INS systems.

The spread of IPM and INS techniques becomes even more important in the case of horticultural crops. The role of women in animal husbandry and aquaculture can also be further Strengthened through appropriate training.

Post-harvest management

There are several possibilities for making considerable improvements in all the links that make up the producer consumer chain. Preparation of value added products is now feasible from every part of plant and animal biomass.

To achieve this, agricultural research and training institutions should integrate the following programme rules

- Job destruction and job creation must be concurrent events;
- The aim of technology should be labour diversification, drudgery reduction, and better energy conversion, but not labour displacement;
- Knowledge and skill transfer mechanisms should be tailored to the socio cultural and socio-economic conditions of rural women, and to the opportunities for remunerative marketing;
- Every effort should be made to increase the active participation of women both in technology development and dissemination;
- Synchronization between knowledge and skill transfer and the supply of the necessary inputs, including credit; for this purpose, the growth of women's organizations and the strengthening of their professional management capacity should be promoted.

New frontiers

Development in the following fields has opened up new opportunities for making rural jobs knowledge-intensive and skilled, and has introduced the dimension of flexibility in the place and duration of work:

- **Biotechnology**, including bio-processing, microbiology, tissue culture, genetic engineering, and applications in animal breeding and disease control;
- **Information technology**, including computer-aided instruction, and extension and mobilization of mass media, particularly radio and television, for awareness generation and knowledge dissemination;
- **Space technology**, including conservation monitoring and supporting ground verification work;
- **Microelectronics**, which lends itself to decentralized production;
- **Management technology**, which would help in integrating all the elements of a technological package in a mutually supportive manner. In particular, the managerial and organizational aspects of decentralized production supported by a few key centralized services need attention. *Also*, the planned strengthening of the beneficial linkages among the primary, secondary and tertiary sectors must be achieved.

Suggested immediate steps are:

- **Technology blending centres**: these should be organized in each agro-ecological zone in order to combine the positive features of traditional and frontier technologies. Such centres should also aim to upgrade the productive role of women. A focus on the needs of women will help both men and women, while the reverse will not be true, as past experience shows. For example, demonstrations in small farmers' fields help big farmers, while the reverse is not always true;
- **Programmes for training rural women in skilled employment**: training should be adjusted to the level of literacy and domestic obligations of the trainees;
- **Women's organizations**: these should be fostered that can serve as links between science and the application of technologies.

Notes

1. F2 and F3 are the second and third generations after a cross is made between two parents

Science and Shaping our Agricultural Future

M.S. Swaminathan

Swaminathan, M.S. (2006). Science and Shaping Our Agricultural Future. Part 1. The K. R. Narayanan Orations 1994 to 2006. *The First Ten K. R. Narayanan Orations: Essays by Eminent Persons on the Rapidly Transforming Indian Economy.* (ed.) Raghbendra Jha. The Australian National University Press, Canberra. 155–183.

It is a privilege to deliver a lecture in honour of Dr K R Narayanan, immediate Past President of India. Dr Narayanan represents all that is best in Indian culture and democratic system of governance. He rose from the lowest to the highest position in Indian Society by virtue of his innate human and professional qualities. Dr Narayanan knows the pangs of hunger and has therefore been on the forefront of the hunger free India movement. He encouraged scientists to work on problems relevant to the alleviation of poverty and eradication of hunger. I have therefore chosen the topic, 'Science and Shaping our Agricultural Destiny' for this lecture.

Introduction

From the beginning of time, technology has been a key element in the growth and development of societies. The spread of technologies has however been uneven throughout history. In food production, we have now reached the age of biotechnology and precision farming. Many of the technologies like improved seeds are scale neutral with reference to their relevance to farms of varying sizes but are not resource neutral. Inputs are needed for output and hence those who do not have access to inputs tend to get bypassed by technological transformation. Synergy between technology and public policy has therefore remained a pre-condition for technologies to confer benefit to all sections of the farming community, irrespective of the size of their holdings and their innate capacity to mobilise capital and take risks. Among factors of production, access to irrigation water has been a major determinant of technological change, since without assured irrigation, it is difficult to apply nutrients in quantities essential for high yields, even if genetic strains capable of high productivity are available.

Today, global agriculture is witnessing two opposite trends. In many South Asian countries, farm size is becoming smaller and smaller and farmers suffer serious handicaps with reference to the cost-risk-return structure of agriculture. Farm size in most industrialised countries is becoming larger and larger and farmers are supported by heavy inputs of technology, capital and subsidy. The recent breakdown of the Cancun negotiations of the World Trade Agreement in the field of agriculture reflects the polarisation which has taken place in the basic agrarian structure of industrialised and developing countries.

In India, average yields of major food crops remained well below 1 metric ton per hectare for centuries, until the introduction of high yielding varieties in the 1960s. To produce one metric ton of rice the rice plant needs at least 20kgs of nitrogen and appropriate quantities of phosphorus, potash and micronutrients. The native soil fertility was often below this level and hence yields tended to remain below a ton.

The steps taken after independence to improve the productivity of food crops fall under the following major categories:

- Package of technology
- Package of services in areas such as input supply and extension
- Package of public policies in areas such as land reforms, rural infrastructure development, investment in irrigation, input and output pricing policies and assured and remunerative marketing.

Improvement of agricultural production through the productivity pathway is essential for both resource poor farmers and consumers. Casual agricultural labourers are the largest in number among the chronically poor and cultivators the second largest group. Most of the chronically poor were either landless or near-landless. The smaller the farm, the greater is the need for increasing productivity, so that the farm family has a higher marketable surplus. Productivity improvement also tends to reduce the cost of the commodity, thereby benefiting resource poor consumers. Above all productivity improvement is essential for safeguarding the remaining forests, since otherwise forest land will get converted to produce food. Thus, the productivity pathway of agricultural advance helps in strengthening ecological, livelihood and food security.

What we need is an evergreen revolution, which can help to increase productivity in perpetuity without associated ecological harm (Swaminathan 1996). Exploitative agriculture offers great dangers if carried out with only an immediate profit or production motive. The initiation of exploitative agriculture without a proper understanding of the various consequences of every one of the changes introduced into traditional agriculture, and without first building up a proper scientific and training base to sustain it, may only lead us, in the long run, into an era of agricultural disaster rather than one of agricultural prosperity (Swaminathan 1968). We need ecotechnologies rooted in the principles of ecology, economics, gender and social equity and employment generation. The vulnerable sections need job-led economic growth and not jobless growth.

Inspite of striking agricultural progress and democratic decentralisation, chronic and transient poverty and poverty induced malnutrition are widespread. International and national media refer to this as the co-existence of 'grain mountains and hungry millions' (Swaminathan 2005). Section 2 outlines the issues in the context of food security and access, sections 3 and 4 the transition from the green to gene to evergreen revolution in rice and wheat, section 5 provides case studies that show how we can bridge the technological divide and section 6 concludes.

Food Availability, Access, Absorption and Threats to Food Security

Food security was formerly considered essentially in terms of production. It was assumed that adequate food production would ensure adequate availability of food in the market as well as in the household. In the seventies, it became clear that availability alone does not lead to food security. It is becoming evident that even if availability and access are satisfactory, the biological absorption of food in the body is related to the consumption of clean drinking water as well as to environmental hygiene. Finally even if physical and economic access to food is assured, ecological factors will determine the long-term sustainability of food security systems. We have to define food as physical, economic, social and ecological access to balanced diet and clean drinking water, so as to enable every child, woman and man to lead a healthy and productive life. The needs of each age group must be addressed (see cycle approach described by MSSRF 2001). Such an approach will involve the following steps (Swaminathan 2002a; 2002b).

Food availability

This is a function of both home production and imports. There is no time to relax on the food production front. The present global surplus of food grains is the result of inadequate consumption on the part of the poor, and should not be mistaken as a sign of over-production.

Food access

Lack of purchasing power deprives a person from access to food even though food is available. Inadequate livelihood opportunities in rural areas are responsible for household nutrition insecurity. For example, India today has over 30 million tonnes of wheat and rice in government godowns; yet poverty induced hunger affects over 200 million persons. It is endemic in south Asia and sub-Saharan Africa (Ramalingaswami *et al.* 1997; WFP 2001). Macro-economic policies, at the national and global level, should be conducive to fostering job-led economic growth based on micro-enterprises supported by micro-credit. Where poverty is pervasive, suitable measures to provide the needed entitlement to food, should be introduced. The State of Maharashtra introduced, nearly 25 years ago, an Employment Guarantee Scheme to assist the poor to earn their daily bread during seasons when opportunities for wage employment are low.

Food Absorption

Lack of access to clean drinking water, poor environmental hygiene and poor health infrastructure, lead to poor assimilation of the food that is consumed. Nutrition security cannot be achieved without environmental hygiene, primary health care and clean drinking water security. Culinary habits also need careful evaluation as some methods of cooking may lead to the loss of vital nutrients.

Threats to food security

The most important among the internal threats to sustainable food security is the damage to the ecological foundations essential for sustained agricultural advance, like land, water, forests and biodiversity. Second, in the areas of farm economics, resource flow to the agriculture sector is declining and indebtedness of small and marginal farm families is rising. Input costs are increasing, while factor productivity is declining. Third, a *technology fatigue* has further aggravated farmers' problems, since the smaller the farm the greater is the need for sustained marketable surplus, in order to have cash income. Linkages between the laboratory and the field have weakened and extension services have often little to extend by way of location, time and farming system specific information and advice (chapter on Wake Up Call in the NCF report).

The external threats include the unequal trade bargain inherent in the WTO agreement of 1994, the rapid expansion of proprietary science and potential adverse changes in temperature, precipitation, sea level and ultra violet ß radiation. Though it is now over ten years since the WTO regime started operating in agriculture, serious attempts are yet to be made to launch in rural areas movements for quality literacy (sanitary and phytosanitary measures and codex alimentarius standards of food safety), trade literacy (likely demand-supply and price situation) legal literacy (IPR, Farmers' Rights) and genetic literacy (genetically modified crops). No wonder the prevailing gap between potential and actual yields even with technologies currently on the shelf is very wide (Table 1).

Table 1: Comparative crop productivity (Kg/hectare).

Crop	USA	China	India
Maize	8900	4900	2100
Paddy	7500	6000	3000
Soybeans	2250	1740	1050
Seed Cotton	2060	3500	750
Tomato	6250	2400	1430

Source: Wake Up Call Chapter in NCF Report

In the area of technology, there is also need to bridge the growing digital and genetic divides. Post-harvest technology is poor and there is little value addition particularly in the case of fruits, vegetables and spices including a wife range of tubers and medicinal and aromatic plants. Sustainable intensification, ecologically, economically and nutritionally desirable diversification and value addition to the entire biomass are important for raising small and marginal farm families above subsistence level. All this will call for initiating an era of knowledge intensive agriculture. Modern information communication technologies (ICT) afford an opportunity for launching a knowledge revolution in rural India.

Technological Transformation of Productivity, Profitability and Sustainability: Rice

Asia grows most of the world's rice output and 90 per cent of rice is produced by small farmers who depend on it for their livelihood and food security. The role of rice in national and global food security systems will increase, not only because of increases in population and purchasing power, but also because of likely changes in climate and sea level rise due to global warming. An immediate task is bridging the gap between potential and actual yields, widely prevalent in several rice growing countries and particularly in different parts of India. This is possible even at currently available levels of technology, through mutually reinforcing packages of technology, services and public policies. In the decades ahead, more rice will have to be produced under conditions of shrinking per capita arable land and irrigation water availability and expanding biotic and abiotic stresses. Due to breeding efforts based on an appropriate integration of Mendelian and molecular techniques, the ceiling to yield is being raised continuously (Figure 1).

Figure 1: Progress in the Yield Potential of Rice

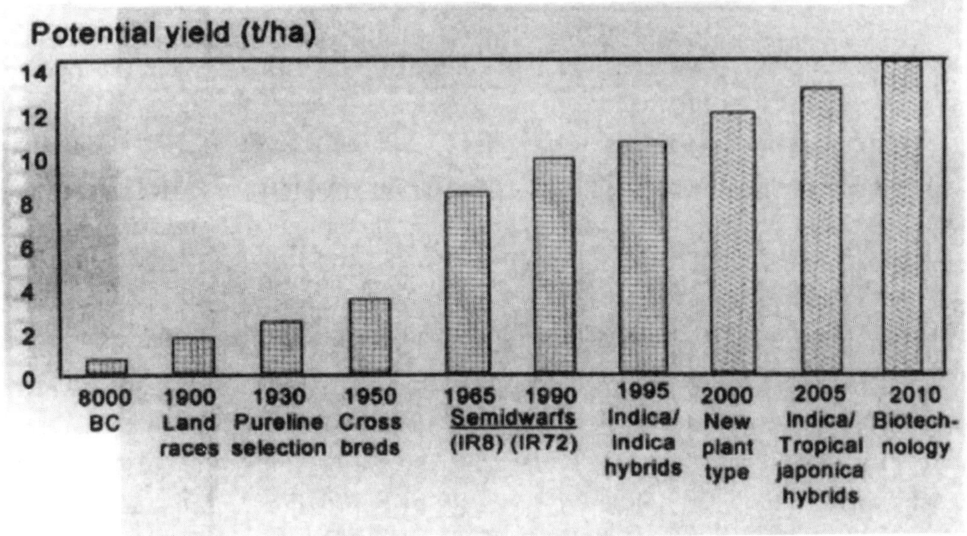

Aided by biotechnology, the greatest potential for productivity gains in yield ceiling in the future lies in rainfed environments (Peacock & Chaudhury 2002). Integrating genetic efficiency with genetic diversity of diverse gene pool through pre-breeding and participatory breeding should be encouraged (Figure 2). Hybrid rice, 'Super rice' and 'Super hybrid rice' are likely to dominate the rice world in the future. What is however important is the initiation of research which can lead to the standardisation of methods of feeding the rice plant for higher yields in an ecologically sustainable manner. Research on breeding and feeding for higher yields should proceed concurrently.

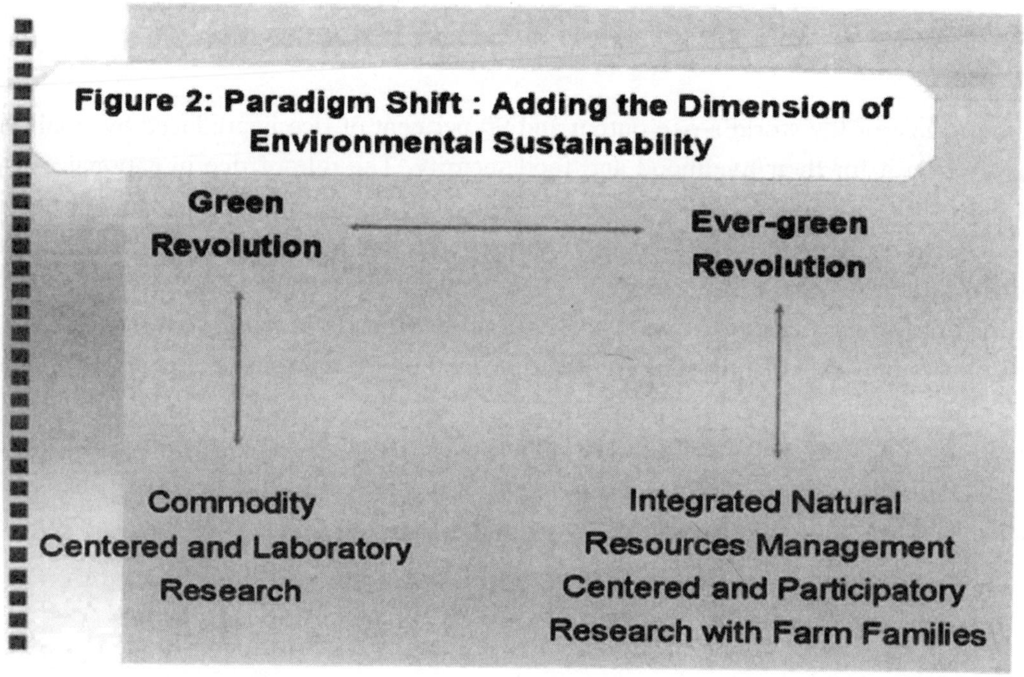

We have several simple and elegant tools that enable us to manipulate the rice genome to elicit desirable responses — tolerance to pests and diseases, moisture stress, salinity-alkalinity, heat, increased photosynthetic efficiency, dry matter accumulation, and source sink partitioning. Rice gene sequence information is widely viewed as an invaluable asset for developing products and technologies. Because of advances in molecular mapping and breeding, there are new opportunities for improving the nutritive qualities of rice, with particular reference to iron, vitamin A and other micronutrients. Under an expanding Intellectual Property Rights (IPR) regime, it is important that research for public good receives the needed support at the national and international levels so that the resource poor can gain from it. Farmers have to achieve revolutionary progress in productivity, quality and value addition. The emerging ecological, economic and social challenges have to be met through partnerships among rice researchers and developmental organizations, committed to the cause of improving the productivity, profitability, sustainability and stability of rice farming systems.

Pingali *et al.* (1997) have described in detail the steps needed to increase rice production in Asia to meet future needs. If global warming and the associated changes in temperature, precipitation and sea level rise do occur, the position of rice in national and global food security systems will increase, since rice has the ability to grow under very diverse environmental conditions. Rice is by far the best-adapted crop to lowland soils that are prone to flooding during wet season. They draw attention to the following challenges facing rice research and development agencies:

- Productivity gains from the exploitation of Green Revolution technologies are close to exhaustion.

- In the absence of further technical change, Asian farmers face increasing costs per tonne of rice produced.
- Adverse agricultural externalities are increasing due to lack of holistic perspective of the farm resource base management.
- Despite an anticipated decline in per capita rice consumption, aggregate Asian demand for rice is expected to increase by 50 to 60 per cent during the 1990–2025 period due to population increase and poverty reduction.
- Economic growth and the commercialization of agricultural systems could reduce the competitiveness of rice relative to other crops and other farm enterprises.
- An upward shift in rice yield frontier is necessary to meet future rice requirements and to sustain farm-level profits.

Compounding these problems, there are potential dangers arising from the diminishing investment in research in institutions devoted entirely to national and international public good and the expanding intellectual property rights (IPR) regime. The question now is how much more improvement can we bring about in productivity without ecological harm? In other words, can we launch an evergreen revolution in rice in the new millennium, marked by sustained advances in productivity, profitability, stability and sustainability of rice farming systems (Swaminathan 1996; 2000; 2002a). How can we also increase the role of rice in the nutritional security of families dependent on it for their dietary energy supply? How can rice production be insulated from the adverse impact of potential changes in precipitation, temperature and rise in sea level? Above all, how can we maintain and strengthen international cooperation in rice improvement?

Increasing Production and Productivity

Bridging the yield gap

Due to imperfect adaptation to local environments, insufficient provision of nutrients and water, and incomplete control of pests, diseases and weeds the present average rice yield is just 40 per cent of what can be achieved even with technologies currently on the shelf. There is considerable scope for further investment in land improvement through drainage, terracing, control of acidification, etc. in areas where these have not already been introduced. While irrigated areas are making good progress, there is need for more attention on intensive research and development in rainfed, low land and upland areas. Ensuring that benefits from technology accrue to resource poor or marginal farmers will require special efforts as outlined in the case studies in section 5.

An integrated approach is necessary to remove the technological, infrastructure and social and policy constraints responsible for the productivity gap and in some cases, productivity decline. Reducing the cost of production through eco-technologies and improving income through efficient production and post-harvest technologies will help to enhance opportunities for both skilled employment and farm income. Public policies should not only pay attention to agrarian reform and input and output pricing, but also to reaching the unreached in technology dissemination through training, techno-infrastructure and trade. A constraints analysis of the type shown in Figure 3 should be undertaken. Public policy on anticipating and avoiding production constrains research on facilitating adoption of new technology by small farmers should receive as much attention as agronomic research.

Future agricultural production programmes will have to be based on a three-pronged strategy designed to foster an evergreen revolution, which leads to increased production without associated ecological and social harm. These strategies include defending the gains already achieved, extending the gains to rainfed areas and making new gains through farming systems diversification and value addition.

Defending the gains already achieved

There is need for stepping up maintenance research for ensuring that new strains of pests and pathogens do not cause crop losses and prevent the introduction of invasive alien species. Water harvesting, watershed development and economic and efficient water use can help to enhance productivity and income considerably. Where water is scarce, high value but low water requiring crops should be promoted. As pulses and oilseeds are important income earning and soil enriching crops, they should be included in rice farming systems.

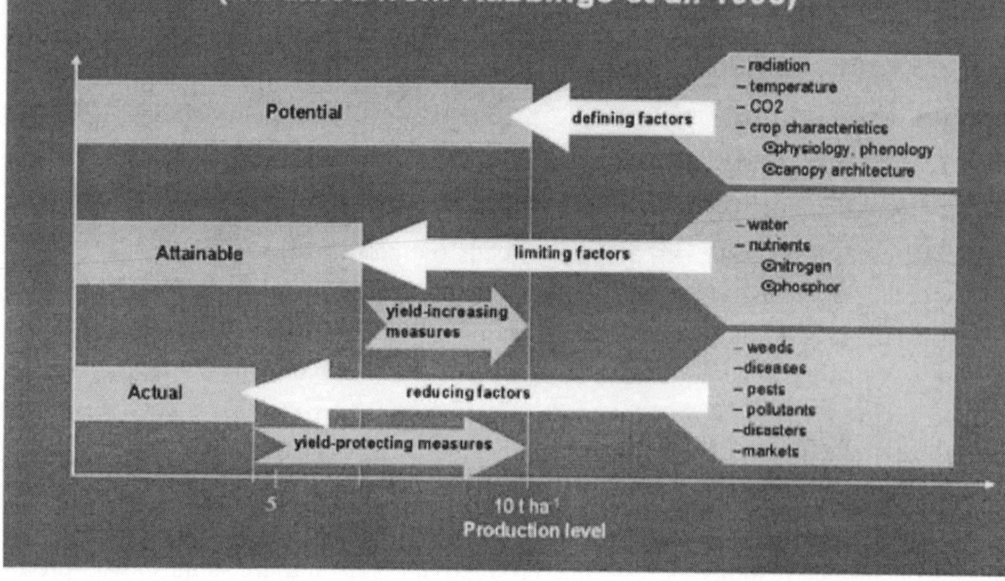

Figure 3: Yield Gap - Constraints Analysis (modified from Rabbinge *et al.* 1993)

Extending the gains

There is need to develop and disseminate eco-technologies for rain-fed and semi-arid, hill and island areas, which have so far been bypassed by modern yield enhancement technologies.

Regional imbalances in agricultural development are growing based largely on the availability of assured irrigation on the one hand and assured and remunerative marketing opportunities on the other. The introduction of eco-regional technology missions that look at all the links in the chain and work towards the stipulated goal, aimed at providing appropriate packages of technology, the specific farm infrastructure services required by the techonology (techno-infrastructure), and input and output pricing and marketing policies, will help to include the excluded in agricultural progress. Technologies for elevating and stabilizing yields are available for semi-arid and dry farming areas (Ryan & Spencer 2001). Therefore the emphasis should be on farming systems that can optimize the benefits of natural resources in a sustainable manner and not merely on cropping systems. Dry farming areas are also ideal for the cultivation of low water requiring but high value pulses and oilseeds.

Making new gains

Farming systems intensification, diversification and value-addition should be promoted. Watershed and wasteland atlases should be used for developing improved farming systems, and designing what crops to grow based on soil structure so as to provide more income and jobs. Value addition to primary products should be done at the village itself. This will call for appropriate institutional structures which can help provide key centralized services to small and marginal farm families and provide them with the power of scale in eco-farming involving techniques like integrated pest management, integrated nutrient supply, scientific water management, precision farming, etc. as well as in marketing. A quantum leap in sophistication of management of all production factors will be required to sustain yield gains from the present levels to the commercially feasible threshold of about 80 per cent yield potential (Swaminathan 2001).

Small farm management

Institutional structures, which will confer upon farm families with small holdings, the advantages of scale at both the production and post-harvest phases of agriculture, are urgently needed. For example, thanks to the cooperative method of organisation of milk processing and marketing, India now occupies the first position in the world in milk production. Strategic partnerships with the private sector will help farmers' organisations to have access to assured and remunerative marketing opportunities.

Vital Areas for Sustainable Advances in Rice Productivity

There are great opportunities for achieving higher yields per unit of land; provision of water at the right time to rice farmers enabled them to shift to precision farming methods. The five vital areas of research, development and extension, which need attention from the point of view of achieving environmentally sustainable advances in rice productivity, are soil health and fertility management; water management; integrated plant health management; energy management; and post-harvest management (for more details see Swaminathan 2004).

Evergreen revolution

As earlier mentioned, this implies improvement of productivity in perpetuity without associated ecological harm. Rice scientists should foster an evergreen revolution in rice through partnerships for the development and dissemination of precision farming technologies. The major goals that were proposed for the FAO sponsored International Network for an Evergreen Revolution in Rice by Swaminathan (2002b) are as follows:

- Initiate an Integrated Gene Management programme.
- Improve productivity per unit of input, particularly of nutrients and water and thereby reduce the cost of production.
- Substitute to the extent possible knowledge and farm produced inputs for capital and market-purchased chemicals.
- Enhance the ecological and social sustainability of high-yield technologies.
- Increase farmers' income and opportunities for skilled employment.
- Establish an information grid and farmer-participatory knowledge system for empowering women and men engaged in rice farming with new knowledge and skills, thereby conferring on rice farmers the strengths of Knowledge Societies.

Research Strategies and Priorities

These Strategies include Integrated Gene Management (IGM), integrated efforts in feeding and breeding rice for high productivity, information empowerment, overcoming hidden hunger caused by micronutrient deficiencies and promoting rice as a substrate for oral vaccines.

The IGM programme in rice should be based on conservation, sustainable use and equity in sharing benefits. The over 100,000 strains available today in rice is the result of the conservation ethics of farm and tribal families. Most of them are from Asian countries. India is the largest contributor to this collection followed by Laos (Appa Rao *et al.* 2002).

Overcoming hidden hunger caused by micronutrient deficiencies

The challenge of micronutrient deficiencies in diet is becoming great especially for the chronically poor. Iodine, Vitamin A and iron deficiencies are serious in many parts of the developing world. Worldwide, iron deficiency affects over one billion children and adults. Recent analyses from the United States Institute of Medicine (Earl & Woteki 1998; Potrykus 1997; Swaminathan 2002a) highlight the effect of severe anaemia in accounting for up to one in five maternal deaths. Maternal anaemia is pandemic and is associated with high MMR; anaemia during infancy, compounded by maternal under-nutrition, leads to poor brain development. Iron deficiency is also a major cause of permanent brain damage and death in children and limits the work capacity of adults (Smith & Haddad 2000; Swaminathan 2002b). There is not enough appreciation of the serious adverse implications to future generations arising from the high incidence of low birth weight (LBW) among newborn babies. LBW is a major contributor to stunting and affects brain development in the child. The new millennium will be a knowledge century, with agriculture and industry becoming more knowledge intensive. Denial of opportunities for the full expression of the innate genetic potential for mental development even at birth is the cruelest form of inequity that can prevail in any society (Smith & Haddad 2000). We must take steps to eliminate as soon as possible such inequity at birth leading to a denial of opportunities to nearly one out of every three children born in South Asia, for performing their legitimate role in the emerging knowledge century.

Wherever rice is the staple, a multi-pronged strategy for the elimination of hidden hunger should be developed by rice scientists. IRRI has undertaken research on enriching rice genetically with iron and other micronutrients. Fortification, promotion of balanced diets, new semi-processed foods involving an appropriate blend of rice and micro-nutrient rich millets as well as genetic improvement, could all form part of an integrated strategy to combat the following major nutritional problems in predominantly rice eating families:

- Protein-energy Malnutrition
- Nutritional anaemia (Iron deficiency)
- Vitamin A deficiency
- Iodine deficiency
- Dietary deficiencies of thiamin, riboflavin, fat, calcium, vitamin C and zinc

Swaminathan (2002a) suggested that the International Rice Commission could include nutrition security aspect as an integral part of the International Network. We must fight the serious threat to the intellectual capital of developing countries caused by low birth weight children and hidden hunger (UN Commission on Nutrition). Some of the research areas worthy of attention in this context are described below.

Breeding for nutritional quality

Nutritive quality is as important as cooking quality for countries in tropical Asia, where rice is the principal source of dietary protein, vitamin (B_1) and minerals (Fe, Ca) (Juliano & Villareal 1993). Rice provides about 40 per cent of the protein in the Asian diet. Among the cereal proteins, rice protein is considered to be biologically the richest by virtue of its high digestibility (88 per cent), high lysine content (+4 per cent) and relatively better net protein utilization. Yet, it is nutritionally handicapped on account of two factors *viz*: (i) its inherently low protein content (6–8 per cent) and (ii) inevitable milling loss of as much as 15–20 per cent. Unlike in other cereals, increased protein content in rice does not result in decreased protein quality as all of its fractions (glutelin 65 per cent, globulin and albumin 15 per cent and lysine-cysteine rich prolamin 14 per cent) are rich in lysine and other essential amino acids. Even a marginal increase of 2 percentage points of protein, therefore, would mean 10–15 per cent increase in the nutritionally rich protein intake in our diet.

Genetic engineering approaches for correcting micronutrient deficiencies

Breeding for Nutritional Improvement was recommended at the 19th Session of the International Rice Commission, which called for an increase in focus on strategies to combat malnutrition (Philip *et al.* 2000; Gopalan 2001). There are four categories of direct interventions believed to be successful in reducing micronutrients malnutrition; supplementation, fortification, dietary diversification and genetic enhancement. Nutritional status of populations will focus on the potential for improving malnutrition, primarily micronutrient malnutrition through genetic improvement.

Golden rice

About 250 million people worldwide are deficient in vitamin A. Over five million children in South and South-east Asia are reported to suffer from the serious eye disease 'xerophthalmia' every year and about 5,00,000 of them eventually become partially or totally blind due to deficiency of vitamin-A. Besides affecting vision, vitamin-A deficiency predisposes children to varied respiratory and intestinal diseases resulting in high mortality. Researchers from Swiss Federal Institute of Technology inserted these genes from daffodil and a bacterium into temperate rice plants to produce a modified grain, which has sufficient β carotene (precursor of vitamin A) to meet total vitamin A requirements in a typical Asian diet (Ye *et al.* 2000). Golden rice technology was made available to developing nations for research. If this technology can be moved to the production stage, it could represent an important contribution to improved human nutrition. In particular, rice fortified genetically with vitamin A and iron will be very useful to improve the nutritional status of pregnant and nursing women.

Iron deficiency

Iron deficiency anaemia (IDA) is the world's most common nutritional deficiency. It affects pregnant and nursing women and young children most commonly. IDA in mothers predisposes to still births, neonatal mortality, anaemia and low birth weight in infants, and increases the risk of maternal mortality (Swaminathan 2002; Earl & Woteki 1994). Regular intake of iron or administration of iron prevents anaemia. Daily supplementation with iron-folic acid tablets is a low-cost and effective intervention.

Technological Transformation of Productivity, Profitability and Sustainability: Wheat

The rediscovery of Mendel's laws of genetics in 1900 opened up a new era in crop breeding in general, and wheat breeding in particular. Although the art of plant breeding is as old as the beginning of agriculture nearly 12,000 years ago, systematic research in the areas of genetics and cytogenetics, which commenced in the early part of the 20th century, created uncommon opportunities for improving the productivity, profitability, stability and sustainability of wheat production. Even before Mendel (1822–1884), plant hybridizers like Kolreuter, Knight, Gartner and Burbank were able to produce improved varieties of crops through careful observation and selection. The concept of sustainability to which we now attach great importance was recognised long ago as essential for sustained agricultural progress.

During the past 100 years, Mendelian genetics has helped not only to exploit naturally occurring genetic variability but has also accelerated the process of generation, manipulation and combination of new variability. We are now in a state of transition from Mendelian to Molecular breeding. Breeder's eye for selection and for spotting the winner will continue to play an important part in successful plant breeding.

It is projected that global demand for wheat will increase by 40 per cent by the year 2020. Also, 67 per cent of the world's wheat consumption will be in developing countries. Between 1961 and 1990, yield increases accounted for 92 per cent of the additional cereal production in developing countries. In the years ahead, there is no option except to produce more from less per capita land and water resources. Can we sustain the yield revolution in wheat? It will be useful to consider this issue in the context of the genetic pathways which led to the wheat revolution of the 20th century.

Progress in yield improvement

Wheat is a crop of great antiquity. We can identify at least 4 major phases in the evolution of wheat breeding during the 20th century.

Phase I (1900–1930): Early days of Mendelian genetics

Soon after the re-discovery of Mendel's laws of genetics in 1900, systematic work on the genetics of resistance to stem, leaf and stripe rusts started. Selection from naturally occurring genetic variability also began.

During the early part of the 20th century, the major breeding challenges were in the area of resistance to rusts and grain quality improvement. A study of the yield improvement achieved between 1900 to 1930 in USA shows only limited progress. The emphasis was more on stability of production through disease resistance than on achieving quantum jumps in yield.

Phase II (1930–1960): Enlarging the base of theory and its application

This period was marked by the introduction of cytogenetic knowledge and tools in wheat improvement. This phase of wheat improvement was characterised by widening the gene pool used by breeders, incorporation of genes for the semi-dwarf plant type, shuttle breeding and breeding to meet the challenge of physiologic specialisation in pathogens.

Phase III (1960–1980): The green revolution phase

This phase is also generally referred to as the *green revolution* era. It was characterised by revolutionary progress in improving wheat production and productivity in several developing countries like India and Pakistan. The introduction of the semi-dwarf plant type enabled the wheat plant to yield well under conditions of good soil fertility and irrigation water management. Farmers who were used to harvesting 1 to 2 tonnes of wheat per hectare started harvesting over 5 tonnes/ha (Swaminathan 1993). In view of the widespread interest in this remarkable transformation in India's agricultural destiny, it will be useful to summarise some of the highlights.

During 1942–43, the Indian sub-continent witnessed a severe famine in Bengal resulting in the death of nearly three million children, women and men. This prompted Jawaharlal Nehru, the first Prime Minister of Independent India to remark in 1948, 'everything else can wait, but not agriculture'.

In 1964, a National Demonstration Programme was started in farmers' fields, both to verify the results obtained in research plots and to introduce farmers to the new opportunities opened up by semi-dwarf varieties for improving very considerably the productivity of wheat. When small farmers, with the help of scientists, harvested over five tonnes of wheat per hectare, its impact on the minds of other farmers was electric. The clamour for seeds began and the area under high yielding varieties of wheat rose from four hectares in 1963-64 to over four million hectares in 1971–72. A small Government programme became a mass movement (Swaminathan 1993). Wheat production in India rose from 10 million tonnes in 1964 to 17 million tonnes in 1968. In 1999, Indian farmers harvested about 72 million tonnes of wheat, taking India to the second position in the world in wheat production.

Greater interdisciplinary collaboration among breeders, plant pathologists, agronomists, physiologists, soil scientists, entomologists, nemotalogists, economists and other social scientists, climatologists and policy makers was the principal factor responsible for the success of the green revolution. The green revolution era can also be termed *the golden age in interdisciplinary and international collaboration* in wheat improvement for sustainable food security. The concept of shuttle breeding transcended continental boundaries and a global college of wheat scientists emerged. Above all, the green revolution showed how to generate synergy between technology and public policy.

Phase IV (1980–2000): Transition from Mendelian to molecular breeding

The last 20 years have witnessed great progress in using sophisticated approaches to wheat breeding. Hybrid wheat is reaching the possibility of large scale commercial cultivation. The use of genetic-cytoplasmic male sterility and of chemical hybridizing agents (CBA) are responsible for progress in the commercial exploitation of hybrid wheat. Different management practices such as lower seed rate, raised bed planting, split nitrogen application and different row width are being tried to enhance the expression of hybrid superiority. The cultivation of hybrid wheat is slowly gaining in momentum in South Africa, Australia (New South Wales), China, Argentina and France. The use of wild relatives in genetic engineering is growing (Khush & Baenziger 1998). The global average yield of wheat is 2.5 t/ha; the low average yield of wheat is because of large areas of wheat being under rainfed conditions. Progress in improving yield is however steady. So far, advances in yield improvement have been associated with increases in harvest index (i.e., grain-straw ratio). Further advances will depend upon greater biomass production and not merely on partitioning the phytosynthates.

Challenges Ahead

At the dawn of the 21st century, we can look back with pride and satisfaction on the revolution, which farm men and women have brought about in our agricultural history during the 20th century. The Punjab farmer, hardworking, skilled and determined, has been the backbone of the revolution.

While we can and should rejoice about the past achievements of farmers, scientists, extension workers and policy makers, there is no room for complacency. We will face several new problems, of which the following are important.

- First, increasing population leads to increased demand for food and reduced per capita availability of arable land and irrigation water.
- Second, improved purchasing power and increased urbanisation lead to higher per capita food grain requirements due to an increased consumption of animal products.
- Third, marine fish production is tending to become stagnant and coastal aquaculture has resulted in ecological and social problems.
- Fourth, there is increasing damage to the ecological foundations of agriculture, such as land, water, forests, biodiversity and the atmosphere and there are distinct possibilities for adverse changes in climate and sea level.
- Fifth, while dramatic new technological developments are taking place, particularly in the field of biotechnology, their environmental, food safety and social implications are still being debated.
- Finally, gross capital formation in agriculture is tending to decline in both public and private sectors during the present decade. The rate of growth in rural non-farm employment has been poor.

Since land and water will be shrinking resources for agriculture, there is no option in the future except to produce more food and other agricultural commodities from less per capita arable land and irrigation water. In other words, the need for more food has to be met through higher yields per unit of land, water, energy and time. It would therefore be useful to examine how science can be mobilised for raising further the ceiling to biological productivity without associated ecological harm. It will be appropriate to refer to the emerging scientific progress on the farms as an evergreen revolution, to emphasise that the productivity advance is sustainable over time since it is rooted in the principles of ecology, economics, social and gender equity and employment generation.

The green revolution based on Mendelian genetics has so far helped to keep the rate of growth in food production above population growth rate. The green revolution was however, the result of public good research supported by public funds. The technologies of the emerging gene revolution based on molecular genetics in contrast, are spearheaded by proprietary science and can come under monopolistic control. How can we take the fruits of the gene revolution to the unreached? This is a challenge, which we need to address.

I would like to list 5 major challenges, which will confront the wheat scientists during this century.

Equity

The Convention on Biological Diversity (CBD) stipulates that plant exploration, collection and introduction should be based on the principles of prior informed consent and equity in benefit sharing. Therefore exchange of wheat genetic resources in the future will be possible only on the basis of Material and Knowledge Transfer Agreements.

Ecology

Ecological sustainability of high productivity will be an important determinant in relation to the choice of technologies. For example, if hybrid wheat can enable us to produce 8 to 10 t/ha, over 300 kg. of nitrogen will be needed by the crop. It is obvious that if the nutrient needs of hybrid or other high-yielding wheat varieties are to be met entirely through mineral fertilizers, there will be serious environmental problems including nitrate pollution of ground water. Hence, success in achieving high productivity on a sustained basis will depend upon our ability to develop new methods of feeding the plant. Research on breeding and feeding should be carried out concurrently by a team of breeders, physiologists, agronomists and soil scientists.

Concerns relating to genetically modified organisms (GMOs)

There are growing public and political concerns relating to GMOs. The concerns relate to food and environmental safety and bioethics. It is essential that these concerns are carefully addressed through a mechanism for risk-benefit analysis, which inspires public confidence. An integrated disease management strategy should be developed to ensure that GMO's with novel genetic combinations for disease resistance do not break down due to the emergence of new physiological strains of pathogens. Also, regulatory procedures should be transparent and should inspire public confidence. There is also need for integrating molecular breeding with organic farming methods.

Expansion of proprietary science

The world is witnessing an expansion of proprietary science governed by Intellectual Property Rights (IPR). Public good research supported from public funds, in contrast, is shrinking. What will be the impact of such a situation on international varietal or other trials organised by CIMMYT? Is the golden age of cooperative research coming to an end? How can we find a balance between public good and private profit?

Climate change and safeguarding genetic diversity

Will molecular breeding resulting in 'super wheats' lead to a high degree of genetic homogeneity in farmers' fields? We know that genetic homogeneity will enhance genetic vulnerability to biotic and abiotic stresses. Hence, we should foster an integrated programme of pre-breeding and participatory breeding. Pre-breeding will help to generate novel genetic combinations, while participatory breeding with farm families will help to combine genetic efficiency with genetic diversity. Numerous location specific varieties can be developed in this manner. This will be the most effective way of meeting challenges arising from potential changes in temperature, precipitation and sea level as a result of global warming arising from the growing imbalance between carbon emissions and absorption.

Sustaining and Strengthening Agricultural Progress

In a predominantly agricultural country like ours, agricultural progress serves as the most effective safety net against hunger and deprivation. There is need for intensifying our efforts to improve agricultural productivity, quality and income. An urgent need in this area is the strengthening of institutional structures, which can help to confer on small and marginal farmers, the ecological and economic benefits of scale at both the production and post-harvest phases of farming. The following are some of the institutional structures whose reach has to be extended.

Without socially relevant and beneficial institutional structures, the extrapolation domain of successful experiences and development efforts will remain limited.

S. No	*Sector*	*Institutional Mechanism*
1.	Dairy	Cooperatives
2.	Poultry	Egg Coordination Council
3.	Integrated on-farm and off-farm employment	Biovillages
4.	Power of scale to small producers	Small Farmers' Agri-business Consortium
5.	Technological upgrading of production and post-harvest sectors	Agri-Clinics Agri-business Centres
6.	Group action for micro-enterprises supported by micro-credit	Market-driven Self-help Groups
7.	Timely and affordable credit	Kisan Credit Cards, Integrated Informal and Formal banking systems
8.	Operation of minimum support price	Food corporation of India and State Food Corporations, as well as assured buy-back arrangement and contract farming by the private sector

Dr K R Narayanan inaugurated the JRD Tata Ecotechnology Centre at the M S Swaminathan Research Foundation (MSSRF) in 1998. In the second part of my lecture, I shall briefly summarise the work done in MSSRF on issues relating to linking ecological security with food and livelihood security in a mutually reinforcing manner.

In 2005, MSSRF whose work has always received encouragement and support from Dr K R Narayanan will be completing 15 years of work in the areas of research, education, capacity building, mentoring, policy advocacy and networking. In retrospect, the decision made in 1990 to choose integrated coastal zone management for priority attention with a view to linking the ecological security of coastal areas and the livelihood security of coastal communities (both fisher and farming families) in a mutually reinforcing manner has proved to be a wise one. The Coastal System Research (CSR) programme of MSSRF was designed to give concurrent scientific attention to sea and land surfaces along the shoreline. The CSR programme was initiated in anticipation of potential adverse changes in sea level as a result of global warming. An early scientific step in this process was the conservation of mangrove genetic resources and the rehabilitation of degraded mangrove wetlands in Tamil Nadu, Andhra Pradesh, Orissa and West Bengal. Such mangrove forests served as 'bio-shields' during the Tsunami attack on 26 December 2004. They had also served a similar purpose during the super cyclone in Orissa in 1999. These observations have helped to generate at both the political and public levels interest in the development of bio-shields along the shoreline.

The following are among the major contributions of the CSR programme during 1990–2005.

- Restoration of degraded mangrove wetlands along the east coast of India: the area restored by MSSRF scientists alone comes to 1475 ha of degraded area restored with the help of 5240 families organized in 33 village mangrove councils. Nearly 7 million saplings were planted.

- Conservation strategy for mangrove genetic resources in the Asia–Pacific Region and the establishment of a Genetic Garden for meeting the challenge of sea level rise at Pichavaram, Tamil Nadu
- Genome mapping of mangrove species (*Avecinnia marina*) and the identification and transfer of genes for sea water tolerance from *A.marina* to rice, mustard and pulses.
- Development of a trusteeship mode of management of the Gulf of Mannar Biosphere and helping to create the first Biosphere Trust in the world, with the support of the Global Environment Facility and the Governments of Tamil Nadu and India
- Development of integrated Bio-shield — Biovillage — Village Knowledge Centre Programmes in coastal areas.
- Development of a Code for the participatory management of mangrove ecosystems, involving cooperative action among fisher and farm communities and Forest and Fisheries departments.
- Development of a comprehensive strategy for the rehabilitation of Tsunami ravaged coastal areas in Tamil Nadu, Pondicherry, Andhra Pradesh, Kerala and the Andaman and Nicobar Islands
- Drawing national and international attention to the urgent need for conserving mangrove wetlands and promoting the adoption of a *Charter for Mangroves*.
- Organising national and international training programmes for creating a cadre of well trained Mangrove forest managers.
- Preparation of comprehensive Atlases of the Mangrove forests of Tamil Nadu, Andhra Pradesh and Orissa
- Standardisation and popularization of mangrove propagation methods based on both vegetative and micro-propagation techniques; and helping local communities to undertake the raising and planting of mangrove forests.
- Preparation of a Tool Kit for raising bio-shields in coastal areas.
- Establishment of the first community developed and managed artificial reef in the Gulf of Mannar area based on the technology developed by the Central Marine Fisheries Research Institute
- Developing eco-agriculture strategies for coastal drylands through participatory evaluation and propagation of varieties of pulses developed at the Bhabha Atomic Research Centre through mutation breeding, and the establishment of green belts and genetic garden of horticultural crops for sustainable food and livelihood security.
- Launching a Fish for All movement at Kolkata in December, 2003 in association with the World Fish Centre (ICLARM) located in Penang, Malaysia, to promote sustainable capture and culture fisheries movements.
- Advocacy for the adoption of aquarian reform measures designed to promote harmony between artisenal and mechanized fisheries as well as aquaculture and agriculture.

The CSR approach helped MSSRF to propose a comprehensive and integrated strategy for launching a ‘Beyond Tsunami’ programme based on concurrent attention to ecological, livelihood, agronomic, psychological and educational rehabilitation. The experience gained by MSSRF in developing integrated coastal zone management procedures helped a National Committee set up by the Ministry of Environment and Forests under my Chairmanship to review the Coastal Regulation Zone Notification of 1991, to propose 12 basic guiding principles for the sustainable and scientific management of the Coastal Zone. Some of these are:

- Ecological and cultural security, livelihood security and national security should be the cornerstones of an integrated coastal zone management policy.
- The coastal zone would include an area from territorial limits (12 nautical miles), including its sea-bed to the administrative boundaries or the biological boundaries demarcated on the landward side of the sea coast. The coastal zone management should also include the inland tidal water bodies influenced by tidal action and the land area along such water bodies. This area should be taken up for an integrated, cohesive, multi-disciplinary and multi-sectoral coastal area management and regulatory system
- Regulation, education and social mobilization should be the three major components of a participatory and sustainable Coastal Zone Management strategy. Panchayati Raj institutions in coastal areas should be fully involved in the educational and social mobilization programmes.
- Coastal regulation needs to be based on sound scientific and ecological principles and should safeguard both natural and cultural heritage. Heritage sites need particular care and should be conserved in their pristine purity. These include areas of environmental significance, rich in biodiversity and scenic beauty. Bird sanctuaries, parks and breeding grounds of migratory birds should be protected.
- The precautionary approach should be used where there are potential threats of serious or irreversible damage to ecologically fragile critical coastal systems and to living aquatic resources. Scientific uncertainty should not be used as an excuse for the unsustainable exploitation of coastal resources — both living and non-living.
- Ecological economics should underpin economic activities, so that present day interests and future prospects are not antagonistic. Significant biological, cultural and natural assets should be considered incomparable, invaluable and irreplaceable and should receive overriding priority in the allocation of resources for coastal area protection and conservation
- Coastal policy and regulations should be guided by the principles of gender and social equity as well as intra-generational and inter-generational equity, (i.e., the interests of future generations). They should be based on Mahatma Gandhi's dictum, 'Nature provides for everyone's needs, but not for anyone's greed'. All stakeholders should be involved in decision making. Precious biological wealth, coming under Marine Biosphere Reserves, should be managed in a Trusteeship mode, with all the stakeholders protecting the unique natural wealth of biosphere reserves as Trustees and not as owners. A case study should be made on how the Gulf of Mannar Biosphere Trust is functioning, so that the Trusteeship pattern of sustainable management by the principal stakeholders can be replicated.

- The regeneration of mangrove wetlands, coral reefs and sea grass beds as well as the promotion of coastal forestry and agro-forestry will confer both short and long term ecological and livelihood benefits. Carbon sequestration through coastal bio-shields will make an important contribution to promoting a balance between carbon emission and absorption, in addition to offering protection during coastal storms and calamities like tsunami. An important lesson taught by the tsunami disaster is that the rehabilitation of degraded mangrove forests and the raising of coastal plantations of salicornia, casuarinas, Vetiver and appropriate species of halophytes will represent a 'win-win' situation both for nature and coastal human habitations. No further time should be lost in initiating a national coastal bio-shield movement along the coasts of the mainland of India as well as islands. This can be a priority task under the National Rural Employment Guarantee and Food for Work Programmes.
- The severe loss of life and livelihoods as well as property caused by tsunami in Andaman & Nicobar Islands and in the coastal regions of Tamil Nadu, Kerala, Andhra Pradesh and Pondicherry teaches us that short term commercial interests should not be allowed to undermine the ecological security of our coastal areas. Human memory tends to be short and neglecting the lessons of tsunami will be equivalent to writing off the future of coastal communities.

Based on the experience gained during the last 15 years, it is proposed to establish in Chidambaram a Resource Centre for Integrated Coastal Zone Management, for the purpose of imparting training in the erection of bio-shields, the development of biovillages and the establishment of Village Knowledge Centres. Tool Kits for these purposes have already been prepared.

In addition to the above, steps have been taken in association with the Tata Relief Committee and the World Fish Centre (ICLARM) to establish a Fish for All Training and Resource Centre at Akkarapettai village near Nagapattinam for imparting training in all aspects of capture and culture fisheries through the principle of learning by doing. The Centre will give attention to capacity building of fisher women and men in every step in the chain of capture / culture to consumption.

During 2004–05, MSSRF's strategic and participatory research to meet the challenges of climate change, which has been so far confined to the coastal zone, was extended to the arid and semi-arid areas of Andhra Pradesh and Rajasthan with financial and technical help of the Swiss Agency for Development Cooperation (SDC) and in partnership with Action for Food Production (AFPRO) and the National Institute of Agricultural Extension Management (MANAGE). This project will help to study vulnerability to adverse changes in temperature and precipitation and develop mitigation and adaptation strategies. Such proactive measures are essential to prevent human suffering resulting from agricultural collapse during drought and flood. The climate change programme will take into account the impact of radiation, carbon dioxide concentration in the atmosphere, temperature and precipitation. It will also help to understand and chronicle traditional coping mechanisms, so that these can be conserved and strengthened. Computer simulation models on the impact of variations in temperature and precipitation will be developed and contingency plans to mitigate the adverse impact of climate change will be introduced.

Besides developing a methodology for conserving the Gulf of Mannar Biosphere Reserve for posterity through a multi-stakeholder trusteeship system of management, MSSRF has evolved during the last 15 years three other major institutional innovations in areas of significance to sustainable food and livelihood security and poverty eradication. These are described briefly below.

Community Nutrition and Water Security System

This system introduced in the Koraput district of Orissa consists of organizing field gene banks (*insitu* on-farm conservation), seed banks, genetic enhancement through participatory breeding, water banks (i.e., water harvesting and saving in farm ponds), and Grain Banks, as shown in Fig. 4.

This system helps to enlarge the food basket by facilitating the inclusion of millets and other under-utilised but nutritious crops in the Community Grain Bank. Such a decentralised Community managed Nutrition Security system helps to foster concurrent attention to conservation, cultivation, consumption and commerce. The tribal community of Koraput pioneering this system was given the Equator Initiative Award by UNDP at the World Summit on Sustainable Development held at Johannesburg in 2002.

Currently there are 2,34,676 village Panchayats in 31 States and Union Territories. In addition, there are traditional councils in Meghalaya, Mizoram and Nagaland. Each of these Panchayats/local bodies can spearhead the Community Food and Water Security movement. This will be the fastest and a sustainable method of making hunger history.

Fostering Job-Led Economic Growth

The most serious challenge facing India is overcoming the famine of jobs or sustainable livelihood opportunities in rural India. MSSRF, whose mandate is imparting a pro-poor, pro-nature and pro-woman orientation to technology development and dissemination, designed and developed the biovillage model of sustainable human well being for this purpose in 1992. The biovillage concept involves the technological upgradation of agriculture and agro-based enterprises in villages through ecotechnologies developed by blending frontier technologies like information and biotechnologies as well as space, nuclear and renewable energy technologies with traditional ecological prudence. Thus, the biovillage based on the economics of human dignity, capitalizes on the benefits conferred by ecotechnology to both the environment and the rural economy. By giving simultaneous attention to on-farm and non-farm employment, the biovillage promotes job-led economic growth and helps to transfer poor families from the primary to the secondary and tertiary sectors of economic activity. This model is now being adopted both in other parts of India and other countries like Bangladesh and Mozambique.

The principal features of the biovillage model of sustainable and equitable rural development are shown in Fig.5

With the help of the Technology Information Forecasting and Assessment (TIFAC) programme of the Govt. of India, a Business Plan was prepared for establishing Rice BioParks. A wide range of economically viable business activities were identified for producing value-added products from rice straw, husk, bran and grain. Business plans were prepared for nearly 28 different enterprises.

Thus, the biomass of cultivated plants can provide opportunities for new enterprises. Similarly, the production and marketing of the biological software essential for sustainable agriculture such as, biofertilizers, biopesticides, vermiculture etc., could help self help groups (SHG) of women and men to enhance their income. MSSRF organized a workshop for sharing experiences on SHGs. It became clear at the workshop that SHGs can become economically sustainable only if they have backward linkages to technology and credit, and forward linkages to markets and management. MSSRF has developed an accounting software for helping SHGs to maintain both accuracy and transparency in accounting.

With the help of the Central Food Technology Research Institute, Mysore, training in post-harvest processing was given to trainees from Ladakh to help them prepare value-added products from apricot and seabuckthorn. Similarly, technical advice was given to the Sher-E-Kashmir University of Agricultural Science and Technology of Kashmir in Srinagar for establishing a Womens' Biotechnology Park at Srinagar on the lines of the one functioning in Chennai.

Under the International Year of Rice Year Programme, consultations were held at Koraput in Orissa, Pattambi in Kerala and Shillong in Meghalaya for reviewing the current status of research on medicinal and aromatic rices. Detailed scientific strategies were developed for the improvement of the *Navara* rice of Kerala and *Kalajeera* rice of Koraput through participatory breeding and knowledge management. In all such programmes the role of women in conservation and enhancement of genetic resources was given specific attention.

Centre

The third major institutional innovation developed by MSSRF for transforming the rural economy is the computer-aided and internet connected Village Knowledge Centre.

The work undertaken by MSSRF in setting up community centered and managed Village Knowledge Centres (VKC) in Pondicherry villages since 1998 based on modern information and communication technologies (ICT) with financial support from IDRC of Canada has shown that ICT helps to improve the timeliness and efficiency of farm operations and enhances income through producer-oriented markets. Also, experience has shown that bridging the digital divide is a powerful method of bridging the gender divide. Knowledge connectivity therefore confers multiple economic and social benefits. The VKC operates on the principles of social inclusion and giving voice to the voiceless. The information provided, which includes location-specific data on entitlements to different Government schemes, is demand-driven and is in the local language.

For example, in Union Territory of Pondicherry there are over 150 schemes designed to help the poor; yet nearly 20 per cent of families are below the poverty line. After the onset of the digital age, knowledge on entitlements and how to access them has grown rapidly. The VKC will be a powerful instrument for operationalising the provisions of the Right to Information Act (2005).

Encouraged by the ability of rural women and men to take to ICT like fish to water, MSSRF initiated in 2003, two major steps to take ICT to every one of the over 600,000 villages in India by 15 august, 2007, which marks the 60th anniversary of 'our tryst with destiny', to quote Jawaharlal Nehru. The first is the organization of a National Alliance for Mission 2007: Every Village a Knowledge Centre which provides a platform for partnership to all committed to the cause of extending the power of ICT to rural India. The National Alliance has now over 150 members comprising Central and State Government agencies, business and industry, academia and non-governmental and mass media organizations.

The second is the establishment of the Jamsetji Tata National Virtual Academy for Rural Prosperity with generous support from the Tata Education Trust. The Internet — community radio combination is a powerful method of reaching the unreached in terms of delivery of dynamic information. Public policy in promoting the use of community radio should be based on the following principle enunciated by the Supreme Court in its judgment delivered in December1995, 'Air waves constitute public property and must be used for advancing public good'. This is the same principle enshrined in the Dandi March movement of Mahatma Gandhi in relation to sea water, which is the basis of MSSRF's programme on sea water farming for coastal area prosperity.

At a recent meeting held at MSSRF, Panchayati Raj leaders have assured that they will provide space, electricity and telephone connection for establishing VKCs in the Panchayat premises. Thus, all the 234676 Village Panchayats in 31 States and Union Territories as well as Traditional Councils in the N.E. States can be brought together under the umbrella of the National Alliance. A hub-spokes model will help to reach all villages from Panchayat VKCs. Such Centres can be operated by ICT-Self-help Groups of rural women and men. MSSRF is assisting NABARD to organize about 10000 ICT — SHGs in 10 States of the country during 2005-06,

Besides connectivity and content, capacity building is essential for ensuring local ownership of VKCs. This is where the Jamsetji Tata National Virtual Academy (NVA) of MSSRF hopes to play a key role. The President of India, Dr APJ Abdul Kalam inducted the first 137 Fellows of the NVA drawn from 15 States on 11 July 2005 at New Delhi. Microsoft is providing generous support for capacity building under its Unlimited Potential programme.

The Fellows of NVA are rural women and men who have studied up to the tenth class or up to the first degree. They serve as Master Trainers and undertake the training of other rural women and men as well as children. These grassroots academicians will be the torchbearers of the rural knowledge revolution. Another significant development in taking the benefits of the space age to the rural poor was the inauguration by the Prime Minister of India, Dr Manmohan Singh on 18 October 2004 of an ISRO–MSSRF joint initiative in setting up Village Resource Centres (VRCs) which can link rural families to the best available sources of knowledge in medicine and health care, education, agriculture, markets and government programmes. This programme which initially linked MSSRF (Chennai) to VRCs in Thiruvayaru, Sempatti and Thangachimadam in Tamil Nadu is being extended to Chidambaram, Pudukottai, Pondicherry, Nagapattinam and Kanyakumari during this year. With the help of the Indian Space Research Organisation, additional centers are being opened in Tsunami affected areas and in Farmers' 'distress hotspots' in Kerala, Andhra Pradesh, Maharashtra and Karnataka. These are areas where suicides by farmers occur. Those operating the computer aided knowledge system at such Centres will be either wives or daughters or sons of those who were driven to take their lives. This will help to provide a sense of realism and urgency in achieving a match between content and the need to save livelihoods and lives. While VKC operates at the village level, the VRC is designed to cover a Block and thereby serve as a resource center for all the villages in the Block. The design of a VKC is on the lines shown in Fig. 6.

The Prime Minister of India has announced a well-funded Bharat Nirman programme to accelerate progress in providing urban amenities in rural areas and to bring an additional ten million hectares under assured irrigation. Knowledge connectivity should be the backbone of the Bharat Nirman programme, since it is fundamental to deriving maximum benefit, in terms of a better quality of life in villages, from the investment on roads, telephone connectivity and other forms of physical connectivity. The involvement of Panchayats and Gram Sabhas in providing the needed logistic and policy support will ensure the efficient functioning of VKCs. To begin with VKC should be tools of information, knowledge and skill empowerment of rural families, particularly of the economically and socially under privileged sections of the society. This is a fundamental responsibility of Government. Hence, the initial expenses should be met from the Bharat Nirman Programme and the Universal Service Obligation (USO) Fund. By the end of this decade (i.e., by 2010), the VKCs will become vibrant centers of economic activity and will provide opportunities for outsourcing of assignments from urban to rural areas. They will then become not only economically self-reliant but will help to create a wide range of skilled jobs for youth in villages. A VKC centred Bharat Nirman will be the most effective method of fostering rural and agrarian prosperity and arresting the unplanned migration of the rural poor to urban areas resulting in the proliferation of urban slums. Therefore, knowledge connectivity through VKCs should be the corner stone of a New Deal for Rural India.

What motivates the scientists and scholars of MSSRF are the words of the Poet Rabindranath Tagore

With your mind intent, cross this sea of chaos
And sail to that shore of new creation

References

Appa Rao, S., Bounphanousay, C., Schiller, J.M. and Jackson, M.T. (2002a). Collection, classification, and conservation of cultivated and wild rices of the Lao PDR, *Genetic Resources and Crop Evolution*, 49, 75–81.

Earl, R. and Woteki, C.E. (eds) (1994). *Iron deficiency anemia: recommended guidelines for the prevention, detection and management among US children and women of childbearing age*. Institute of Medicine.

Food and Nutrition Bulletin, 21(3), (September 2000). Supplement, Ending Malnutrition by 2020: An Agenda for Change in the Millennium, Final Report to the ACC/SCN by the Commission on the Nutrition Challenges of the 21st Century, United Nations University Press.

Gopalan, C. (2001). Combating Vitamin A Deficiency and Micronutrient Malnutrition through dietary improvement, MSSRF, Chennai (mimeographed).

Juliano, B.O. and Villareal, C.P. (1993). *Grain quality evaluation of world rices*. IRRI: Manila.

Khush, Gurdev S. and Baenziger, P. Stephen (1998). Crop improvement: emerging trends in rice and wheat, V.L. Chopra, R.B. Singh and A. Varma (eds), *Crop Productivity and Sustainability — Shaping the Future*, Proceedings of the 2nd International Crop Science Congress, pp. 113–25.

MSSRF (M.S. Swaminathan Research Foundation). (2001). *Community Grain Bank: An Instrument for Local Food Security*. MSSRF, Chennai.

Peacock, K. and Chaudhury, A. (2002). The impact of gene technologies on the use of genetic resources, J. M. M. Engels, V. Ramanatha Rao, A.H.D. Brown and M.T. Jackson (eds), *Managing Plant Genetic Diversity*, 23–31. IPGRI.

Philip, James *et al.* (2000). Ending malnutrition by 2020: an agenda for change in the millennium, *Food and Nutrition Bulletin*, 21(13): 88.

Pingali, P.L., Hossain, H. and Gerpacio, R.V. (1997). *Asian Rice Bowls: The Returning Crisis*, CAB International in association with IRRI, p. 341.

Potrykus, I. (1997). Transgenic rice (Oryza sativa) endosperm expressing daffodil (Narcissus pseudonarcissus) phytoene synthase.

Ramalingaswami, V., Johnson, U. and Rohde, J. (1997). Malnutrition: A South Asian Enigma, Stuart Gillespie (ed.), *Malnutrition in South Asia*, Rosa Publication 5. Nepal: UNICEF.

Ryan, J.G. and Spencer, D.C. (2001). *Future Challenges and Opportunities for Agricultural R&D in the Semi-Arid Tropics*. Patancheru: ICRISAT.

Smith, L.C. and Haddad, L. (2000). *Overcoming Child Malnutrition in Developing Countries: Past Achievements and Future Choices*. Washington D.C.: International Food policy Research Institute.

Swaminathan, M.S. (1968). The Age of Algeny, Genetic Destruction of Yield Barriers and Agricultural Transformation, Presidential Address, Agricultural Science Section, Fifty-fifth Indian Science Congress, Varanasi, January.

Swaminathan, M.S. (ed.), (1993). *Wheat Revolution: A Dialogue*, Madras: Macmillan India Ltd, pp. 164.

Swaminathan, M.S. (1996). *Sustainable Agriculture: Towards an Evergreen Revolution*, pp. 232. Delhi: Konark Publishers Pvt. Ltd.

Swaminathan, M.S. (2000). Bridging the nutritional divide, *The little magazine*, II(6), available at http://www.littlemag.com/hunger/swami.html.

Swaminathan, M.S. (2000). An evergreen revolution, *Biologist*, 47(2): 85–9.

Swaminathan, M.S. (2001). Food Security and sustainable development, *Current Science*, 81, 948–954.

Swaminathan, M.S. (2002a). Building a National Nutrition Security System, paper presented at India-ASEAN Eminent Persons Lecture Series 11, Jan 2002, FAO, Bangkok.

Swaminathan, M.S. (2002b). Nutrition in the Third Millennium: Countries in Transition, Plenary Lecture 17th International Congress on Nutrition Vienna, 27–31 August 2002.

Swaminathan, M.S. (2004). Technological change and food production: implications for vulnerable sections, Working Paper 20, CPRC-IIPA, New Delhi.

Swaminathan, M.S. (2005). *India's Greatest Living Industry: Hundred Years Later*, IARI New Delhi, IARI Centenary Lecture. 16 March.

WFP (2001). *Enabling Development: Food Assistance in South Asia*, New Delhi: Oxford University Press, pp. 316.

Ye, X., Al-Babili, S., Kloti, A., Zhang, J., Lucca, P., Beyer, P. and Potrykus, I. (2000). Engineering provitamin A (β-carotene) biosynthetic pathway into (carotenoid-free) rice endosperm, *Science*, 287: 303–305.

SECTION III

FOOD SECURITY AND ECOLOGICAL BALANCE — HOW THE GAINS OF GREEN REVOLUTION ARE IMPACTED BY CLIMATE CHANGE, HOW SCIENCE WILL BE HELPFUL IN ENSURING SUSTAINABLE FOOD SECURITY, GREEN REVOLUTION TO EVER-GREEN REVOLUTION — ROADMAP

Section III

Food Security and Ecological Balance

How the gains of green revolution are impacted by climate change, how science will be helpful in ensuring sustainable food security, green revolution to ever-green revolution a roadmap

Commentary by M.S. Swaminathan

Conservation and enhancement of the ecological foundations essential for sustainable agriculture are the prerequisites for achieving the goal of Food for All and Forever. There is need for concurrent attention to soil health care and improvement, and water conservation and efficient use. There is also need for the conservation of agro-biodiversity in order to meet the needs of nutrition security now, and yield security in an era of climate change in the future. We have to conserve genetic diversity in all living organisms ranging from microbes to man. For this purpose, we should develop a conservation continuum beginning with in-situ onfarm conservation in farmers' fields and ending with ex-situ preservation under permafrost conditions in the Svalbard Global Seed Vault. It is a matter of satisfaction that the plea made in my Presidential Address at the International Congress of Genetics in 1983 for establishing an ex-situ genetic conservation centre under permafrost conditions has come true with the establishment by the Government of Norway of the Svalbard Global Seed Vault. The loss of every gene and species limits our options for the future and hence the conservation and sustainable and equitable use of agro-biodiversity must receive overriding priority.

An Evergreen REVOLUTION

1968 – The beginning of the green revolution

M.S. Swaminathan, *M. S Swaminathan Research Foundation, India*

The term 'Green Revolution' was coined in 1968 to indicate revolutionary improvements in crop yield in several Asian countries. Many of these improvements came at the cost of adverse environmental effects in areas subjected to intensive farming. However, where population pressure is high, there is no option except to produce more food. Productivity must increase, but in ways which are environmentally safe, economically viable and socially sustainable. This has been christened an 'Evergreen Revolution'.

The Green Revolution

The Food and Agriculture Organization of the United Nations (FAO) was established before the end of World War II, since combating food insecurity was recognised as a major challenge of the postwar era. A severe famine in Bengal in 1942–43 resulted in nearly three million deaths and prompted Jawaharlal Nehru, the first prime minister of independent India, to remark, 'Everything else can wait, but not agriculture.' This priority was reflected in land reform, expansion of irrigation facilities, and greater support for research and the production of inputs, particularly seeds and fertiliser.

Fertiliser trials with wheat and rice in the 1950s revealed that the varieties then cultivated lodge (fall over) when about 20 kg of nitrogen is applied per hectare. As well as making the crop difficult to harvest, lodging also makes it difficult to provide irrigation during grain development, when the crop benefits most from water availability. Thus, with these varieties, it is difficult to get an enhanced response to additional nutrients and water.

India's search for crops that could respond to higher levels of nutrition started in 1952, when a programme was initiated for crossing fertiliser-responsive japonica rice with standard indica strains. The goal was to select lines with the ability to use about 100 kg of nitrogen, producing about five tonnes of rice, per hectare. Several genetic problems including semi-sterility arose, rendering the search difficult. With the advent in the early 1960s of semi-dwarf, non-lodging varieties, interest in japonica crosses waned, and semi-dwarfed indica rices provided the material for breeding high-yielding varieties.

When I joined the Indian Agricultural Research Institute (IARI) in late 1954, I started a research programme for developing non-lodging, fertiliser-responsive varieties of wheat,

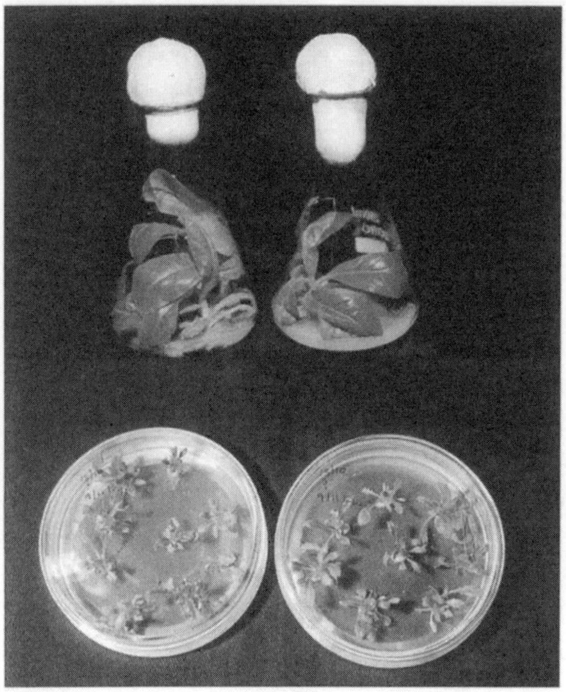

Transgenic system

along the lines of the work already begun in rice. The research strategy had three components: crosses between cultivated bread wheat and the semi-dwarf, stiff-straw compactum and sphaerococcum subspecies; induction of stiff-straw mutants through the use of radiation and chemical mutagens; and increasing straw stiffness through chemical treatment. Unfortunately, in all three approaches, short and stiff straw was always associated with short panicles (flower clusters) and fewer grains, and yields stagnated at less than one tonne per hectare.

During the late 1950s, publications began to appear on the work of Dr Orville Vogel in the United States on the transfer of dwarfing genes to North American winter wheats. Dr Vogel was kind enough to send seeds of Gaines, a semi-dwarf winter wheat, and further indicated that Dr N.E. Borlaug in Mexico had semi-dwarf varieties of spring wheat, which would grow better under the short days of India's winter. In March 1962, a few dwarf spring wheat strains were grown in IARI fields. Their phenotype was most impressive: they had reduced height but long panicles, unlike the earlier hybrids and mutants. In 1963, Dr Borlaug sent a wide range of semi-dwarf wheat, which provided the material for accelerated advances in productivity.

In 1964, a National Demonstration Programme was started in farmers' fields, both to verify the results obtained in research plots and to introduce farmers to the new opportunities for improving their productivity. When small farmers harvested over five tonnes of wheat per hectare, the impact on the minds of other farmers was electric. The clamour for seeds began, and the area under high-yielding wheat varieties rose from four hectares in 1963–64 to over four million in 1971–72. The small government programme had become a mass movement.

Wheat production in India rose from 10 million tonnes in 1964 to 17 million in 1968, and similar results were obtained with semi-dwarf varieties of rice. In 1968, the Indian government released a 'Wheat Revolution' postage stamp to commemorate the progress that had been made. In the same year, Dr William Gaud of the United States coined the term 'Green Revolution' to stress that what was happening in the wheat and rice fields of Asia was revolutionary, not just evolutionary, progress.

As early as 1967, I had observed that farmers in northwest India with relatively large holdings tended to use large quantities of fertilisers and grow single genetic strains in large, contiguous areas. In my presidential address to the Agricultural Sciences Section of the Indian Science Congress, I stressed the need for considering ecological sustainability in efforts to improve yield: 'The initiation of exploitative agriculture without a proper understanding of the various consequences of every one of the changes introduced into traditional agriculture, and without first building up a proper scientific and training base to sustain it, may only lead us, in the long run, into an era of agricultural disaster rather than one of agricultural prosperity.' (Swaminathan, 1968).

Agriculture in the 21st century

At the dawn of the 21st century, we can look back with pride and satisfaction on the revolution in agriculture during the 20th century. In 1969 I wrote in the *Illustrated Weekly of India:* 'Brimming with enthusiasm, hard-working, skilled and determined, the Punjab farmer has been the backbone of the revolution.... Farmers, young and old, educated and uneducated, have easily taken to the new agronomy. It has been heartwarming to see young college graduates, retired officials, ex-army men, illiterate peasants and small farmers queuing up to get the new seeds. At least in the Punjab, the divorce between intellect and labour, which has been the bane of Indian agriculture, is vanishing.'

While we can and should rejoice about past achievements, there is no room for complacency, as we face several new prob-

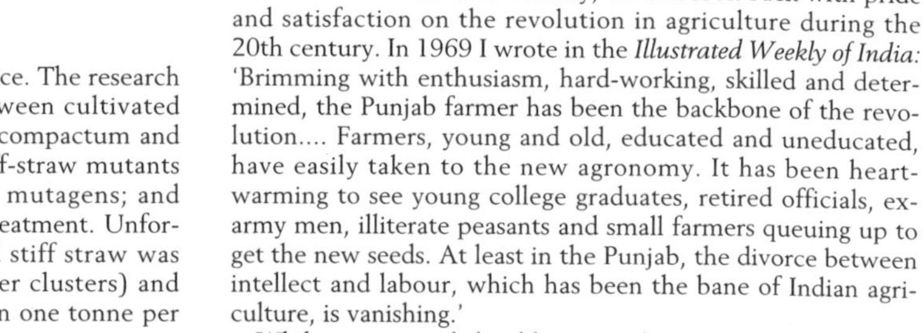

Green Revolution

Wheat Production – India now occupies the Second Position in the World

lems. An increasing population leads to increased demand for food but reduced per capita availability of arable land and irrigation water. Improved purchasing power and increased urbanisation also lead to higher per capita grain requirements, due to increased consumption of animal products. At the same time, there is increasing damage to the ecological foundations of agriculture (land, water, forests, biodiversity, atmosphere) and distinct possibilities for adverse changes in climate and sea level. While dramatic new technological developments are taking place, particularly in the field of biotechnology, their environmental, safety and social implications are yet to be fully understood. Finally, gross capital formation in agriculture is declining in both public and private sectors.

Since land and water will be diminishing resources, there is no option in the future except to produce more food and other agricultural commodities from less arable land and irrigation water. It will therefore be useful to examine how science can be mobilised to further raise the ceiling of biological productivity without associated ecological harm. The emerging scientific progress on the farms can be called an 'Evergreen Revolution', emphasising that the advance in productivity will be sustainable over time because it is rooted in the principles of ecology, economics, social and gender

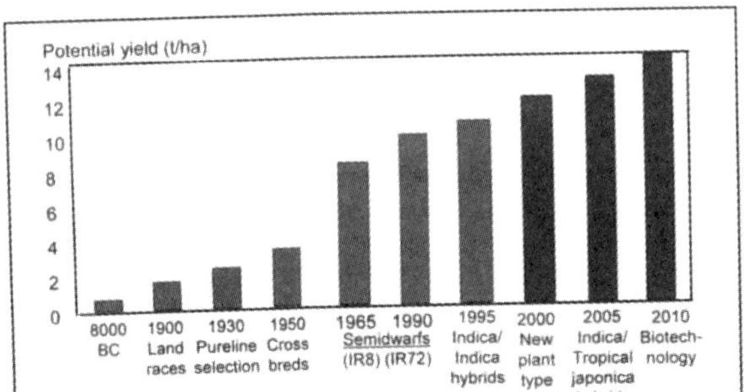

Figure 5. Progress in the yield potential of rice

equity, and employment generation.

The Green Revolution, based on Mendelian genetics, has so far helped to keep the rate of growth in food production above the rate of population growth. It was, however, the result of public-good research, supported by public funds. The technologies of the emerging gene revolution based on molecular genetics, in contrast, are spearheaded by proprietary science and can come under monopolistic control. How can we take the fruits of the gene revolution to the unreached?

The Gene Revolution

The past ten years have seen dramatic advances in our understanding of how biological organisms function at the molecular level, and in our ability to analyse, understand, and manipulate DNA. Thus the new scientific discipline of genomics has arisen. This has contributed to powerful new approaches in agriculture and medicine and has helped to promote the biotechnology industry. Several commercial corporations in Europe and the United States have made major investments in adapting genomics and its new offshoot, proteomics, to produce new plant varieties of agricultural importance for large-scale commercial agriculture. These technologies have equally important potentials for addressing food security in the developing world.

For example, it is now clear that the 21st century may witness changes in temperature, precipitation, sea level and ultraviolet radiation. Such changes in climate are expected to affect India and sub-Saharan Africa adversely. (All human-induced calamities have especially adverse effects on poor nations and the poor among all nations.) This led us to initiate in 1992 an anticipatory research programme to breed salt-tolerant varieties of crop plants in coastal areas, in order to be prepared for seawater intrusion into farmland as a result of a rise in sea level. The donor of genes for salt tolerance was a mangrove tree species of the family Rhizophoraceae. Transferring genes for salt tolerance from mangrove to rice, mustard or tobacco is an impossible task without recourse to recombinant DNA experiments. The immense benefits that can accrue from genomics and molecular breeding are clear.

What then are the principal concerns? In industrialised countries, the main concerns relate to the impact of genetically modified organisms (GMOs) on human health and the environment; these food and environmental safety concerns have been well documented and are widely known. The professional, public and political leaders of developing countries are equally concerned about the food and environmental safety aspects of GMOs. But what are the additional issues which agitate the minds of the public and professionals in the developing world?

The first issue relates to biosafety. Fortunately, negotiations relating to the formulation of an internationally agreed biosafety protocol (as called for under Article 19 of the Convention on Biological Diversity) were concluded at Montreal in January 2000. After this agreement is ratified by 50 nations, it will come into force.

Another issue relates to the potential impact of GM foods on biodiversity. The modernisation of agriculture has resulted in a narrowing of the base of food security, both in terms of the number of species constituting the 'food basket' and the number of genetic strains cultivated. Local cultivars have often been the donors of many useful traits, including resistance to pests and diseases. Under small farm conditions, every farm is a genetic garden, containing several crops and several varieties of each. The present need is to enlarge the food basket, not to shrink it further.

A final aspect of GM foods and biodiversity relates to the equitable sharing of benefits between biotechnologists and the primary conservers of genetic resources and holders of traditional knowledge. At present, the primary conservers remain poor, while those who use the knowledge and material become rich, resulting in accusations of bio-piracy. It is time that genetic engineers promoted genuine partnerships with the holders of indigenous knowledge and conservers of genetic variability, based on principles of ethics and equity in benefit sharing. The invaluable contributions of tribal and rural people to genetic resource conservation and enhancement also deserve recognition and reward (Swaminathan, 1996).

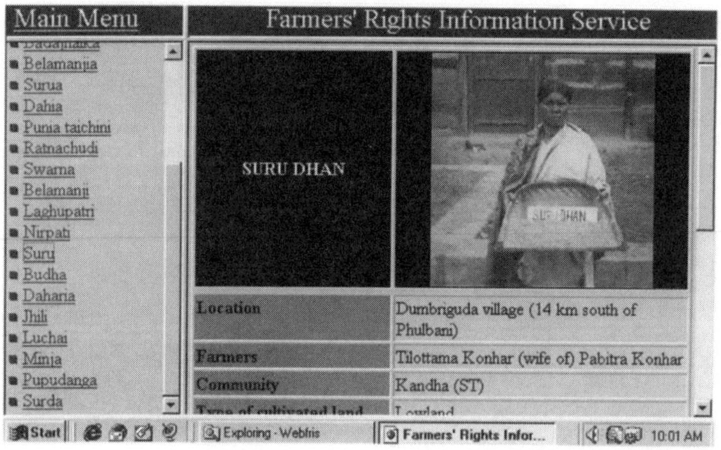

Women: Seed selectors and savers

Unless research and development of GM foods is based on principles of ethics, safety, biodiversity conservation and partnership, there will be serious public concern about the ultimate nutritional, social, ecological and economic consequences of replacing numerous local varieties with a few GMOs. Also, where the market is the dominant factor in determining research priorities, 'orphan' problems (those unlikely to be profitable enough to attract private investment) will remain orphans unless the public sector steps in.

The ecotechnology revolution

Knowledge is a continuum. There is much to learn from the past in terms of ecological and social sustainability of technology. At the same time, new developments have opened up new opportunities for the development of technology that can lead to high productivity without adverse impact on natural resources. Blending traditional and new technologies leads to a combined strength in terms of economics, ecology, equity, employment and energy.

For example, in the area of water use, there are many lessons to be learnt from the past. In the desert area of Rajasthan, India, drinking water is available even in areas with 100 mm annual rainfall, largely because women continue to harvest water in simple structures called kunds. In contrast, drinking water is scarce during summer months in some parts of northeast India that have an annual rainfall of 15 000 mm. This demonstrates the need to conserve traditional wisdom and practices, which are tending to become extinct (Agarwal and Narain, 1998).

The decision of the World Intellectual Property Organization (WIPO) to explore the needs, rights and expectations of holders of traditional knowledge, innovations and culture is an important step in widening the concept of intellectual property. Principles of ethics and equity demand that this invaluable component of intellectual property rights is included when the TRIPS agreement (Trade-Related Intellectual Property Rights) of the World Trade Organisation comes up for review in 1999–2000. The FAO has been a pioneer in promoting the concept of farmers' rights, and the Union for the Protection of New Varieties of Crops should also undertake the task of preparing an integrated concept of breeders' and farmers' rights.

Ecotechnologies are knowledge-intensive. Fortunately, modern information technology provides opportunities for reaching the unreached. Computerised, networked 'virtual colleges', which link scientists to people living in poverty, can be established to launch a knowledge and skill revolution. This will help to create better awareness of the benefits and risks associated with GMOs, so that both farmers and consumers get better insights into the processes leading to the creation of novel genetic combinations.

The future of small farms lies in precision agriculture, which involves the use of the right inputs at the right time and in the right way. Biotechnology will play an important role in the following major components of integrated resource management and precision farming: integrated gene management (involving concurrent attention to conservation, sustainable use and equitable sharing of benefits); soil health care; water conservation and management; integrated nutrient supply; integrated pest management; and efficient post-harvest management, resulting in value addition to primary products. Ecotechnology-based precision farming can help cut costs, enhance marketable surplus and eliminate ecological risks.

The yield revolution

Productivity improvement will be possible only if greater attention is paid to improving the efficiency of input use,

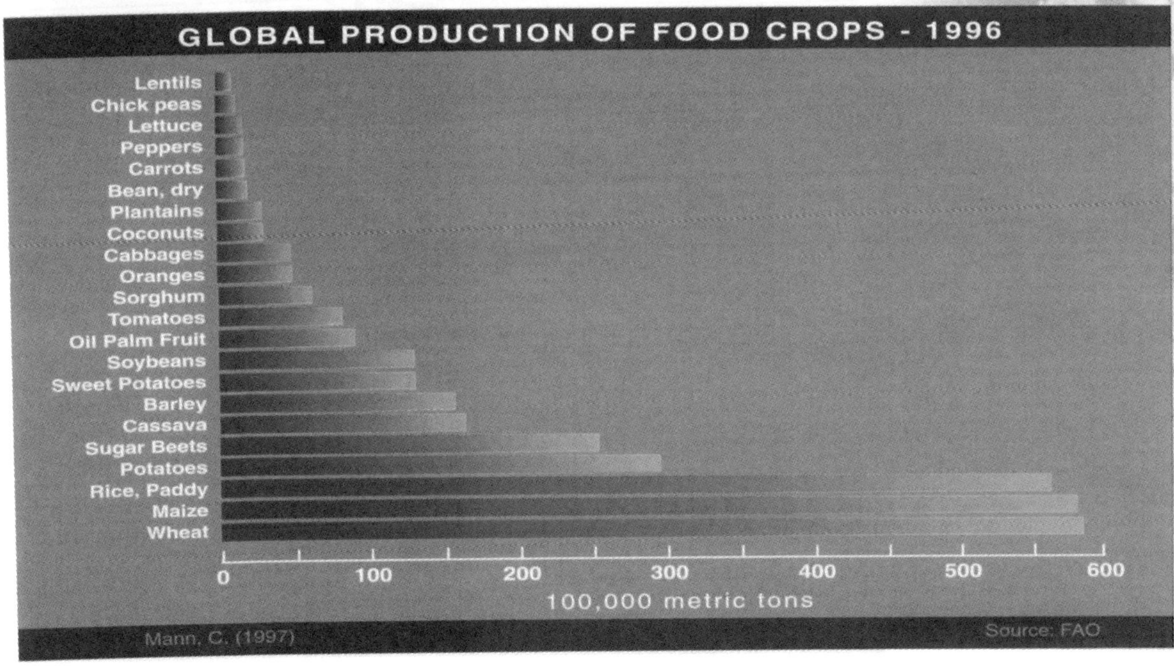

Figure 6. Global Production of Food Crops – 1996

particularly the use of nutrients and water. To cite just one example, cotton yields in India are less than 20% of the yields achieved in several other countries, such as Egypt and the USA, yet Indian farmers use 25 times more water to raise a ton of cotton than farmers in California.

To bridge the gap between actual and potential yields, a multidisciplinary constraints analysis will have to be undertaken in different regions, covering different farming systems. In the short term, the highest priority should go to exploiting the present untapped production potential. In the longer term, the prospects for further improving yield without associated ecological harm will have to be explored.

Lester Brown (1995) has analysed the prospect for increasing food production in China and India, and has concluded that both countries will need to import substantial quantities of food grains by the year 2030. The proposed import needs, about 240 and 40 million tonnes for China and India respectively, exceed the current total world trade in food grains. Therefore, the question is being asked, who will feed China and India by 2030?

China has taken several steps to improve agricultural production, including channelling surplus water from the Yangtze to the north of the country. In the case of India, there is fortunately great opportunity to raise productivity because of the gap between potential and actual yields. This yield gap can be bridged through an integrated package of technology, services and public policies.

Another area which needs attention is enlarging the food basket. There are considerable opportunities for increasing the production of under-utilised or minor crops. With increasing urbanisation, the demand for processed food increases, and there is much scope for including minor crops in the manufacture of processed and semi-processed foods. Farming systems intensification, diversification and value addition are all important for achieving the goal of food for everyone.

The Green Revolution provided a breathing spell, allowing countries to achieve a balance between population growth and food production. However, in the future, the production technologies adopted must be both environmentally and socially sustainable. Achieving sustainable advances in the productivity of major farming systems and the wellbeing of farming families is the pathway towards an Evergreen Revolution in agriculture.

References and further reading

Agarwal A. and Narain S. (1997) *Dying wisdom: rise, fall and potential of India's traditional water harvesting systems.* Faridabad, Thomson Press (India) Ltd.

Brown L.R. (1995) *Who will feed China? Wake-up call for a small planet.* New York, W.W. Norton.

National Research Council, US Board on Science and Technology for International Development (1996) *Lost crops of Africa.* Washington, DC, National Academy Press.

National Research Council, US Board on Science and Technology for International Development (1989) *Lost crops of the Incas: little-known plants of the Andes with promise for worldwide cultivation.* Washington, DC, National Academy Press.

Swaminathan M.S. (1968) *The age of algeny, genetic destruction of yield barriers and agricultural transformation.* Presidential Address, Agricultural Science Section, 55th Indian Science Congress, January 1968. Proceedings Indian Science Congress, Varanasi.

Swaminathan M.S. (ed.) (1996) *Agro-biodiversity and farmers' rights.* Konark Publishers Pvt. Ltd.

Swaminathan M.S. (ed.) (1993) *Wheat revolution: a dialogue.* Madras, Macmillan India Ltd.

Swaminathan M. S. (1999) *I predict: a century of hope. Towards an era of harmony with nature and freedom from hunger.* Madras, East West Books (Madras) Pvt. Ltd.

Professor Swaminathan, FRS, is Chairman of the M. S. Swaminathan Research Foundation, 3rd Cross Street, Institutional Area, Tavamani, Chennai 600 113, India.

Agriculture and Food Systems

by M. S. Swaminathan *(IUCN)*

1. The Changing Scenario

The unfavourable weather of 1972 and 1974 in the middle latitudes and the failure of monsoons in the sub-Saharan countries provided the backdrop to the discussions on the impact of climate on global food security at the first World Climate Conference in 1979. The eighties witnessed both uncomfortable food gluts in some parts of the world and acute food scarcity in others, particularly in the Sahelian region of Africa. According to the State of Food and Agriculture Survey of FAO (1989), the years 1987 and 1988 witnessed a marked turnaround in the world agricultural and food security situations. Some important agricultural commodity markets shifted from having a global surplus to a situation of relative scarcity, and international prices increased significantly, after having fallen to their lowest levels in many years. FAO's March 1990 Food Outlook predicts that the cereal supply situation will remain tight in 1990/1991. Even assuming normal weather, 1990 production is unlikely to be large enough to meet trend consumption in 1990/1991 and allow stock replenishment. With stocks at their lowest level for many years, adverse weather would have serious consequences (Figure 1).

The past 10 years have witnessed great progress in methods of both monsoon forecasting and climate impact assessment. In addition, several basic shifts have occurred in our approach to the analysis of the inter-relationships between climate and food production systems. First, while in the past the focus was on the ***impact of climate on human activity***, the current concern is more on the ***impact of human activity on climate***. Second, while studies on the possible impact of a cooling trend in the world's climate on crops like maize and soybeans attracted interest in the seventies (Thompson, 1975), the interest now is more on interactions among CO_2 concentration in the atmosphere, temperature, precipitation and biological productivity. The feed-back linkages among climate change, crop yields and spatial shifts of crop potential are being investigated in several countries and in several major crops (Parry, Carter and Konijn, 1989). There have been apprehensions about a possible increase in the warming of the El-Niño current, thought to be a major contributory factor to droughts in Brazil, India, Australia and the Sahelian region of Africa during 1982–1983, in case of a rise in world temperature.

Third, with an increase in human population, precipitation patterns are assuming great significance not only for stability of crop production but also for drinking water security for human and animal populations. In spite of a growing awareness of the need to stabilise human population, a global annual population growth of 1.6 to 1.7% is still occurring. This would lead to a net increase of 90 million more people to feed each year. The course of fertility decline during the current decade will largely determine whether the world's ultimate population reaches 10 billion, 15 billion or some other plateau. Demo-

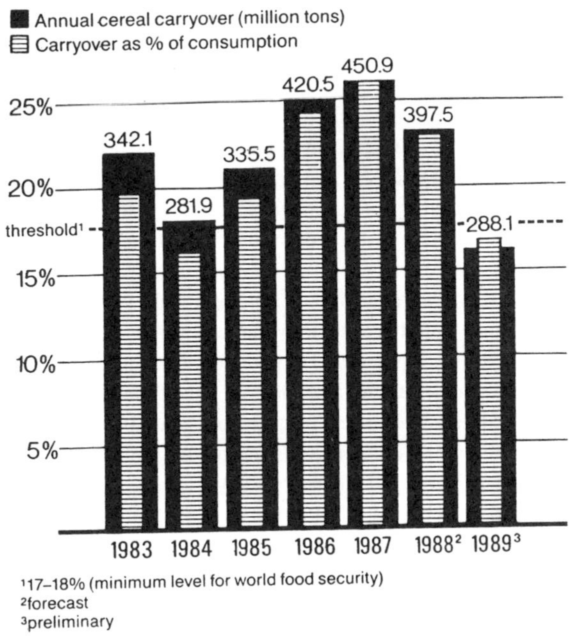

Figure 1: World carry-over of cereal stocks, 1983-1989 (million tons). Source: FAO (1989).

***Table 1**: Changes per person staple food production, imports and consumption in developing countries 1970–1986/1987*

	Production	Imports	Consumption
	(1970–1987) Average annual percent change		
Low-income food-deficit countries	0.81	2.89	1.39
Africa	-1.25	4.82	0.37
Near East	-1.23	4.82	0.78
Far East	1.22	0.75	1.59
Latin America	-0.66	3.53	0.88
Oceania	-0.86	2.71	2.99
Developing countries	0.71	4.32	1.24

* 69 countries with GNP/per caput of up to $940 in 1987, net cereal importers on average during 1983/84–1987/88
Source: Food and Agriculture Organization of the United Nations. AGROSTAT

graphically the industrialised and developing countries will differ not only in the absolute size of the population but also in the age composition. For example, in India about 70% of the population will be below the age of 35 in the year 2000, while in the United States a similar percentage will be above the age of 50. In many developing countries, over 60% of the population will be rural, depending for their livelihood security on crop husbandry, animal husbandry, fisheries, forestry and agro-industries. While at present about 75% of the world's population lives in developing countries, the proportion will increase to 79% in 2000, to 81% in 2010 and to 83% in 2025, according to UN data. FAO (1989) predicts that although agricultural production will continue to increase in most developing countries, there will be little improvement on a per capita basis.

About a half a billion people are believed to be at nutritional risk because they have limited access to a balanced diet. FAO (1989) estimates that the number of people with consumption levels below the critical threshold value is likely to increase from the present 510 million, to 630 million in 2000. A recent report from the World Bank defined poverty as the inability to attain a minimum standard of living (World Bank, 1990). By this standard, it has been estimated that 1.1 billion people are struggling to survive on less than $370/year. About three quarters of these poor, about 800 million people, live in Asia, primarily in Bangladesh, China and India. The largest absolute number of people affected by potentially critical shortage of land is also in Southeast Asia. It is in this context that the further complication likely to arise in relation to national and global food security systems will have to be viewed. Finally, an important development between 1979 and 1990 is the addition of the dimensions of ecological sustainability and equity to the goal of enhancing productivity and profitability in research and development programmes designed to improve major farming systems. Equity is now defined in two time dimensions: (a) intra-generational equity safeguarding the interests of those living today, and (b) inter-generational equity safeguarding the interests of the generations yet to be born. Agricultural scientists are thus now faced with the task of combining ecological sustainability, economic viability and intra- and inter-generational equity in technology development and dissemination.

2. Nature of Food Security Challenge

During the seventies, the major food security challenge was quantitative adequacy or physical access to food supplies. During the eighties also, physical access continued to be an important problem in some years and in some regions. For example, the 1988 harvest worldwide was about 5% less than the 1985 harvest. The Sahelian food famine of the mid-eighties also underlined the precarious food situation in many African countries where ***per capita*** food production was declining steadily (Table 1).

On the other hand, the spread of modern technologies, particularly irrigation, fertilizer and high yielding seeds of cereals helped the densely populated countries of South and South-East Asia to keep growth rates in food production above that of the population. In such countries the food security challenge became one of economic access to food or one of lack of entitlements, to use the terminology of Amartya Sen (1981). Sen (1987) has also emphasized that a public distribution system geared to the needs of the vulnerable sections of the community can bring the essentials of livelihood within easy reach of people whose lives may remain otherwise relatively untouched by the progress of real national income (Table 2 and 3).

Swaminathan (1987) has stressed that while dealing with food systems the following three evolutionary steps should be recognized.

a) ***Food Self-sufficiency*** which implies adequate supplies in the market
b) ***Food Security*** which involves both physical and *economic* access to food, and
c) ***Nutrition Security*** which implies physical and economic access to ***balanced diets and safe drinking water*** for all people at all times.

Many developing countries including India which have experienced a "green revolution" are still in the first stage of this evolutionary process.

Kates et al (1989) have calculated that the biosphere at current production levels could support about 6 billion people, if the diet of all human beings were vegetarian. If 15% of calories were derived from animal products, planet earth could support about 4 billion people. If 25% of calories are derived from animal products, the total population which could be supported would fall to 3 billion. Fish production is the most efficient from the point of view of conversion of plant calories into animal protein. In this context, FAO's (1989) report that the world fish catch levelled off in 1987, after ten years of steady growth is a matter for concern. The world harvest of fish in 1987 was 92.7 million tons, as compared to 92.4 million tons in 1986. El Niño warm currents in the Southeastern Pacific substantially reduced catches of some pelagic species in South America. In contrast to this trend, consumer demand for fish is likely to increase by 28.4 million tons by the year 2000. Therefore, an integrated approach to coastal and inland aquaculture and capture fisheries will be important.

3. Challenges Ahead

Sustainable advances in biological productivity are essential for meeting the needs of both enhanced agricultural production and greater agricultural diversification. For this purpose we need new agricultural technologies capable of raising the population-carrying capacity of land and water. Such techniques will have to be tailored to the following major land use systems:

a) ***Mountain ecosystems***, where damage upstream would have serious repercussions on downstream agriculture (e.g. Himalayas and the Indo-Gangetic agricultural region)
b) ***Coastal ecosystems***, where it will be necessary to promote the integrated management of land and sea surfaces,
c) ***Sustainable intensification areas*** (could also be referred to as "green revolution" areas) which, with appropriate support from maintenance and anticipatory research, can sustain intensive output of crops and livestock at high and rising levels of productivity
d) ***Semi-arid areas***, where inadequate or unreliable rainfall coupled with over-exploitation leads to chronic land degradation
e) ***Arid areas***, belonging to both the hot and cold desert categories, where sylvi-pastoral and sylvi-horticultural systems of land use are ideal, and
f) ***Island ecosystems*** where changes in sea-levels could have important implications.

Unsustainable management of the resource and environmental systems associated with such regions is having serious repercussions, as is evident from the following visible signs in many developing countries.

- Precipitous drying up of drinking water resources
- Vanishing forests, flora and fauna
- Intensifying drought and floods
- Loss of grazing lands and growing degradation of land
- Deterioration of the quality of air and water
- Explosive growth of rural and urban unemployment
- Mushrooming of urban slums

It is the poor and the marginalised who suffer most from such environmental breakdown.

The nexus among people, resources, environment, technology and agricultural development is thus a closed one. FAO's estimates of total harvested land of different potentials is given in Figure 2. FAO's study, World Agriculture: Toward 2000, concluded from an analysis of 93 developing countries, excluding China, that nearly 60% of harvested land in 1982–84 belonged to the high potential category. Unirrigated arid and semi-arid areas of developing countries accounted in 1983–88 for only about 9% of total cereal production and 6% of root and tuber production. Thus for achieving the production goals of the future, it will be essential to maintain and further enhance the production potential of "sustainable intensification areas" and upgrade degraded land. In this context, it will be useful to consider briefly the present state of global land and water resources management.

i) ***Land***: It is estimated that 30 to 50% of the earth's land area is degraded due to improper management. In particular, both the conversion of forest land for agriculture and exploitative agricultural practices

***Table 2**: Opulence, Life and Death*

	GNP Per Head (Dollars) 1985	Life Expectancy at Birth 1985	Under-5 Mortality Rate (Per Thousand) 1985
Oman	6,730	54	172
South Africa	2,010	55	104
Brazil	1,640	65	91
Sri Lanka	380	70	50
China	310	69	48

Source: World Development Report, 1987; The State of the World's Children, 1987

***Table 3**: Mortality rates of children under the age of 5 (U5MRs) per 1,000 live births, 1987*

Per capita GNP (US$) 1986	Lowest U5MRs		Highest U5MRs	
300–400	China	45	Afghanistan	304
	Sri Lanka	45	Sudan	184
			Haiti	174
401–600	Guyana	39	Angola	288
			Yemen Democratic Republic	202
			Yemen	195
			Ghana	149
601–800	Dominican Republic	84	Nigeria	177
	Papua New Guinea	85	Bolivia	176
			Côte d'Ivoire	145
801–1000	Jamaica	23	Cameroon	156
			Botswana	95
1000+	Hong Kong	10	Gabon	172
	Singapore	12	Peru	126
	Costa Rica	23	Saudi Arabia	102
	Trinidad and Tobago	24		
	Chile	26		
	Mauritius	30		

Source: Based on United Nations Children's Fund. State of the World's Children 1989

have led to an increase in soil erosion over the past 25 years. The rate of soil erosion is almost imperceptible (1mm of soil loss in a storm amounts to 15 tonnes per hectare) and significantly exceeds its floor renewal rate (2.5 cm/500 years and at best 1 tonne per hectare per year). The rate of soil erosion in temperate countries is 10 to 20 times the soil renewal rate, while in the tropics it is almost 20 to 40 times.

FAO's estimates suggest that approximately 6 million ha/year are becoming unfit for agriculture. In some areas, the productivity of eroded soils cannot be restored even at enormous costs (equivalent to the application of 2000 tonnes of quality soil per hectare, or 50 tonnes dry rotted cattle manure per hectare).

Soil loss also leads to nutrient depletion. One tonne of good agricultural soil may contain a total of

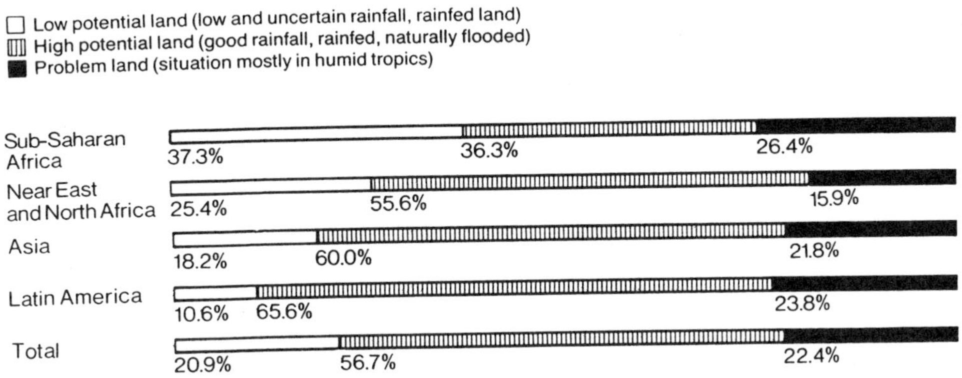

Figure 2: Shares of total harvested land of different potentials, 1982-1984 in 93 developing countries. Source: FAO, World Agriculture: Toward 2000.

***Table 4**: Current Status of Threatened Species*

	EX	E	V	R	I	Globally Threatened Taxa
Plants	384	3325	3022	6749	5598	 19,078
Fish	23	81	135	83	21	 343
Amphibians	2	9	9	20	10	 50
Reptiles	21	37	39	41	32	 170
Invertebrates	98	221	234	188	614	 1,355
Birds	113	111	67	122	624	 1,037
Mammals	83	172	141	37	64	 497

Key: EX = Extinct (post–1960), E = Endangered, V = Vulnerable, R = Rare, I = Indeterminate
Source: Reid and Miller, 1989; WCMC, unpublished data, Jan. 1989

4 kg. of nitrogen, 1 kg of phosphorous, 20 kg of potassium and 2 kg of calcium. Further, soil erosion results in a loss of organic matter which plays a pivotal role in improving infiltration, water retention, soil structure and cation exchange capacity.

Organic matter and soil micro-flora and micro-fauna (including earthworms) are interdependent in maintaining soil quality and in promoting recycling of nutrients and degradation of wastes.

ii) ***Water***: The World Commission on Environment and Development in its report, "Our Common Future" (1987) has drawn attention to the serious state of global water resources.

Global water use has doubled between 1940 and 1980 and it is expected to double again by 2000, with two thirds of the projected water use going to agriculture (WCED, 1987). Yet 80 countries, with 40 percent of the world's population, experience serious water shortage even now. There will be growing competition for water for irrigation, industry and domestic use. River water disputes will multiply between nations and within nations. No easy solution is in sight unless solar desalination of sea water becomes an economic proposition.

Concurrently steps to increase national efforts in saving and sharing water in the rainfed areas are important. The most serious component of potential changes in climate is precipitation. Efforts in enhancing the efficiency of water harvesting and on-farm water management are therefore essential.

In areas of intensive agriculture, problems of salinity, sodicity and water logging as well as the incidence of malaria, schistosomiasis and other waterborne diseases are becoming important. Ground water resources are being adversely affected qualitatively by the excessive use of mineral fertilizers and pesticides. Most countries are yet to develop policies for regulating ground water use in accordance with the recharge capacity of the aquifer.

Figure 3: Population experiencing a fuelwood deficit, 1980 and 2000 (millions). Source: FAO. 1983.

1980	Latin America	Africa	Near East and North Africa	Asia and the Pacific	Total	
Total population	26	55		31	112	Acute
Rural population	18	49		29	96	Scarcity
Total population	201	146	104	832	1283	
Rural population	143	131	69	710	1052	Deficit
2000						
Total population	512	535	268	1671	2986	Acute scarcity
Rural population	342	464	158	1434	2398	or deficit

Source: FAO, Fuelwood Supplies in Developing Countries, 1983

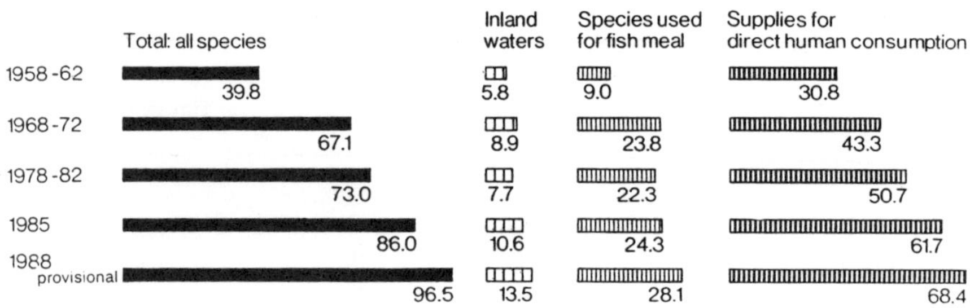

Figure 4: Annual landings (million tons) of aquatic resources (excluding mammals and seaweeds).

iii) ***Living aquatic resources***: The world catch of marine living resources is now growing only 1 to 2 percent annually and is approaching its estimated maximum sustainable yield of 100 million tonnes (Figure 4). Since as much carbon is fixed in the sea as on land, it is important that coastal zones are subjected to an ocean capability analysis, on the model of land classification systems currently in use.

iv) ***Biological diversity***: Biological diversity is the foundation upon which the edifice of sustainable advances in biological productivity can be built. Recent advances in molecular biology and genetic engineering, which render the transfer of genes across sexual barriers possible, have further enhanced the economic and ecological value of global biological wealth. Unfortunately, serious losses are now occurring at all the three levels in which biological diversity manifests itself, namely intra-species and ecosystem levels, due largely to the destruction of habitats rich in genetic resources, such as tropical rain forests. Recent estimates put the rate of deforestation in the tropics at 17 million ha or 1% per annum (TFAP, 1990). The disappearance of forests also reduces the extent of carbon absorption on the earth. The carbon emission-absorption balance is thus upset.

Many plant and animal species are now under threat of extinction (Table 4). Some scientists predict that if present trends in habitat destruction continue, at least 25 percent of the world's species will be lost in the next several decades. Such a loss of bio-diversity has profound implications for development. Biological resources are renewable; forests, fisheries, wildlife and crops reproduce themselves and even increase when managed properly. Further, the highly diverse natural ecosystems which support this wealth of species also maintain hydrological cycles, regulate climate, build soils, cycle essential nutrients, absorb and break down pollutants and provide for recreation, research and richer quality of life.

Marine protected areas are yet to receive the same attention as their counterparts on land. The area of sea and seabed is more than two and a half times as great as the total area of land masses of the world but less than one percent of that marine area is currently within established protected areas. This compares with about 3 percent of area which is protected in the terrestrial environment. Conservation of biological diversity under wetlands and aquatic conditions should receive greater support.

Conserving biological diversity, therefore, is urgent because diversity provides the raw material for human communities to adapt to change. The loss of each additional gene, species or habitat reduces the available options.

Loss of the biological potential of the soil, adverse changes in water availability and quality, increasing biotic and abiotic stresses and the biological impoverishment of the earth are all occurring at a time when on the one hand, human population is expanding and on the other, the pathways of economic and industrial development chosen so far have built-in potential for climatic alterations.

4. Energy and Fuelwood

FAO's (1989) estimate of fuelwood shortages is given in Figure 3. In India, it has been calculated that a minimum of 1.4 percent growth rate in energy availability will be needed for achieving a 1% growth rate in GNP. However, the energy-mix made available is itself often the cause of both short-term and long-term environmental damage. The deficit of firewood is growing day by day. At the present level of consumption of forest produce and on the current productivity of forests, India needs a minimum of 0.47 ha of forest land for every individual. The existing forest area on this basis would be adequate only for a population of 150 million (in contrast to the present population of 850 million). The task of improving the productivity of forests and mobilising alternative sources of energy is thus urgent. A similar situation obtains with regard to fodder for animals. The population of domesticated animals is increasing, while the area under grasslands and pastures is shrinking. Thus, we will need land-saving agricultural technologies, grain-saving animal rearing methods and energy and cost saving methods of enhancing biological productivity, if we are to face the challenges on the food production and distribution front.

5. Challenges on the Economic Scene

The financial and technical resources needed to promote ecologically sustainable agricultural and food production systems and to meet possible future changes in temperature, precipitation and sea levels are enormous. For example, the participants of the Keystone Dialogue on Plant Genetic Resources held at Madras, India, in January 1990, concluded that a minimum of US $500 million per year of new money will be needed to undertake the tasks essential for conserving for posterity a sample of the genetic variability existing in crop plants through *ex situ* conservation techniques. The tasks associated with sustainable agriculture such as upgrading degraded land, conserving water and *in situ* protection of biological diversity need considerable additional resources. It is hence unfortunate that there is today a net outflow of resources from developing to developed countries. (WCED, 1987). For developing countries as a whole, external debt increased 4 percent in real terms in 1987, reaching US $1,218,000 million by the end of the year. High debt-servicing payments coupled with the low level of commercial lending and new investment, resulted in growing net transfers of resources from the poor nations to the rich (by World Bank estimates, no less than US $43,000 million in 1988, compared to US $38,100 million in 1987).

International agricultural trade is in disarray. The policies adopted by industrialised countries with regard to trade and domestic subsidies have contributed to surplus production and subsidized exports of agricultural products. Developing countries, already burdened by the debt crisis, fluctuations in exchange rates, economic recession and volatile oil and commodity markets, are also experiencing a deterioration of terms of trade.

Even the slogan "trade and not aid" is losing its meaning. Trade barriers are growing. The scope of intellectual property rights is expanding, while there is little effort to give economic recognition to the informal innovation system which is the very foundation of agricultural evolution and of the conservation of biological diversity. The GATT-TRIPS (Trade related intellectual property rights) negotiations should give serious attention to methods of recognising informal innovations (Keystone Dialogue, 1990).

The concept of "dependence" now being introduced in the revised draft of the UPOV (International Union for the Protection of New Varieties of Plants) should recognise the dependence not only on varieties covered by the PBR (Plant Breeding Rights) System but also on land races and the genetic material arising from informal innovation. A better common future for all will not be possible without a better common present.

6. Implication of Potential Changes in Climate

The impacts on agriculture could be of two major types. First, by altering production adversely in the main food-producing areas, climate change could enhance food scarcities. The location of main food-producing regions could change. Second, there could be profound impacts on the physiological mechanisms regulating plant and animal productivity. The greatest impact is likely to come from changes in precipitation patterns.

Spatial Impact

North America continues to be the principal grain surplus area today. The heavy dependence on North America for world grain reserves has increased the sensitivity of world food supply to the weather and climate of that region. If unfavourable growing conditions occur simultaneously in the major mid-latitude regions of North America, the USSR and Australia, the global food security system will be under severe stress. Fortunately, detailed studies are now in progress on the potential effects of global climate change on US agriculture.

Oram (1985) has drawn attention to four vulnerable producer groups who may be affected severely by adverse changes in temperature and precipitation. The first is located in humid tropics, in lowland areas of Asia and in the Pacific and Caribbean. These areas, normally prone to excessive rains and flooding, may be less severely affected by climatic change. The second group located in the arid and semi-arid areas of the tropics in Africa and South Asia and in the Mediterranean climate of West Asia and North Africa will be extremely vulnerable. A third group comprising farmers at high altitudes may experience both favourable and unfavourable effects. The fourth group consisting of farmers located at the cold margins at higher latitudes may also experience diverse effects. Parry, Porter and Carter (1990) feel that overall levels of production can be maintained through a combination of shifts of agricultural zones and adjustments in technology and management. They have identified the countries in the lower middle and lower latitudes of Africa, South America and elsewhere as the areas most at risk from the effects of climate change.

During the last 10 years, special attention has been given to the agriculture of countries in sub-Saharan Africa. This region is ecologically diverse and covers an area of 22,245 sq. kms. An estimated 30% of the area can sustain production of rainfed crops. The World Bank in a long-term perspective study titled "Sub-Saharan Africa: From Crisis to Sustainable Growth" (World Bank, 1990) has proposed a series of measures which can enhance productivity and reduce vulnerability to ecological and economic factors. An important recommendation related to arresting soil erosion. Soil erosion, widespread in all areas of sub-Saharan Africa, is perhaps most serious in Ethiopia, where top soil losses of up to 290 metric tons a hectare have been reported for steep slopes. The report lays specific stress on developing farmer's associations and recognising the role of women.

Physiological Impact

The Goddard Institute of Space Studies (GISS) and the University of Birmingham, UK, initiated in 1989 a three-year study of the impact of climate change on global agricultural output and food trade. To date, only three comprehensive regional or national assessments of the consequences of climate change for agriculture have been completed. The International Institute for Applied Systems Analysis has conducted several case studies with support from the United Nations Environment Programme (Parry *et al.* 1989).

Generally, it is assumed that increased atmospheric CO_2 would enhance growth rates of certain types of crop plants and that changes in temperature and precipitation would affect livestock, crops, pests and soils. They will also affect ground water replenishment patterns and evapo-transpiration rates. Normally, increased CO_2 in the atmosphere can help to increase the rate of photosynthesis, if water and nutrients do not become limiting factors. C3 and C4 plants (i.e. those which have a 3-carbon or 4-carbon path for photosynthesis) respond differently. C3 crops like wheat, barley, rice and potatoes could respond positively to CO_2 enrichment. For wheat and barley, yield increases of as much as 40% have been suggested (Cure, 1985).

Sinha and Swaminathan (1990) examined the integrated impact of a rise in temperature and in CO_2 concentration on the yield of rice and wheat in India. The study showed that for rice, increasing mean daily temperature decreases the period from transplantation to maturity. Such a reduction in duration is often accompanied by decreasing crop yield. There are, however, genotypic differences in per-day yield potential. Breeders can consciously select strains with a high per-day productivity. Increasing levels of CO_2 increase photosynthetic rate and hence dry matter production but an increase in temperature reduces crop duration and thereby yield.

In the case of wheat, there will be an adverse impact on yield if mean temperatures rise by 1 to 2°C. For each 0.5°C increase in temperature there would be a reduction of crop duration of 7 days, which in turn would reduce yield by 0.45 t per ha. According to Parry *et al* (1990), a 1°C increase in mean annual temperature would tend to advance the thermal limit of cereal cropping in mid-latitude northern hemisphere regions by about 150–200 km, and to raise the altitudinal limit to arable agriculture by about 150–200 m.

For India as a whole, rice may become even more important than now in the national food security system, since rice can give high yields under a wider range of growing conditions than wheat. International Testing Networks like those operated by the International Rice Research Institute (IRRI) and the International Centre for Maize and Wheat Research (CIMMYT) provide scope for identifying genotypes suitable for diverse growing conditions (Table 5).

Warrick (1988) has investigated the sensitivity of crop yields to changes in climate variables using several approaches - crop impact analysis, marginal-spatial analysis and agricultural systems analysis. Such investigations suggest that for the core mid-latitude cereal

Table 5: Rice - Varietal Response to Climatic Stresses in IRTP Trials

Drought tolerance	
• From rainfed lowland:	IR 13240–39–3–3–P1 IR19274–26–2–3–1 IR21015–80–3–3–1–2
• From upland nurseries:	ADT31 CR222–MW10 CR289–1208 IR9729–76–3
Submergence tolerance	FR13A IR26933–16–2–3 IR31338–30–2–1 IR31429–14–2–3 IR31432–8–6
Low temperature tolerance	JKAU9KO450–126–10 No. 11 (Takanenishiki) Suweon 303, Akiyudaka

Source: International Rice Testing Program (IRTP).
IRRI Annual Report, 1988

regions, an average warming of 2°C may decrease potential yields by 3 to 17 percent.

It is obvious that more studies are needed at the micro-level in order to understand fully the inter-relationships among CO_2 concentration, temperature and precipitation with reference to their impact on crop yields. The position relating to world food security was characterised by FAO's Committee on World Food Security, at its Session held in Rome from 26–30 March 1990, as "critically dependent on the outcome of the 1990 crops". This underlines the urgency of intensifying on-going research on crop-weather interaction.

Implications of Sea-level Rise

The magnitude of a likely rise in sea level is not yet clear, because of inadequate knowledge of the processes of ice-melt and thermal expansion of ocean water. The definition of sea-level at shoreline positions also presents difficulties, as a result of variations arising from tectonic or isostatic movement of the land mass. Several projections indicate a rate of sea-level rise of 0.5 cm per year or 50 cm per 100 years over the next half century.

Rise of mean sea-level has an immediate and direct effect on the ecosystems of the intertidal zone. It is likely that species with specific tolerances within the tidal spectrum will migrate landwards. Several studies have been conducted on the potential effect of sea-level rise on mangrove swamps (Ellison, 1989). Mangroves generally withstand tropical storms. However, if the intensity and frequency of tropical storms increase as a result of global warming, mangrove ecosystems will be under greater stress.

Coral reefs, sea grasses and mangroves are being destroyed in the coastal areas of several countries due to expansion of tourism, coastal aquaculture, industrial exploitation and pollution. It will be difficult to adapt to new situations in the future if the existing genetic variability in coastal fauna and flora is lost. Hence efforts to conserve the mangrove genetic resources are needed.

7. Consolidating Gains and Adapting to Change

Improving Consumption

Considering the delicate balance between current food production and consumption, every effort will have to be made to improve the productivity, profitability, stability and sustainability of major farming systems. This will have to be done through integrated packages of technology, services and public policies. Agriculture is by and large a location-specific vocation and hence these packages will have to be tailored to match the prevailing agro-ecological, socio-economic and socio-cultural situations.

An area of widespread concern is the persistence of chronic hunger in many countries of the world, in spite of progress in improving agricultural production. The twin deficiencies of past developmental efforts are, first, failure to eliminate chronic hunger (over 600 million children, women and men still go to bed hungry each day) and secondly, failure to arrest the degradation of the environmental capital stocks upon which sustained advances in productivity improvement depend. Most often, hunger arises from inadequate purchasing power. This in turn is due to lack of opportunities for gainful employment.

Experience from employment guarantee programmes organized in India and elsewhere, as well as from the programmes sponsored by the World Food Programme, have shown that such intervention methods are very helpful in mitigating chronic hunger. What is needed urgently is a Social Security Programme for the poorest billion of the human population consisting of the following three components:

a) **Work**: Employment guarantee for ecological security, which integrates food for work and similar programmes with measures to strengthen the ecological foundations of a productive agricultural system
b) **Wage**: Enforcement of minimum wages, with special attention to women workers, since when women have independent access to income, child nutrition is better, and

c) **Welfare**: Steps to ensure that all children have access to better nutrition, health care and education.

A Social Security system for the economically underprivileged sections of the human family involving concurrent attention to work, wage and welfare will help to eliminate chronic hunger, reduce infant mortality and create conditions favourable for the stabilisation of human population and promotion of ecological security. This will call for a bold and imaginative initiative on the part of the affluent nations, since development which is not equitable will not in the long run be sustainable. The World Summit for Children held at the United Nations issued on September 30, 1990, a charter for the very young. If implemented by all nations, the global capacity to meet disasters arising from future changes in climate would be greatly enhanced owing to the availability of the necessary social infrastructure for adaptation to change.

Improving Production

During the decade of the eighties, there has been rapid progress in several areas of science and technology as applied to the improvement of terrestrial and aquatic farming systems. Advances in biotechnology and micro-electronics have enhanced our agricultural capability and improved our early warning and monitoring systems. The new technologies, if integrated in an appropriate manner with traditional technologies, can help to improve the efficiency of management of land and water, energy and post-harvest operations. Improved post-harvest technologies can not only help to save what is produced but also to prepare value-added products from every part of the plant and animal biomass. Biomass refineries can be established in rural areas. This will help to provide more jobs in the off-farm sector. What is needed is the stepping up of agricultural research and extension efforts in developing countries through the establishment of **Technology Blending Centres** for deriving maximum benefit from existing and emerging technologies.

Adapting to Change

Every effort should be made to arrest deforestation and to promote upgrading of degraded lands through agro-forestry and other appropriate forms of land-use. This will help to increase carbon fixation on the earth. Along with efforts to prevent the continued increase of CO_2 in the atmosphere, it is necessary to initiate anticipatory measures which can help us to enhance our ability to adapt to new climatic patterns. The following are some of the steps needed urgently.

a) Establishment of Crop-Weather Watch Groups

Special groups consisting of meteorologists, agricultural research and extension workers, developmental administrators and mass media representatives should be established in every agro-ecological region of a country to continuously monitor the weather situation and analyse its implications for crop growth and pest incidence. Such **crop-weather watch groups** can benefit at the national level from the World Climate Programme of WMO and the early warning systems of FAO. This will help to equip farmers and fisher families with location-specific information to increase their preparedness for floods, tropical cyclones and drought. Mass media and computer-aided extension techniques can help to take the latest information derived from weather satellites to every village.

b) Contingency Planning

Computer simulation models facilitate the preparation of contingency plans to meet different weather probabilities. Alternative land and water use plans can be prepared to adapt to different precipitation patterns. For example, if it is too late to sow a traditional late duration **Sorghum**, an early maturing pearl millet or grain legume crop can be taken. For implementing such contingency plans at the field level, reserves of seeds of the alternative crops will be needed. Fortunately, modern plant breeding has helped to incorporate genes for photo- and thermo-insensitivity in many crop varieties. Thus there is built-in resilience concerning sowing dates. It will be essential that the crop-weather watch groups identify suitable alternative cropping patterns. Just as grain reserves are important for food security, seed reserves are essential for the security of crop production.

c) Ocean Capability Analysis

Over 40% of the global population live in coastal areas. We should initiate in each country an ocean capability analysis through an integrated coastal systems research program (CSR) involving concurrent attention to capture and culture fisheries and coastal forestry and agroforestry. While land capability classifications exist we have yet to initiate similar studies in the coastal zones. We must derive greater benefit from the enormous amount of carbon fixed in the oceans.

d) Genetic Resource Centres for Adaptation to Climate Changes

Gene pools occur in nature for adaptation to drought, floods and sea level changes. Unfortunately, gene erosion, through a variety of causes, is leading to the loss of valuable genetic material. Every effort should be made to establish specialised genetic resources centres for collecting and conserving species and genotypes with the desirable genes. For example, adaptation to sea level rise will be facilitated by the conservation of mangrove species, sea grasses and deep-water or floating rices. A centre for the conservation of genetic resources of plants for adaptation to sea level rise is being established in the State of Tamil Nadu in India. The Centre for Research on Sustainable Agricultural and Rural Development, Madras and the International Tropical Timber Organization (ITTO) are organising in Madras in January 1991 an International Workshop for organising a global network of genetic resource centres for adaptation to climate change. These centres will identify and maintain candidate genes for use in recombinant DNA experiments.

e) Establishment of Genetic Enhancement Centres

Tools of molecular biology and recombinant DNA techniques make the transfer of genes across sexual barriers possible. For example, theoretically the capacity to withstand sea-water conditions can be transferred from mangroves to other species. The "elongation" gene from rice which enables it to grow to the needed height for remaining above flood water levels, can be transferred to other crops. To achieve such goals, it will be necessary to establish **Genetic Enhancement Centres** consisting of experts in molecular biology and genetic engineering. An integrated global grid of Genetic Resources Conservation Centres and Genetic Enhancement Centres will be of immense help in the breeding of crop varieties possessing greater tolerance to drought, floods and sea water intrusions.

f) National Studies

Adams *et al* (1990) have undertaken a detailed analysis of the likely impact of global climate change on the agriculture of the United States of America using the NASA/Goddard Institute of Space Studies (GISS) and the Princeton Geophysical Fluid Dynamics Laboratory (GFDL) models. They found that climate change could lead to a slight reduction in total US cropped acreage. For GISS scenarios, aggregate acreage is reduced due to increased yields. For the GFDL scenarios, acreage is reduced owing to shifts in production between regions and associated changes in resource availability. Because of the possible change in domestic and foreign production assuming the GFDL climate scenario, the role of the USA in agricultural exports may change. Also, patterns of agriculture in the USA are likely to shift as a result of changes in regional crop yields and in crop irrigation requirements. Such potential changes have implications both for public policy and research priorities. **It will be desirable that as a follow-up of the Second World Climate Conference, an International Research Network on Climate Change and Food Security** is organised to conduct at critical locations around the world studies of the kind undertaken by Adams *et al* (1990) in the United States.

g) Specific Measures for Reducing Ggriculture's Contribution to the Build-up of Greenhouse Gases

Reilly and Bucklin (1989) have estimated the contribution of the farm sector to increases in carbon dioxide, methane, nitrous oxide and other greenhouse gases in the atmosphere (Figure 5). It would be useful to standardize methods of reducing the contribution of agriculture to greenhouse gas accumulation in the

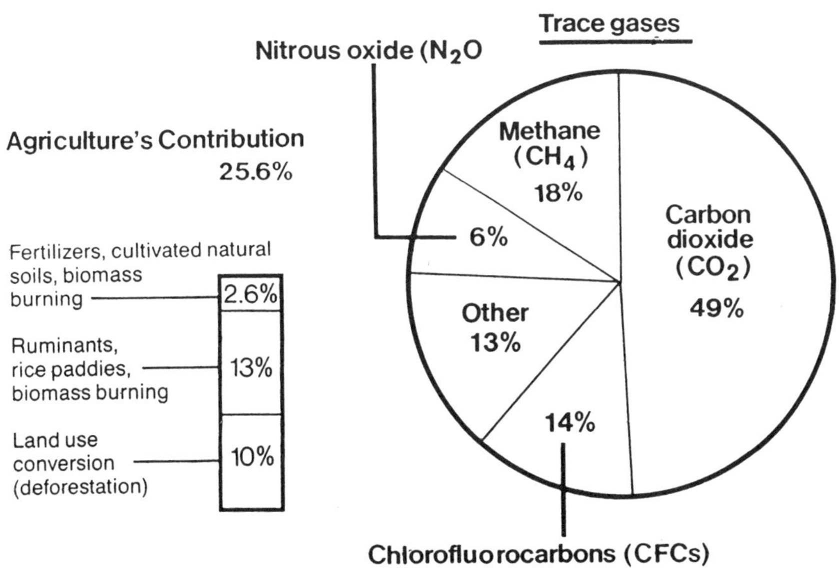

Figure 5: Contributions to increased radiative forcing in the 1990's. Source: Reilly and Bucklin (1989).

atmosphere. Some of these measures are indicated below.

i) CO_2 emissions
Developing countries contribute 10 to 20% of global CO_2 emissions due to deforestation. This contribution can be reduced or eliminated through the restoration of damaged ecosystems and through extensive afforestation.

ii) Methane
Irrigated rice fields, swamps and bogs are believed to be important sources of methane. There is a need for more extensive studies of methane fluxes in rice growing countries. Microbiological and irrigation management methods can help to reduce the release of methane. Methane from the decomposition of organic refuse can be captured and utilized in biogas plants.

iii) Nitrous oxide
Nitrogenous fertilizer as well as the exposure of soils after deforestation could contribute to the emission of nitrous oxide. We need more research on the agricultural sources of nitrous oxide and of methods to reduce such emissions. At the same time, research on the use of biofertilizers, green manure crops and other substitutes for mineral fertilizers needs to be stepped up.

iv) Chlorofluorocarbons (CFC)
Besides contributing to global heating, CFCs also deplete the stratospheric ozone layer which shields the earth from ultraviolet rays. In tropical and sub-tropical areas, more facilities for refrigerated food storage are essential for reducing the spoilage of perishable commodities like meat, milk, fish, vegetables and fruits. There is therefore need for the intensification of research on alternative methods of refrigerated storage of food products, which will not damage the ozone layer. Solar refrigeration methods will be particularly valuable in the tropics, provided they become more economical and efficient.

Every country should develop an integrated package of greenhouse gas free agricultural technologies which will minimise and, where possible, eliminate the contributions of farm operations to the increase of CO_2, CH_4 and NO_2 in the atmosphere and find substitutes for CFC based refrigeration methods.

Conclusions

National and global food and nutrition security systems can be insulated from potential adverse impacts of alterations in temperature, precipitation, sea levels and incidence of UV radiation through well planned anticipatory and participatory research and development programs. This will call for mutually reinforcing packages of technologies, services and public policies. The organization of an International Research Network on Climate Change and Food Security would help to promote the needed degree of multi-disciplinary and multi-institutional cooperation. The speedy development and effective dissemination of avoidance and adaptation strategies will then become possible.

The major components of such a global research network which can be jointly sponsored by WMO, UNEP, FAO and CGIAR should be the following.

a) Estimation and monitoring of the emission of greenhouse gases as a result of farm operations under different agro-ecological and technological conditions
b) Standardisation of techniques both for minimising agriculture's contributions to the accumulation of greenhouse gases and for withstanding additional biotic and abiotic stresses on crops and farm animals arising from climate change
c) Promotion of research and training in the field of restoration ecology and in the afforestation of degraded forests and organisation of demonstration and training programmes for spreading sustainable management procedures for tropical rain forests
d) Anticipatory research and development measures in coastal areas to avoid or minimize the adverse impact of potential changes in sea levels
e) Standardisation of methods for deriving benefit from higher CO_2 concentration in the stratosphere
f) Standardisation of post-harvest technologies for perishable commodities like vegetables, fruits and animal products based on non-CFC dependent refrigeration methods, and
g) Stimulating policy research designed to strengthen the public policy back-up for both avoidance and adaptation.

References

Adams, R.M., Rosenzwerg, C., Pearl, R.M., Ritchie, J.J., McCarl, B.A., Olyer, D.J., Curry, B.R., Jones, J.W., Boote, K.J., and Allen, H.L., 1990. Global Climate change and US Agriculture. Nature Vol. 345: 219–224 pp.

Cure K.D., 1985. Carbon dioxide doubling response: a crop survey. In: Strain, B.R. and Cure, J.D. (Eds.) Direct effects of increasing carbon dioxide on vegetation. US Department of Energy, Washington. 19–116 pp.

Ellison, Joanna C., 1989. The Effect of Sea-Level Rise on Mangrove Swamps. Review for the Commonwealth Secretariat Expert Group on Climatic Change and Sea-Level Rise. 22 pp.

FAO, 1989. The State of Food and Agriculture Food and Agriculture Organization of the United Nations, Rome. 171 pp.

Human Development Report, 1990. United Nations Development Programme and Oxford University Press, New York, 189 pp.

IRRI, 1988: International Rice Research Institute, Annual Report, Los Banos, Philippines, 646 pp.

Kates, R.W., Chen, R.S., Downing, T.E., Kasperson, J.X., Messer, E. and Millman, S.R., 1989. The Hunger Report: Updated 1989. The Alan Shawn Feinstein World Hunger Program, Brown University, Providence, R.I., USA.

Keystone Dialogue, 1990. International Dialogue on Plant Genetic Resources. Proceedings of Madras Plenary Session, Keystone Center, Colorado, USA.

Oram, P.A., 1985. Sensitivity of Agricultural Production to climatic Change. Climatic Change 7: 129–152 pp.

Parry, M.L., Carter, T. and Konijn, N. (Editors), 1989. The Impact of Climatic Variations on Agriculture. Volume 1—Assessments in Cool Temperate and Cold Regions. Volume 2—Assessments in Semi-Arid Regions. Kluwer Academic Publications.

Parry, M.L., Porter, J.H. and T.R. Carter, 1990. Climatic change and its implications for agriculture. Outlook on Agriculture 19: 9–15.

Reilly, J. and Bucklin, R., 1989. Climate Change and Agriculture. World Agriculture Situation and Outlook Report. Washington, D.C.: USDA/ERS, WAS-55.

Reid, W., and K. Miller, 1989: Programme for the Conservation of biodiversity. World Resources Institute, Washington, D.C.

Sen. Amartya, 1981. Poverty and Famines. Oxford, Clarendon Press, 79 pp.

Sen. Amartya, 1987. Food and Freedom. Sir John Crawford Memorial Lecture, Washington, D.C., 25 pp.

Sinha, S.K. and Swaminathan, M.S., 1990. Climate Change and Food Production. Climate Change (under publication).

Swaminathan, M.S., 1987. Building a National Nutrition Security System. In Global Aspects of Food Production, Tycooly Press.

TFAP (1990). Taking Stock: The Tropical Forestry Action Plan after five years. World Resources Institute. Washington, D.C., USA.

Thompson, L.M., 1975. Weather Variability, Climatic Change and Grain Production. Science 188: 535–541 pp.

Warrick, R.A., 1988. Carbon Dioxide, Climatic Change and Agriculture. The Geographical Journal 154: 221–233 pp.

World Bank, 1990. Sub-Saharan Africa. From Crisis to Sustainable Growth. The World Bank, Washington, D.C., 300 pp.

World Commission on Environment and Development. Our Common Future, 1987. Oxford University Press. 400 pp.

World Development Report, 1987. The State of the World's Children. World Bank.

World Development Report: Poverty, 1990. Published for the World Bank by Oxford University Press. 260 pp.

World Food Council, 1989. The Cyprus Initiative against Hunger in the World. 23 pp. (Mimeographed).

World Resources (1990–91)—1990 A Guide to the Global Environment, Oxford University Press, Oxford & New York. 388 pp.

PHILOSOPHICAL TRANSACTIONS OF THE ROYAL SOCIETY A | MATHEMATICAL, PHYSICAL & ENGINEERING SCIENCES

Managing extreme natural disasters in coastal areas

P.C Kesavan and M.S Swaminathan

Phil. Trans. R. Soc. A 2006 **364**, 2191-2216
doi: 10.1098/rsta.2006.1822

PHILOSOPHICAL TRANSACTIONS OF THE ROYAL SOCIETY A

Phil. Trans. R. Soc. A (2006) **364**, 2191–2216
doi:10.1098/rsta.2006.1822
Published online 27 June 2006

Managing extreme natural disasters in coastal areas

BY P. C. KESAVAN* AND M. S. SWAMINATHAN

M.S. Swaminathan Research Foundation (MSSRF), Third Cross Street, Taramani Institutional Area, Taramani, Chennai 600 113, India

Extreme natural hazards, particularly the hydro-meteorological disasters, are emerging as a cause of major concern in the coastal regions of India and a few other developing countries. These have become more frequent in the recent past, and are taking a heavy toll of life and livelihoods. Low level of technology development in the rural areas together with social, economic and gender inequities enhance the vulnerability of the largely illiterate, unskilled, and resource-poor fishing, farming and landless labour communities. Their resilience to bounce back to pre-disaster level of normality is highly limited. For the planet Earth at crossroads, the imminent threat, however, is from a vicious spiral among environmental degradation, poverty and climate change-related natural disasters interacting in a mutually reinforcing manner. These, in turn, retard sustainable development, and also wipe out any small gains made thereof. To counter this unacceptable trend, the M.S. Swaminathan Research Foundation has developed a biovillage paradigm and rural knowledge centres for ecotechnological and knowledge empowerment of the coastal communities at risk. Frontier science and technologies blended with traditional knowledge and ecological prudence result in ecotechnologies with *pro-nature*, *pro-poor* and *pro-women* orientation. The rural communities are given training and helped to develop capacity to adopt ecotechnologies for market-driven eco-enterprises. The modern information and communication-based rural knowledge centres largely operated by trained semi-literate young women provide *time-* and *locale-specific* information on weather, crop and animal husbandry, market trends and prices for local communities, healthcare, transport, education, etc. to the local communities. The ecotechnologies and time- and locale-specific information content development are need-based and chosen in a 'bottom-up' manner. The use of recombinant DNA technology for genetic shielding of agricultural crops for coastal regions against abiotic stress (induced by the water- and weather-related natural disasters), strengthens the foundations of sustainable agriculture undertaken by the resource-poor small farm families.

Keywords: natural disasters; developing countries; disaster management; sustainable development; technological and knowledge empowerment; rural communities

* Author for correspondence (pckesavan@mssrf.res.in).

One contribution of 20 to a Discussion Meeting Issue 'Extreme natural hazards'.

1. Perspectives on extreme natural disasters

Life on planet Earth is accustomed to encounters with natural disasters. An astro-physical event (i.e. impact of a meteorite on Earth) approximately 65 Myr ago is said to have wiped out dinosaurs and many other species. However, such astro-physical hazards are very rare. Geophysical hazards (i.e. earthquakes, volcanic eruptions) originate from within the Earth's interior and are common. Powerful earthquakes resulting in vertical displacement of the tectonic plates at the bottom of the sea can produce destructive tsunami waves as happened on 26 December 2004 (Bilham 2005, Mishra & Rajasekhar 2005). The hydro-meteorological hazards consist of tropical cyclones, landslides, floods and droughts. In the 1990s, more than 90% of those killed in natural disasters lost their lives in hydro-meteorological events (mainly droughts, floods and wind storms, GEO-3). While earthquakes accounted for 30% of the estimated risk, they caused just 9% of all fatalities due to natural disasters. In contrast, hunger caused by famines worldwide killed 42% of people in the affected regions, but accounted for just 4% of damage over the past decade (GEO-3; IFRC 2001). Further, the number of geophysical disasters has remained fairly steady during the past two decades (1980s and 1990s), but the number of hydro-meteorological disasters has increased substantially (GEO-3). Emerging events such as melting of ice in the Polar Regions and the Himalayas, rise in sea level and increasing intensity and frequency of floods, hurricanes and typhoons suggest that global warming is influencing climate change. Experts at the National Snow and Ice Data Center in Boulder, Colorado at the National Aeronautics and Space Administration of USA have recently announced that the floating cap of sea ice on the Arctic Ocean shrank during summer, 2005 to its smallest size in a century (Revkin 2005). Trenberth (2005) argues that higher sea surface temperatures in the Atlantic Ocean and increased water vapour in the lower atmosphere—caused by global warming—are to blame for the past decade's intense storms. Recently, there were newspaper reports associating the hurricanes 'Katrina' and 'Rita' in August and September 2005, respectively, which devastated New Orleans and neighbouring towns in Texas, USA, with global warming. Emanuel (2005) reports that global warming may lead to an upward trend in tropical destructive potential. He also points out that with an ever increasing coastal population, a substantial increase may occur in hurricane-related losses in the twenty-first century. Growing anthropogenic pressures are exacerbating the risks.

A scientific consensus on the potential impact of climate change is yet to emerge, although there is agreement that anthropogenic factors are influencing global climate systems. Against this background, the declaration by Lord May of Oxford, President of the Royal Society, UK, that 'the world leaders can no longer use uncertainty about aspects of climate change as an excuse for not taking urgent action to cut greenhouse gas emissions' is quite significant. It is indeed becoming increasingly clear that human-associated climate change is real. The Anniversary Address (2004) by the President of the Royal Society dealt in part with climate change (Sunday Herald, London, 19 June 2005).

The majority of natural disasters affect poor nations and the poor in all nations the most, since their coping capacity is limited. Climate change directly affects sensitive sectors like agriculture, forestry and fishery and thereby the livelihoods of millions of coastal communities (Sinha & Swaminathan 1991;

PHILOSOPHICAL TRANSACTIONS OF THE ROYAL SOCIETY A MATHEMATICAL, PHYSICAL & ENGINEERING SCIENCES

Swaminathan 1996). The adverse impacts of climate change, in the form of declining rainfall alternating with heavy downpours causing floods, and long dry spells leading to droughts, severely threaten livelihoods, economy and food security. To add to this, poor infrastructure facilities, weak institutional mechanisms, lack of financial resources and vast sectoral and regional disparities adversely affect the adaptive capacity of developing countries to climate change. Thus, climate change exerts additional stress on the ecological and socio-economic systems that are already facing tremendous pressure due to rapid industrialization, urbanization and globalization of economies.

Hydro-meteorological hazards would exert even greater hardship to the people as well as damage to the natural resources in small island state developing countries and countries with long coastlines, such as India. It is reported (UNEP 1989) that India, with a low-lying densely populated coastline extending to over 7000 km, is highly vulnerable to sea-level rise. Most of the paddy fields of coastal regions are highly vulnerable to inundation and salinization (Gupta 2005). The small island nations, such as Kiribati, Seychelles and Maldives, have more than 80% of the land area at less than a metre above the present sea level. Such low-lying islands may have to pay a heavy price in terms of loss of human lives, and property if sea level rises to the extent forecast by global climate models (UNEP/GRID; Arendal 2005). Further, migration of coastal communities to inland areas will swell the number of environmental refugees (Myers 2000, 2002). Earthquake risk is high in several parts of India (Rao 2005). There is also concern about future *large magnitude* earthquakes along the Sumatra Fault southeast of the 2005 event rupture and the adjacent region that last failed in 1833, which is likely to have accumulated substantial strain (Lay *et al.* 2005; Sieh 2006). International efforts to improve the tsunami-warning capabilities are warranted, given the inevitability of future great thrust earthquakes along the Sumatra zone. Consequently, the Indian Ocean countries are creating a framework for a region-wide warning system and girding for the next killer wave (Stone & Kerr 2005). The growing concentration of population along India's coastline will have to be prepared particularly for much higher intensity of hydro-meteorological hazards. For these reasons, small island developing countries and the countries with long coastlines need to focus on disaster management particularly in their coastal regions. Discussion of these issues with special reference to the coastal regions in the developing countries is the main thrust of this paper. The need of the hour is to promote concerted efforts for preserving natural ecosystems and diversifying coastal economies, which can enhance recovery from disasters and resilience to their effects (Adger *et al.* 2005; Allenby & Fink 2005).

2. Vulnerability

Vulnerability to disasters has social, gender, ecological and economic dimensions in addition to the well-acknowledged technological capacities of the countries. It describes the degree to which a socio-economic system or physical assets are either susceptible or resilient to the impact of natural hazards. Development practices which do not take into account the susceptibility to natural hazards result in enhancing the vulnerability. There is a close correlation among the trends of increased demographic pressure in developing countries, enhanced environmental degradation, higher human vulnerability and the intensity of the natural disasters. For instance, both the intensity of damage and loss of human lives were more severe where the mangrove forests had been degraded at the time of the Orissa cyclone in 1999. The degradation of the environment, particularly the forest cover, is often a critical factor that transforms a natural hazard, or a climatic extreme such as heavy downpour, into a disaster (Myers 1989).

Poverty, environmental degradation and hazard vulnerability form a vicious spiral, and are mutually reinforcing. The poor are compelled to exploit environmental resources for survival and also to live dangerously close to the high tide zones, and thus become highly vulnerable to cyclones, tsunamis, etc. Furthermore, among the poor, women and children are particularly vulnerable. Apart from the physical constraints caused by pregnancy, delivery, nursing care, etc., social and economic handicaps also contribute to their enhanced vulnerability. A generally poor reproductive health care system, particularly for women in these rural areas, combined with increasing incidence of tuberculosis, malaria and HIV/AIDS substantially weaken their health status. In India, economic globalization has accelerated the pace of migration of young men from rural to urban areas, where new employment opportunities have emerged. While this trend may be good in one sense, the detrimental consequence is the 'feminization of agriculture' in the rural areas. With much of the agriculture in the rural areas remaining as 'subsistence farming', the real effect is the 'feminization of poverty'.

Already handicapped by gender-related disadvantages, the worsening of economic status makes women much more vulnerable socially. Aglionby (2005) reported that four times as many women died in the tsunami as men on 26 December 2004. In Cuddalore, Tamil Nadu, India, 391 women were killed, compared with 146 men. According to the United Nations Children's Fund (UNICEF), children account for a third of the casualty. In some districts of Tamil Nadu, the casualty among children was an overwhelming 50–60% ('UNICEF Lays Out Four Point Programme for Tsunami children', Press Trust of India, 4 January 2005). The young girls and women face the danger of being pushed into prostitution by the anti-social elements. The trafficking of young women and girls in the human trade was noted as a serious dimension of gender-specific vulnerability resulting from natural disasters. What matters in the end, however, is that both sudden and extremely violent as well as the not-so-aggressive natural hazards with considerable lead time exert their adverse effects differentially on urban and the rural environments, the rich and poor and males and females. The need is to bridge the gaps in technological, social and gender divides.

PHILOSOPHICAL TRANSACTIONS OF THE ROYAL SOCIETY A MATHEMATICAL, PHYSICAL & ENGINEERING SCIENCES

3. Vicious spiral between poverty, environmental degradation and natural disasters

The United Nations Department of Economics and Social Affairs developed a background document in February 2002 on Sustainable Development for the *World Summit on Sustainable Development* held during August 2002 in Johannesburg, South Africa. One of the background papers submitted by the United Nations relating to an International Strategy for Disaster Reduction discusses natural disasters and sustainable development. The question posed in it was whether sustainable development along with the international instruments aiming at poverty reduction and environment protection, can be accomplished without taking into account the risk of natural hazards and their impacts. The answer was negative. Disaster reduction policies and measures should enable societies to become resilient to natural hazards while ensuring that development efforts do not increase their vulnerability to these hazards.

Way back in 1972, at the UN Conference on the 'Human Environment' held in Stockholm, Sweden, late Ms Indira Gandhi, the then Prime Minister of India, stated that without addressing the problems of poverty and deprivations, it will not be possible to establish harmony between humankind and nature. Dr Maurice Strong, then Adviser to the UN Secretary-General also drew attention to the vicious spiral between poverty and environmental degradation. Resource-poor and landless rural women and men migrate to neighbouring urban areas in order to eke out a living and end up as 'environmental refugees' (Myers 2002). The M.S. Swaminathan Research Foundation (MSSRF) that was set up in Chennai, India about 15 years ago, therefore, laid stress on linking the livelihood security of landless and resource-poor rural women and men with ecological security in a mutually reinforcing manner. MSSRF's research programmes are also designed to transform the 'Green Revolution' referred to as 'exploitative agriculture' (Swaminathan 1968) into an 'Evergreen Revolution', a term coined by Swaminathan (1996) to denote farming systems that do not cause degradation of the ecological foundation of agriculture and would, therefore, be able to enhance productivity in perpetuity. The vicious spiral between environmental degradation and ever increasing frequency and intensity of hydro-meteorological disasters has received the foundation's attention. It is now widely acknowledged that anthropogenic pressures leading to degradation of mangrove forests (FAO 2003) in the coastal regions and non-mangrove forests throughout the developing world reduce carbon dioxide absorption and generation of oxygen. Soil erosion is yet another serious problem. Degradation of forests directly and indirectly also enhances human vulnerability to natural disasters. The mutually reinforcing vicious linkages among environmental degradation, poverty, destruction by natural disasters and sustainable development are illustrated (figure 1).

The strategy to break the vicious linkages requires concurrent attention to reduction of threats to sustainable rural livelihoods, restoration, conservation and enhancement of ecosystems and arresting environmental degradation and sustainable use of natural resources. The dictum, 'good ecology is good business' (Swaminathan 1999*a*) should underpin all development activities. Developing countries like India, must launch vigorous family planning programmes and contain the unsustainable population explosion. In India, poor economic conditions and the craving to have a son account for a large number of children per couple in many households. Technologically advanced countries, which also have high levels of human resource development, must not only reduce their consumption of energy from fossil fuels, but also share their technologies for 'clean energy' production technologies and energy-saving strategies with the developing countries. The transboundary nature of the harmful effects of climate change leaves humankind with no other option. It is equally important to integrate disaster management with policies, programme of activities and support for sustainable development. The Brundtland Commission Report (1987) was titled 'Our Common Future' to stress that irrespective of political frontiers, the fate of humankind is ecologically intertwined.

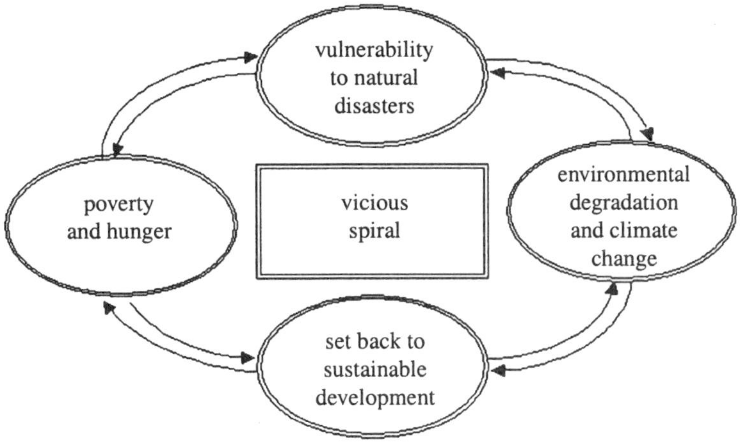

Figure 1. Feedback relationships.

4. Mainstreaming disaster management into sustainable development

(a) *Bottom-up approach to sustainable use of natural resources*

(i) *Sustainable development*

By the 1980s, humankind started facing serious ecological and social crises. The ecological problems consisted of serious damage to the basic life support systems of land, water, forests, biodiversity and atmosphere. Renewable energy sources also became a matter of great concern. The social dimension of crises arose from increasing poverty and hunger, economic, social and gender inequities, rapid growth in human population resulting in reduced *per capita* availability of arable land and irrigation water. The 'famine of employment' (i.e. livelihood) largely accounts for food insecurity at the individual household level, because of population explosion on the one hand, and rapid technological advances leading to automation and jobless economic growth, on the other. All these factors led to MSSRF defining its research agenda in terms of *sustainable development*, rooted in the principles of ecology, social and gender equity, employment (particularly rural livelihood) generation and economic viability and energy efficiency (A Social Vision for Science 2000). If technology was an important factor in the past in increasing economic and social disparities and causing ecological harm, the MSSRF's approach has been to enlist appropriate blends of traditional and frontier technologies as allies in the movement for economic and ecological well being and gender equity. A social contract between science and society (Swaminathan 1999*b*, 2000*a*) has been underpinning the programme of activities of the MSSRF. In the field of agriculture, the MSSRF's goal is to spread an Evergreen Revolution (Swaminathan 2000*b*). Wilson (2002) in his analytical book 'Future of life' refers to the significance of Swaminathan's concept of Evergreen Revolution in the following words: 'The problem before us is how to feed billions of new mouths over the next several decades and save the rest of life at the same time, without being trapped in a Faustian bargain that threatens freedom and security. No one knows the exact solution to this dilemma. The benefit must come from an Evergreen Revolution'.

Sustainable development as defined by the Brundtland Commission (1987) is 'development that meets the needs of the present without compromising the ability of future generations to meet their own needs'. It is clear then that sustainable development that takes the impact of human activities on the environment into account and tries to minimize environmental damage is the key to poverty reduction, ecological security and mitigation and management of weather- and water-related natural disasters. The programme of research and development initiatives of the MSSRF provides replicable models for realizing sustainable development (Swaminathan 2005*a*). Focusing on the resource-poor and landless communities in the rural, and particularly in the coastal, regions is urgent, imperative for reasons of population density, work and income security, rapidly declining natural resources and high level of vulnerability to natural disasters. The Gulf of Mannar Biosphere Reserve in the State of Tamil Nadu, India, is an example of a precious biological heritage affected by the vicious spiral of poverty, degradation of natural resources, and natural disasters (M.S. Swaminathan Research Foundation, 1998). Their marine and adjacent land resources and biodiversity have started to decline rapidly because of a combination of anthropogenic pressure, pollution and hydro-meteorological disasters. Consequently, sea cows (dugong) and several kinds of marine organisms notably disappeared and these necessitated the MSSRF to initiate appropriate remedial action (Twelfth Annual Report 2001–2002, of the M.S. Swaminathan Research Foundation). This action is described below.

(ii) *Operationalizing sustainable development*

With a firm conviction that the vicious spiral of poverty and environmental degradation could be broken by technological and knowledge empowerment of the illiterate, unskilled and resource-poor rural women and men, the MSSRF initiated in 1998 a programme for fostering sustainable livelihoods in the Gulf of Mannar area with support from the Global Environment Facility and the United Nations Development Programme. Earlier, similar work had been initiated in about 20 villages in Pondicherry and great deal of experience had been gained. The thrust in these projects is the blending of frontier technologies with traditional knowledge in order to provide a *pro-nature*, *pro-poor*, *pro-women* and *pro-employment* orientation to technology development and dissemination. The resultant technologies, known as 'ecotechnologies' are readily adopted by rural families, because of their economic and ecological advantages. Local natural resources are sustainably managed. The question whether the largely illiterate and semi-literate rural women and men could develop knowledge and skills and become capable of handling eco-enterprises had already been answered in the 1970s. Swaminathan (1972) had coined the term 'techniracy' to describe the pedagogic methodology of 'learning by doing'. When learning is through work experience, the poor are able to master new technologies within a short span of time.

Groups of women and men separately or jointly organize themselves into self-help groups (SHGs) and undergo training and capacity building in one or more ecotechnologies based on the resources available in the region. Whole villages are organized into biovillages, where concurrent attention is paid to the conservation and enhancement of natural resources and to on-farm, non-farm and off-farm livelihood opportunities. In biovillages, the extreme poor (i.e. those earning one

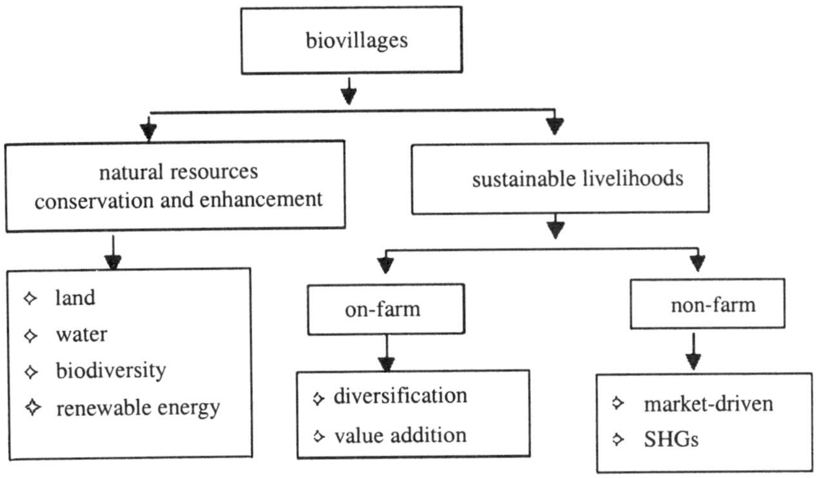

Figure 2. Biovillage paradigm.

US dollar or less per day) are enabled to take to multiple enterprises based on market demand so that the total income of the family is raised to an adequate level. The biovillage paradigm, designed to foster sustainable and equitable rural development and job-led economic growth (Swaminathan 2005*a*), is represented in figure 2.

Location of biovillages in three states of India, numbering 100 at present, is shown in figure 3.

In order to build the capacity of rural families to manage the various enterprises on their own, Biocentres are established as integral part of the biovillages. The organization and functions of the Biocentre are presented in figure 4. The eco-enterprises shown in figure 4 are just illustrative and not exhaustive. The aim is to provide key centralized services to promote economically viable, decentralized production, thereby combining the benefits of 'mass production' and 'production by masses' approaches to economic activity.

Demystification of technologies, especially relevant to sustainable rural development, training and capacity building of the rural communities, provision of microcredit for the microenterprises and establishing market linkages, are all integral components of a biovillage paradigm.

The five E_s of ecotechnology in the biovillage paradigm are economics, ecology, equity (gender and social), energy and employment. Without job-led economic growth, the poor and the marginalized people will not be able to come out of the poverty trap. Renewable energy is essential, particularly in the context of an ever increasing price of petroleum-based fuels.

Further, the sustainability factor would not be firmly anchored unless social and gender equities are mainstreamed into all the development activities. The actions to be taken today for sustainable development, keeping in view the need to manage climate change, are outlined by Swaminathan (2002).

The coastal communities can access both the land and marine resources for developing ecotechnologies and eco-enterprises. The coastal biovillage paradigm, therefore, takes into its account both the marine- and the land-based natural

PHILOSOPHICAL TRANSACTIONS OF THE ROYAL SOCIETY A MATHEMATICAL, PHYSICAL & ENGINEERING SCIENCES

PHILOSOPHICAL TRANSACTIONS OF THE ROYAL SOCIETY A MATHEMATICAL, PHYSICAL & ENGINEERING SCIENCES

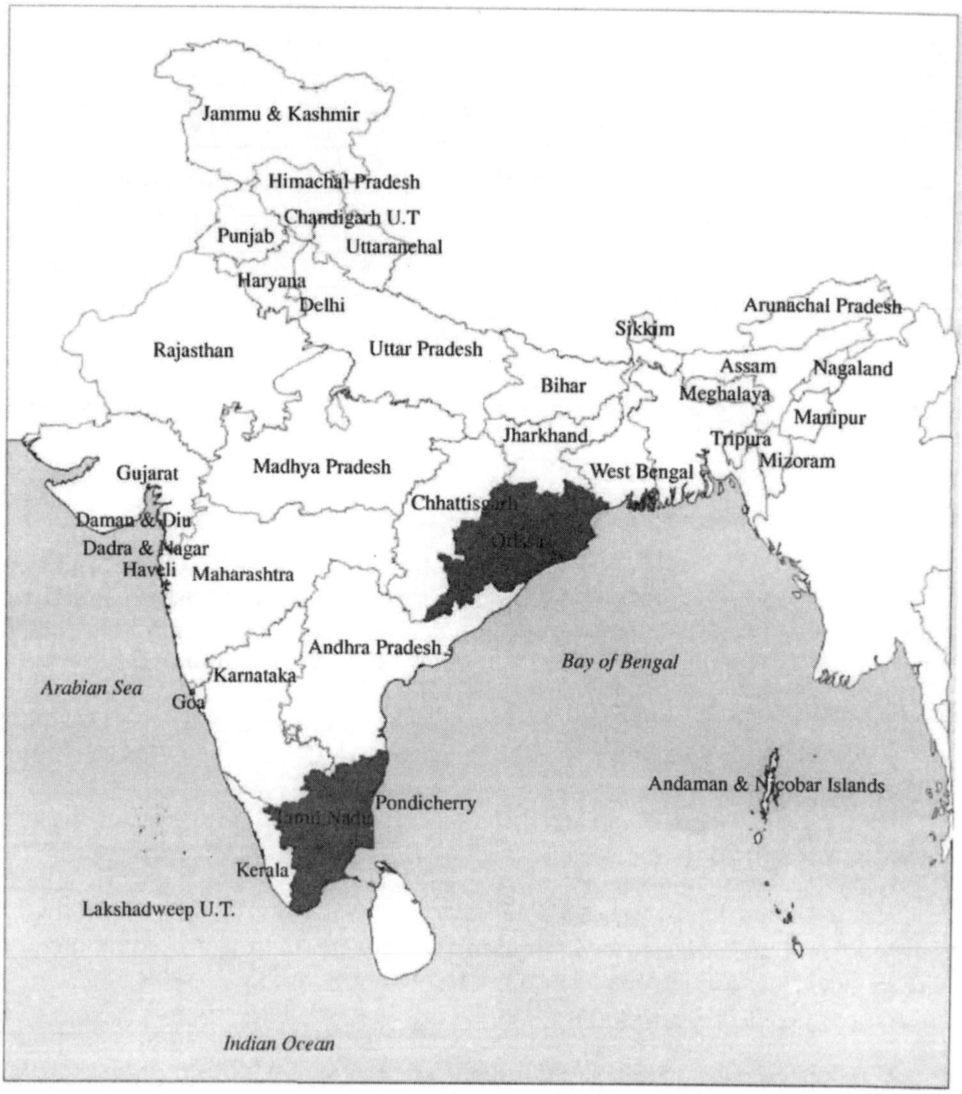

Figure 3. Biovillages are located in states shown in green colour.

resources for developing eco-enterprises, as well as training and capacity building of the local communities. The MSSRF has prepared a toolkit describing the various ecotechnologies and how to develop these. Wherever feasible, 'aquaculture estates' are also developed. Individual fishermen can bring the catch of the day to these estates, which are equipped to process and market the fish, crabs, prawns, etc. While individual fishermen may not have time and resources to process and market their commodities, the centralized aquaculture estates are designed for this purpose. The power of scale is thus provided to the resource-poor fisher communities. A 'fish for all' movement was also launched in 2003 jointly with World Fish Centre in Penang in order to ensure integrated attention to all steps in the capture to consumption chain.

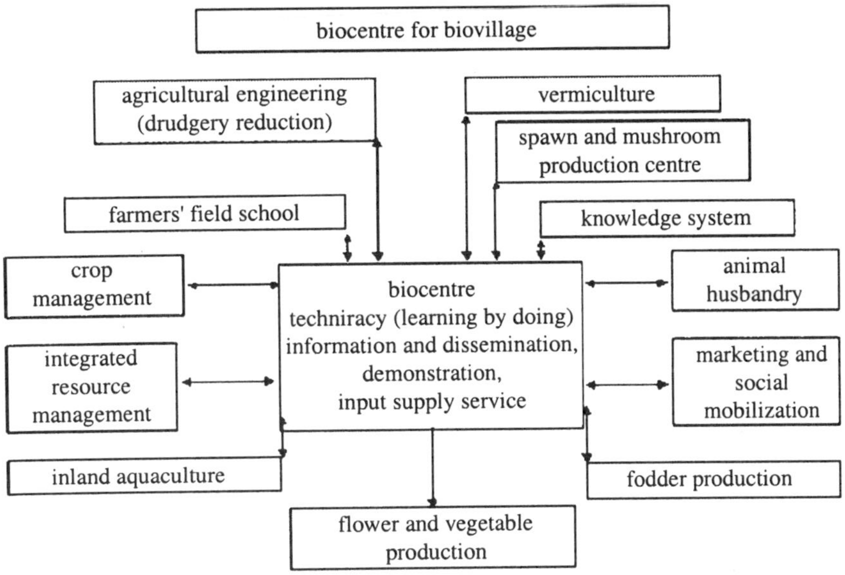

Figure 4. Biocentre for biovillage.

Marine- and land-based natural resources in the coastal areas for developing on-farm and non-farm enterprises are wide ranging. For example, the brackish water area, especially mud flats and saline-affected areas, largely remain barren or with limited biological productivity due to low soil organic content, low level of nutrients and hypersaline conditions for most part of the year. In the estuaries, mangroves provide detritus that nourish crabs, prawns and fish (Mumby *et al.* 2004). Developing an enterprise of mud crab fattening is pro-nature, pro-poor, pro-women and pro-livelihood oriented (figure 5*a*). Organic shrimp aquaculture is another community-centred eco-enterprise. Culturing ornamental fish is very remunerative since it has an export market and landless women are given training and initial resources to take it up. Fish pickle is also a very promising enterprise. Since many of the coastal villages do not have cold storage, the fisher women are forced to sell the fish catch of the day to the middlemen at low prices. Introduction of fish pickle technology has resulted not only in value-addition, but also in reducing the involvement of middlemen and others who exploit the fishers. Pearl culture is yet another market-driven enterprise. In the Gulf of Mannar area, the MSSRF introduced local communities to the science and art of artificial coral reefs. This has helped to revive fisheries in small areas. Coral reefs are an important ecosystem, both in terms of biodiversity and for the invaluable goods and services they provide to millions of coastal dwellers at tropical latitudes (Moberg & Folke 1999).

Paddy is a major agricultural crop in the eastern coastal areas of India. The landless women are trained to use the paddy straw as substrate to culture oyster mushrooms (figure 5*b*). Making paper and boards from banana waste is yet another pro-nature, pro-women and pro-poor enterprise.

The new paradigm of Evergreen Revolution envisages agriculture to be more biology- than chemistry-based. What this means is that biofertilizers and biopesticides would replace the chemical fertilizers and chemical pesticides, respectively. SHGs of women are trained to produce vermicompost/worm compost (i.e. the process of using earthworms to digest kitchen and garden waste to create a faster than normal composting; earthworm castings contain five times more nitrogen, seven times more phosphorus and 11 times more potassium to enrich the soil (figure 5*c*) and biofertilizers and biopesticides such as the *Trichogramma* egg parasitoid (figure 5*d*). It must be emphasized that demystification of science and successful standardization of the laboratory-based technologies into ecotechnologies leads to the shifting of assetless families from unskilled to skilled work, thereby adding economic value to their time and labour.

The experience with biovillage paradigm shows that people-centric sustainable management of local land, forest and marine resources establishes mutually reinforcing linkages between development and environment. The environmental degradation ascribable to anthropogenic pressures is greatly minimized. As of now, there are 100 biovillages located in Tamil Nadu, Orissa and union territory of Pondicherry (figure 3).

(*b*) *Bottom-up participatory management of bioshield in the coastal regions*

During the decades preceding the 1990s, degradation of mangrove forests in India was unbridled. The current mangrove area worldwide has fallen below 15 million ha, down from 19.8 million ha in 1980 although the rate of mangrove deforestation has decreased in the 1990s from that of the 1980s (Mayaux *et al.* 2005). With substantial financial support from the India–Canada Environmental Facility (ICEF), the MSSRF took up a massive restoration of the mangrove forests in the coastal areas of West Bengal, Orissa, Andhra Pradesh and Tamil Nadu. Systematic studies on the causes of degradation whether natural or human-influenced were first carried out, and then suitable methods to ensure successful restoration were followed. A very substantial number of nursery material of different mangrove species was required. This provided an opportunity to train local women and men SHGs for raising nursery material. In doing so, they became aware of the role of mangrove trees and shrubs in providing nutrients to fish, prawns and crabs. The meaning of the traditional saying by fishermen in the Andaman sea, 'mangroves are the roots of the sea' became clear. The realization that their livelihood security was closely entwined with the security of the mangrove ecosystem was profound. Emergence of new harmony between the mangrove-dependent rural communities and the mangrove forests, prompted the MSSRF to develop a model for joint mangrove management (JMM), to be adopted and replicated by the Forest Department which manages mangrove wetlands in India. JMM aims at joint participation and sharing of experiences in mangrove management by the Forest Department, the mangrove user community (particularly the women), related government departments and non-governmental organizations in all mangrove management functions—resource mapping, planning, regeneration, protection and benefit-sharing (The Mangrove Decade and Beyond 2002).

The role of mangrove species acting as a bioshield to reduce the destructive potential of cyclones and tsunami became evident during the Orissa

Figure 5. Eco-enterprises for sustainable livelihood: (*a*) mud crab production, (*b*) mushroom production (oyster mushroom), (*c*) vermicompost and (*d*) trichogramma egg parasitoid production.

supercyclone in October 1999 and the Indian Ocean tsunami in December 2004. In tsunami-affected Pichavaram in Tamil Nadu, the hamlets located between the sea and the mangroves were severely affected and many were destroyed, with a few deaths and substantial damage to houses, boats and fishing nets. On the other hand, the hamlets within the physical cover of the mangrove forest were largely protected, and there were no tsunami-associated deaths at all. Remote sensing (figure 6), ground-based data (Selvam *et al.*

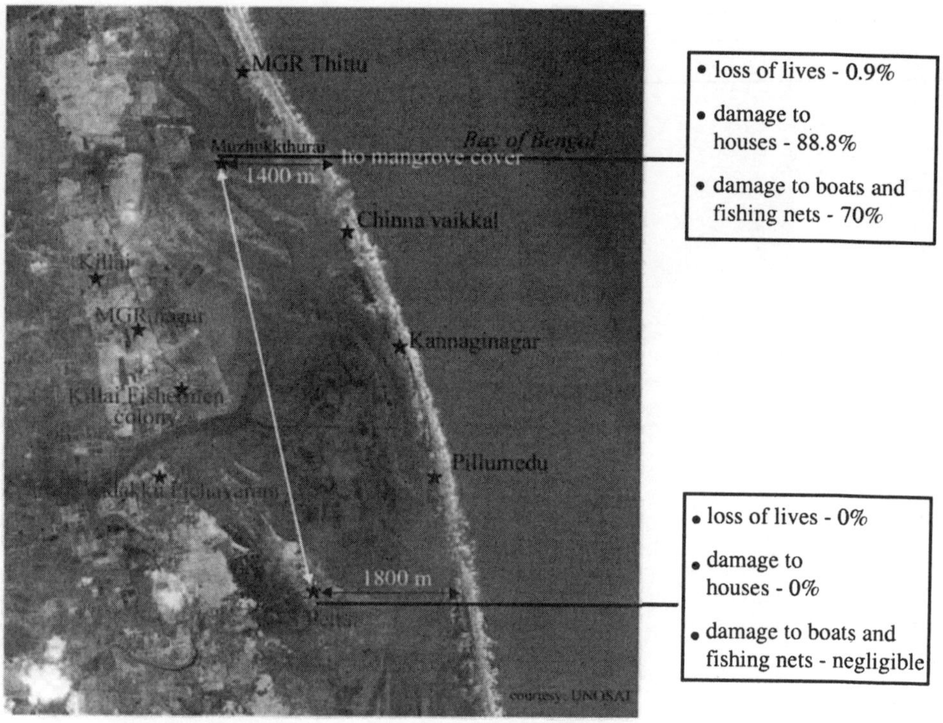

Figure 6. Protective effect of mangrove bioshield.

PHILOSOPHICAL TRANSACTIONS OF THE ROYAL SOCIETY A MATHEMATICAL, PHYSICAL & ENGINEERING SCIENCES

in press) and eyewitness accounts of the survivors of the two natural disasters have convinced the Indian government, the non-governmental organizations and coastal communities of the protective role of the mangrove forest. Wherever these have been degraded, plans are afoot for planting mangrove, *Casuarina* and other suitable species along the shore line. Danielsen *et al.* (2005) have demonstrated the protective role of the mangrove bioshield in a tsunami-affected region in Tamil Nadu. These post-tsunami observations are largely in accordance with the analytical models showing that 30 trees per 100 m^2 in a 100 m wide belt may reduce the maximum tsunami flow pressure by more than 90% (Hiraishi & Harada 2003). In addition, coral reefs are also as important as the bioshield. For island state countries like Maldives, which are just a metre above the sea level, coral reefs provide ecological and economic security. The corals and their ecosystem are a major tourist attraction (Moberg & Folke 1999).

In areas, where the edaphic and hydrological conditions are not favourable for putting up a mangrove bioshield, the plans are to go for non-mangrove species like *Casuarina*, followed by salt-tolerant species such as *Pandanus*, coconut, cashewnut, etc. Intercropping of these with low input and high value crops (pulses, spices, fruit crops depending upon soil, climate, etc.) would combine ecology (bioshield) and economy (livelihood). The strong and harmonious relationships among ecology, economics and development not only reduce the death and devastation by natural disasters, but also enhance the coping capacity

Figure 7. Community (*a*) grain and (*b*) water banks.

of the local rural communities. Mangrove species are an effective carbon sink and so would contribute to remedying the imbalance between carbon emissions and carbon absorption.

(*c*) *Community-centred cyclone/tsunami shelters, and food, fodder and water banks*

In areas highly vulnerable to cyclones, tsunamis, surge storms, etc., the MSSRF's model of Community Seed, Grain, Water Banks (figure 7*a*,*b*; i.e. banks with a difference) can help in enhancing the coping power of the people during a natural hazard. Often the transport and communication are severely disrupted by natural disasters, and the worse-hit areas are almost completely cut-off from rest of the world. The initial relief measures invariably consist of food packets dropped from aircrafts and helicopters. The least developed countries may not have capacity to do even this. In several hydro-meteorological disaster-prone areas, such relief measures are an annual feature. A decentralized, community-centred grain and water (food) bank would be able to provide relief almost immediately, and avoid 'transient hunger'. Built into this 'bottom-up' model is the suitable storage system.

The storage facilities could be traditional, conforming to traditional knowledge and ethos, but incorporating modern scientific inputs to withstand the impacts of natural disasters. The MSSRF facilitated putting up two multi-purpose cyclone shelters in the cyclone-prone coastal villages of Orissa. These are a safe haven during extreme natural hazards such as cyclones, windstorms and floods. These could also be linked in the neighbourhood with food and water banks, as well as first-aid facilities. Modern information and communication technology (ICT)-based village knowledge centres (VKCs) provided in these shelters will also help internet facilities for communication during emergency (§4*d*). Women, children and the infirm should be evacuated to these shelters as soon as early warning is received. The availability of first-aid, food and water would enhance the coping capacity of the communities during the disaster. The details of the community seed, grain, water and fodder banks, as well as their role in ensuring food security, are found in the book titled, 'Community Grain Bank' (2001) jointly published by the M.S. Swaminathan Research Foundation and the World Food Programme, New Delhi.

(*d*) *Modern ICT-based VKCs for extreme natural hazard management*

With a firm conviction and foresight that knowledge empowerment of rural women and men is crucial for sustainable development and poverty reduction, the MSSRF set up in 1997 what were then referred to as village information centres. These are now called VKCs. Wherever telephone lines are not available, connectivity is established through a wired-wireless hybrid technology. Particular attention is given to creating content that is time- and locale-specific and demand-driven. The time- and locale-specific and demand-driven information pertain to weather, crop and animal husbandry, integrated pest management, market trends and prices of local commodities, health care, immunization of children, poverty alleviation schemes of the government, details of eco-enterprises, transport, education, etc. A good example of value-addition to generic information that has been acclaimed world over is the 48 h advance information on the sea wave heights in the Bay of Bengal provided round the clock for the benefit of the fishermen in a small village, Veerampattinam, in the union territory of Pondicherry. In this case, the MSSRF downloads the data on the sea wave heights from the US Naval website, which is unclassified and generic. The MSSRF enhances the value of this information by pinpointing the sea wave heights close to Veerampattinam. The value-added information so gained from the use of high technology is announced in Tamil, the local vernacular, through low-tech loudspeakers. The need to use loudspeakers arises from the fact that some of the fishers cannot read and most of them are totally engrossed in their work. Over the years, the fishers have acknowledged how useful this timely, locale-specific and value-added information has been in avoiding risk to lives on the sea in their country rafts (Catamaran). The fisher women are very relieved that their men would not be caught in a severe cyclonic storm, and drown in the sea.

The village knowledge revolution launched by the MSSRF in a few villages in 1997 involved integrated use of the internet, cable TV, radio and the vernacular press. The experience reveals that the internet-radio combination is particularly powerful for 'reaching the unreached' and 'voicing the voiceless' in rural India (figure 8).

The success of the VKCs set up by the MSSRF is not only because of their time- and locale-specific, value-added, demand-driven information content, but also the basic principles of these being user-controlled, and user-managed with strong social and gender equities. In fact, in almost all the VKCs, young women who have studied up to seventh or eighth class and have learnt computer literacy are the operators and managers. The experience of the MSSRF is that given a sense of ownership and training, the rural youth take to making use of the internet like a fish to water.

On the morning of 26 December 2004, the VKC with its loudspeaker system saved many lives in the village Veerampattinam. It so happened that on that fateful Sunday morning, a few young men sitting on the seashore noticed strange behaviour of the sea. The sea first receded and at the very rear, as far eyes could see, the waves were rising alarmingly high. Guessing that something was drastically wrong, they rushed to the VKC and used the loudspeakers to ask the women, children and men to run to higher grounds away from the seashore. This saved the lives of all those who were minding their business in their huts and outside close to the shoreline. Damage to property was inevitable but precious human lives were saved.

PHILOSOPHICAL TRANSACTIONS OF THE ROYAL SOCIETY A MATHEMATICAL, PHYSICAL & ENGINEERING SCIENCES

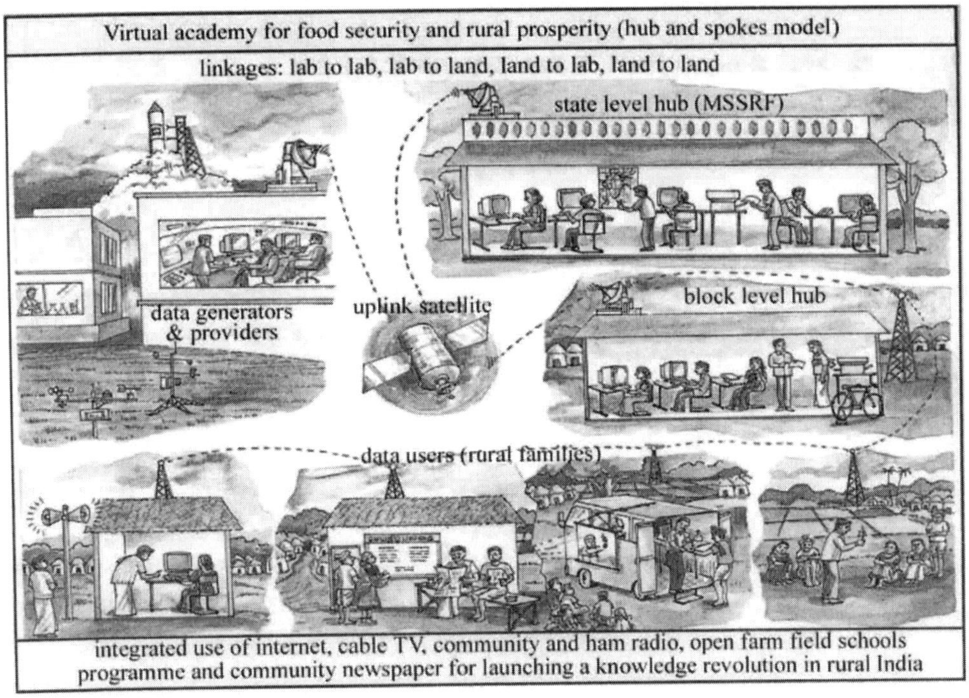

Figure 8. Modern ICT for knowledge revolution in rural India.

Figure 9. Mission 2007: every village a knowledge centre.

With the initiatives of the MSSRF and support of the Government of India, there is now a vigorous movement, by mobilizing the power of partnership, of National Alliance for Mission 2007: Every Village a Knowledge Centre (figure 9). The goal of Mission 2007 is to have a knowledge centre in each of the 600 000 plus villages of India by the year 2007, which marks the 60th anniversary of India's Independence.

At present, 40 VKCs have been set up in various states as indicated by arrows (figure 10), in addition to 13 Village Resource Centres (hubs) each of which has connectivity with several villages falling within a radius of approximately 25 km. The work has picked up momentum involving scores of organizations each concentrating on a particular region/state of the country.

PHILOSOPHICAL TRANSACTIONS OF THE ROYAL SOCIETY A MATHEMATICAL, PHYSICAL & ENGINEERING SCIENCES

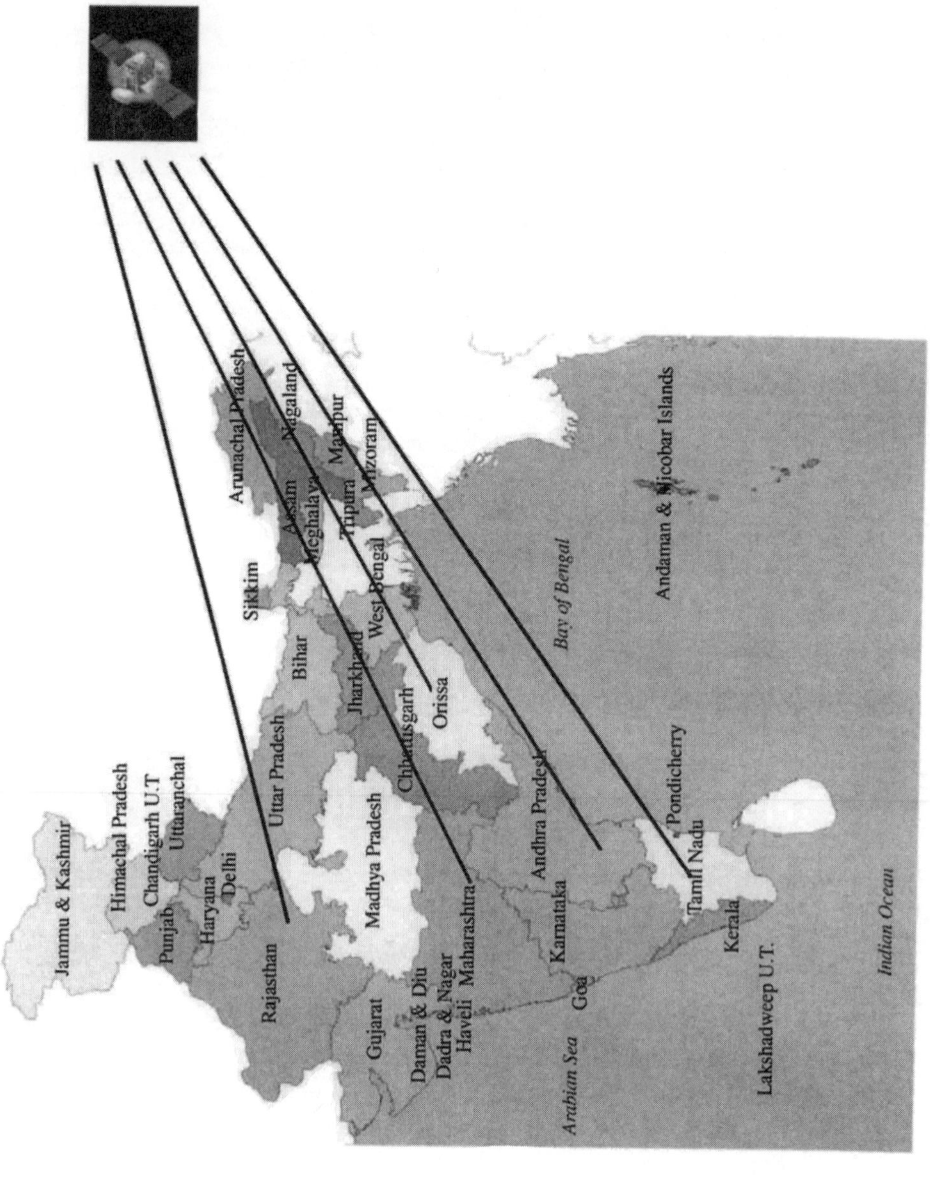

Figure 10. Location of 13 village resource centres and 40 village knowledge centres in India.

In the disaster-prone rural areas, at least one woman and one man will need to be trained to make use of the internet, GIS, remote sensing and communication systems. GIS can help in improving the quality and power of analysis of mitigation measures and in the implementation of emergency preparedness and response action. Identification of hazardous areas and monitoring the planet for its changes on a real time basis and to give early warning about many impending disasters is now possible with remote sensing. Practical ways of integrating disaster management with sustainable development are summarized by Swaminathan (2005*b*). As medium- and long-term rehabilitation strategies, he has suggested three steps of action all along the coast. These are: (i) strengthening the ecological foundations (bioshield) of sustainable human security; (ii) fostering sustainable livelihood security; and (iii) putting in place a network of rural knowledge centres. Swaminathan (2005*b*) has also described the principles and methodologies of mainstreaming disaster management into sustainable development.

(*e*) *Genetic shields in coastal areas vulnerable to seawater inundation*

In his Anniversary Address (2004), Lord May refers to the rapidly growing water shortage as a global problem that could lead to armed conflicts among neighbouring countries which share rivers, etc. He also suggested appropriate use of the recombinant DNA technology to produce crops that are drought-resistant, and/or salt-tolerant, thus moving in the direction of producing crops that are adapted to their environment (figure 11).

Based on the suggestion of Prof. M.S. Swaminathan as early as 1988 at a meeting on climate change in Kyoto, Japan, the scientists at the MSSRF identified several salt-tolerant genes from the mangrove species, *Avicennia marina* (Mehta *et al.* 2005) and transferred these through the recombinant DNA technology to rice. It is an important crop in the coastal areas, which are threatened by increasing sea level, and frequent storms and cyclones. The salinity-resistant rice developed with the help of genes transferred from unrelated mangrove species (*A. marina*) also emphasizes the urgent need to prevent loss of valuable genes through conservation and enhancement of biodiversity in the developing countries. The transgenic rice tolerates salinity up to 150 mM. The development of transgenic salt-tolerant rice (Prashant & Parida 2005) is scientifically fascinating and socially relevant. MSSRF is currently engaged in transferring drought-resistant genes from *Prosopis juliflora*, a common desert tree, to water-thirsty rice. Such genetic shielding of crop plants against salinity, submergence and drought would sustain the coastal agriculture and the livelihoods of millions of resource-poor farming families, even if the frequencies and intensities of extreme hydro-meteorological disasters increase. Recombinant DNA technology provides an opportunity to design and develop genetic shields against adverse changes in temperature, precipitation and sea level as a result of global warming.

(*f*) *Guiding principles for integrated coastal zone management*

For integrating disaster management strategies with sustainable development, the developing countries need to adhere to certain guidelines. An Expert Committee chaired by Prof. M. S. Swaminathan has proposed 12 basic guiding principles for the sustainable and scientific management of the coastal zone (Swaminathan 2005*a*,*b*). The first and the foremost one is that ecological security, cultural security, livelihood security and national security should be the cornerstones of an integrated coastal zone management policy. The coastal zone would include the area from territorial limits (12 nautical miles), including its sea bed, to the administrative boundaries or the biological boundaries demarcated on the landward side of the sea coast. The coastal zone management should also include the island water bodies influenced by tidal action and the land area along such water bodies. The *precautionary approach* should be used where there are potential threats of serious or irreversible damage to ecologically critical coastal systems and to living aquatic resources. Scientific uncertainty should not be used as an excuse for the unsustainable exploitation of coastal resources—both living and non-living. Ecological economics should underpin economic activities, so that present day interests and future prospects are not antagonistic. Significant biological, cultural and natural assets should be considered incomparable, invaluable and irreplaceable and should receive overriding priority in the allocation of resources for coastal area protection and conservation.

5. Tsunami devastation of Andaman and Nicobar Islands: converting calamities into opportunities

(*a*) *Nature of devastation*

According to Government of India reports (see the document of United Nations Country Team India, March 2005 on 'Recovery Framework in Support of Government of India' for a Post-tsunami Rehabilitation and Reconstruction Programme), 10 749 people in India lost their lives, and 5640 persons were still missing, after the tsunami of 26 December 2004. Approximately 7000 people escaped with injuries (Tsunami Rehabilitation Programme Planning Commission Government of India March 2005, Link: http://www.icsf.net/jsp/english/flashnews/rehabDocs/ind0236.pdf).

The island of Andaman and Nicobar archipelago consists of 572 emerald islands, islets and rocks. It stretches approximately 750 km from the northern most Andaman to the Southern most Indira point in Great Nicobar. Located in the Bay of Bengal, these islands stretch from approximately 6 to 14° north latitude, and approximately 92 to 94° east longitude.

The Sumatra–Andaman earthquake (M 9.3) on 26 December 2004 released 4.3×10^{18} J, equivalent to the energy of a 100 Gton bomb. Shifts in the sea floor displaced more than 30 km^3 of seawater (Bilham 2005), generating powerful tsunami waves. The rupture process was initially slow, but picked up speed at approximately 2.5 $km\,s^{-1}$ and proceeded north northwest approximately 1200–1300 km along the Andaman trough (Ammon *et al.* 2005).

From the point of rehabilitation and preparation of an action plan to integrate disaster management with sustainable development, the geological and geomorphological changes induced by the earthquake and tsunami need to be taken into account. Denyer (2005) has reported that the Andaman and Nicobar Islands appear to have shifted southwest by around 4 m (13 feet), according to GPS data. The northernmost inhabited island Diglipur in the north Andamans

PHILOSOPHICAL TRANSACTIONS OF THE ROYAL SOCIETY A MATHEMATICAL, PHYSICAL & ENGINEERING SCIENCES

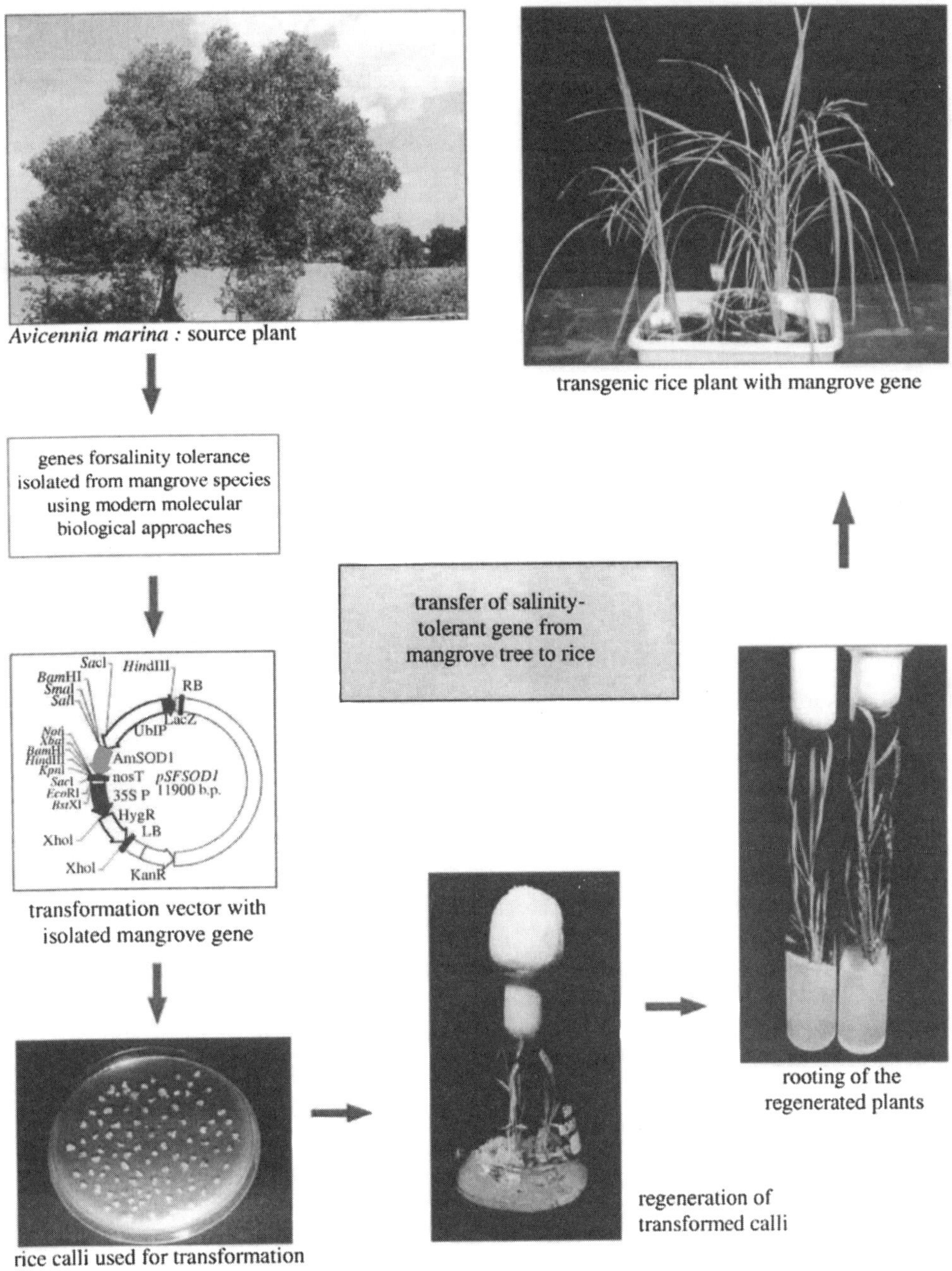

Figure 11. Genetic engineering of crop plants against abiotic stress.

has risen by 0.5–0.8 m (1.7–2.6 feet), whereas the southernmost Indira point (Great Nicobar) has sunk by approximately 1.4–1.8 m (4.6–5.9 feet). Many of the islands suffered extensive damage, as is shown for Katchal islands before (figure 12*a*) and after the tsunami (figure 12*b*).

PHILOSOPHICAL TRANSACTIONS OF THE ROYAL SOCIETY A MATHEMATICAL, PHYSICAL & ENGINEERING SCIENCES

(*b*) *Integrating disaster management with sustainable development of post-tsunami Andaman and Nicobar island archipelago*

The extensive geographic and geological changes caused by the extreme natural disaster have necessitated a new outlook and action plan for the Andaman and Nicobar islands. The past unsustainable practices should be brought to an end so that a new era, in which humans and nature are in a harmonious and mutually reinforcing relationship, can begin.

Keeping all these in view, the MSSRF has recommended locale-specific, sustainable solutions. The main focus is to integrate ecological security with livelihood security largely based on forestry and fisheries and put in place rural knowledge centres with the internet and GIS for early warning and disaster mitigation. The blue print for integrating disaster management into sustainable development was presented by Swaminathan (2005*b*).

The agriculture recommended is *organic* agriculture with *low input, low volume* and *high value crops*. The first and foremost is the erection of bioshield or the shelterbelt with mangrove or non-mangrove tree species based on ecological, edaphic and the hydrological conditions. After several rows of these, the other salt-tolerant, and economically useful species (e.g. coconut, *pandanus*, bamboo, canes, cashewnut, arecanut, etc.) which are all capable of reducing the velocity of cyclones, tsunamis, and the harmful effects of seawater rise due to global warming, should be planted. In between the rows of coconut or arecanut trees, spices (especially pepper, clove and nutmeg) can be grown organically as intercrop. This approach will effectively protect the lives and enhance the livelihoods.

The forest cover of Andaman and Nicobar islands is approximately 86%; the biodiversity is rich, and it includes a few unique fauna and flora which are endemic. Particular care needs to be taken to protect the uniquely rich biodiversity not only from human encroachment (i.e. subsistence farming with low value grain crops, polluting industries, etc.)

Ecological Security—a Prerequisite for Food and Livelihood Security

M. S. Swaminathan

Professor Mostafa Tolba has been a crusader for ensuring the ecological security of our planet. His contributions to the conservation and sustainable management of our land, water, forests and biodiversity resources are truly monumental. His untiring efforts led to several significant international protocols and conventions like the Montreal Protocol for protecting the ozone shield and the Biodiversity, Climate and Desertification Conventions.

In a Festschrift commemorating his lifelong contributions to protecting our life support systems, I have dealt briefly with the critical importance of ecological security for the food, health and livelihood security of present and future generations of humankind. I have also dealt with some ethical issues which should govern North-South relationships in sharing benefits from biological diversity, since this is an aspect on which Professor Tolba laid considerable emphasis.

The Green Revolution

In 1798, Thomas Malthus warned humankind about the dire consequences of unchecked population growth. The French mathematician, Marquis de Condorcet, a contemporary of Malthus, pointed out that population will stabilise itself if children are born for happiness and not just for mere existence. Thanks to dramatic advances in preventive and curative medicine, particularly since IUCN was established in 1948, human numbers have been increasing at a rapid rate. A billion persons are now being added to the human population every 12 to 13 years. This will call for an increase in annual food grain production by 250 million tons once in every 10 years. When Malthus published his essay in 1798 the global population was 980 million. The population of India alone is now 990 million. In spite of such a rapid rise in population, food production has on the whole kept pace with the needs of the

---

Ahmad K.Hegazy (ed.),
Environment 2000 and Beyond.
Printed by Horus for Computer and Printing, Cairo, Egypt. 1999.

expanding population during recent decades. How did this happen?

There were four important factors which helped to keep Malthusian fears at bay. These are:

- Rapid advances in science and technology, particularly in the area of breeding new varieties and hybrids of food crops which can respond well to irrigation water and good soil fertility.
- Services like the production and distribution of good seeds, fertilisers and pesticides, as well as the organization of efficient credit and extension services.
- Public policies in the areas of land reform, rural infrastructure development, input and output pricing and marketing.
- The hard work of farm women and men, who have demonstrated that whether they live in industrialised or developing countries, they will respond to technological progress and opportunities for assured and remunerative marketing.

Mutually reinforcing -packages of technology, services and public policies led to the birth of what was termed, in October 1968 by Dr William Gaud of the United States of America, the "green revolution." In January of the same year (i.e. 1968), I had made the following points in my Presidential address to the Agricultural Science Section of the Indian Science Congress held at Varanasi, India.

"Exploitive" agriculture offers great possibilities if carried out in a scientific way, but poses great dangers if carried out with only an immediate profit motive. The emerging exploitive farming community in India should become aware of this. Intensive cultivation of land without conservation of soil fertility and soil structure would lead, ultimately, to the springing up of deserts. Irrigation without arrangements for drainage would result in soils becoming alkaline or saline. Indiscriminate use of pesticides, fungicides and herbicides could cause adverse changes in biological balance as well as lead to an increase in the incidence of cancer and other diseases, through the toxic residues present in the grains or other edible parts. Unscientific tapping of underground water will lead to the rapid exhaustion of this wonderful capital resource left to us through ages of natural farming. The rapid replacement of numerous locally adapted varieties with one or two high-yielding strains in large contiguous areas would result in the spread of serious diseases capable of wiping out entire crops. Therefore, the initiation of exploitive agriculture without a proper understanding of the various consequences of every one of the changes introduced into traditional agriculture, and without first building up a proper scientific and training base to sustain it, may only lead us, in the long run, into an era of agricultural disaster rather than one of agricultural prosperity.'

The above analysis resulted in the intensification of research on integrated pest management (IPM) and the initiation of whole village operational research projects involving IPM procedures in rice and cotton (Swaminathan, 1975). Also, in 1974, the mineral fertiliser industry was requested to adopt integrated nutrient supply systems (INS) involving organic manures, compost, green manures, biofertilisers and cereal-legume crop rotations (Swaminathan, 1974). Through the integration of INS and IPM procedures, technologies were developed to make the excessive use of mineral fertilisers and chemical pesticides unnecessary.

In subsequent years, the significance of my 1968 analysis has been widely realised. Genetic homogeneity enhances genetic vulnerability to pests and diseases. Excessive use of mineral fertilisers and chemical pesticides have caused soil degradation, ground water pollution and the spread of pests resistant to pesticides in several green revolution areas. This has led to warnings of an impending food crisis in the coming millennium. In the mid-1990s, Lester Brown and his associates, for example, have concluded that population-rich but land hungry countries like China and India may have to resort to substantial food imports in another 30 years due to the following reasons.

- First, increasing population leads to increased demand for food and reduced per capita availability of arable land and irrigation water.

- Second, improved purchasing power and increased urbanisation lead to higher per capita food grain requirements due to an increased consumption of animal products.
- Third, marine fish production is tending to become stagnant.
- Fourth, there is increasing damage to the ecological foundations of agriculture, such as land, water, forests, biodiversity and the atmosphere and there are distinct possibilities for adverse changes in climate and sea level.
- Finally, no dramatic technological breakthroughs are on the horizon, which can help to halt the fatigue of the green revolution.

Since land is a shrinking resource for agriculture, there is no option except to produce more food and other agricultural commodities from less per capita land. In other words, the need for more food has to be met through higher yields per units of land, water, energy and time. It would therefore be useful to examine how science can be mobilised for raising further the ceiling to biological productivity without associated ecological harm.

It will be appropriate to refer to the emerging scientific progress on the farms as an "ever-green revolution," to emphasise that the productivity advance is sustainable over time since it is rooted in the principle of ecology, economics, social and gender equity and employment generation (Swaminathan, 1996).

To give an idea of the dimensions of challenges faced by all involved in developing scientific strategies and public policies for sustainable food security, I would like to refer to the situation prevailing in my country, India.

In global terms, India today has:

- 16 per cent of human population
- 15 per cent of farm animal population

- 2 per cent of the geographical area
- 1 per cent of rainfall
- 0.5 per cent of forests
- 0.5 per cent of grazing land

Such a situation also prevails in many other countries in Asia and Africa. The demand for food and water will increase every year. At the same time, the ecological foundations essential for sustained advances in biological productivity, such as land, water, forests, biodiversity and the atmosphere are experiencing degradation or depletion. *The Living Planet Index* published recently by the World Wide Fund for Nature (WWF) underlines the seriousness of the situation.

Meeting the challenges ahead

Julian Huxley wrote in 1957, 'It is as if man has been suddenly appointed managing director of the biggest business of all, the business of evolution, whether he wants it or not he is determining the future direction of evolution on the earth.' It is clear that if children are to be born for happiness in the coming millennium, nature, humanity and technology must work in harmony. The present pattern of development based on the destruction of environmental capital stocks and life support systems, gross economic, social and gender inequity and jobless economic growth cannot lead to a better common future. We need a paradigm shift in our approach to technology development. Nature resources management concerns must be integrated in genetic improvement methodologies. Sir Francis Bacon once said, 'It would be an unsound fancy to expect that things which have never yet been done can be done except by methods which have never been tried.'

What are the new methods available to meet the first among the hierarchical needs of humankind, namely food? Fortunately, as we approach the new millennium, we are experiencing three major revolutions in science and technology, which will influence agricultural technology in a fundamental manner. It will therefore be appropriate to make a brief reference to them.

(i) The gene revolution—which provides a molecular understanding of the genetic basis of living organisms, as well as the ability to use this understanding to develop new processes and products for agriculture, the environment and human and animal health.

(ii) The information and communication revolution—which allows a very rapid growth in the systematic assimilation and dissemination of relevant and timely information, as well as a dramatically improved ability to access the universe of knowledge and communicate through low cost electronic networks.

(iii) The ecotechnology revolution—which promotes the blending of the best in traditional knowledge and technology with frontier technologies such as biotechnology, space and information technologies, renewable energy and new materials.

In principle, these three types of advances—when coupled with improvements in management science and governance—greatly increase the power of a scientific approach to genetic improvement, agronomics, the integrated management of natural resources and ecosystems, and the management of local and regional development policies. Eco-technologies enable the adoption of ISO 9000 and ISO 14000 standards of environmental management. These scientific revolutions seem to be proceeding at an ever increasing pace, with most of the action occurring in a few places in industrialised nations. Also, new discoveries of great relevance to sustainable food security are coming under the purview of proprietary science.

(i) *The gene revolution*

The past ten years have seen dramatic advances in our understanding of how biological organisms function at the molecular level, as well as in our abilities to analyse, understand and manipulate DNA molecules, the biological material from which the genes in all organisms are made. The entire process has been accelerated by the Human Genome *Project, which* has poured substantial resources into the development of new technologies for working with human genes. The same technologies are directly applicable to all other organisms, including plants. Thus, a new scientific discipline of genomics has arisen. This discipline has contributed to powerful new approaches that can be used in agriculture as well as in medicine and has helped to promote the biotechnology industry.

Several large corporations in Europe and the United States have made major investments in adapting these technologies to produce new plant varieties of agricultural importance for large-scale commercial agriculture. The same technologies have equally important potential applications for addressing food security in the developing world.

The key technological developments in this area are:

(a) *genomics*, the molecular characterisation of species;

(b) *bioinformatics*, data banks and data processing for genomic analysis;

(c) *transformation,* introduction of individual genes conferring potentially useful traits into plants, trees, livestock and fish species;

(d) *molecular breeding,* identification and evaluation of useful traits by use of marker assisted selection, which greatly speeds up traditional breeding processes;

(e) *diagnostics,* identification of pathogens by molecular characterisation; and

(f) *vaccine technology,* use of .modern immunology to develop recombinant DNA vaccines for improved disease control against lethal diseases of animal and fish.

There are widespread concerns about the potential adverse impact of genetically modified organisms (GMOs) on human health and the environment. Some of these concerns are genuine. In order to take advantage of recombinant DNA technologies without associated harm to human or ecological health, it is important that every country has in place suitable institutional structures and regulations for biosafety, bioethics and biosurveillance. A recent statement, issued by the Royal Society of London on 'Genetically modified plants for food use,' provides guidelines for the safe handling of biotechnological applications (Royal Society, 1998). Also, the basic feedstock for the biotechnology industry is biodiversity.

(ii) *The information technology revolution*

New communication and computing technologies will have profound implications in everyday research activities.

(a) Access to the Internet will soon be universal, and it can provide unrestricted low-cost access to information, as well as highly interactive distance learning. The Internet will not only facilitate interactions among researchers, but also greatly improve their ability to communicate effectively with the potential users of their research knowledge.

(b) Computing makes it possible to process large-capacity databases (libraries, remote sensing and GIS data, gene banks) and to construct simulation models with possible applications in ecosystem modelling, preparation of contingency plans to suit different weather probabilities and market variables.

(c) The software industry is continuously providing new tools that increase research productivity and create new opportunities for understanding complex systems of growing conditions.

(d) Remote sensing and other space satellite outputs are providing detailed geographic information useful for land and natural resources management.

(iii) *The ecotechnology revolution*

Knowledge is a continuum. There is much to learn from the past in terms of the ecological and social sustainability of technologies. At the same time, new developments have opened up new opportunities for developing technologies which can lead to high productivity without adverse impact on the natural resources base. Blending traditional and frontier technologies leads to the birth of ecotechnologies with combined strength in the following areas:

- Economics
- Ecology
- Equity
- Employment
- Energy

For example, in the area of water harvesting and sustainable use, there are many lessons to be learnt from the past. In the desert area of Rajasthan, India, drinking water is available even in areas with 100 mm annual rainfall, largely because women are continuing to harvest water in simple structures called *kunds*. In contrast, drinking water is scarce during summer months in some parts of North-East India with an annual rainfall of 15000 mm. Thus, there is need to conserve traditional wisdom and practices, which are often tending to become extinct (Agarwal and Narain, 1996). The decision of the World Intellectual Property Organization (WIPO) to explore the intellectual property needs, rights and expectations of holders of traditional knowledge, innovations and culture is hence an important step in widening the concept of intellectual property. Principles of ethics and equity demand that this invaluable component of IPR gets included when the TRIPS agreement (Trade-related Intellectual Property Rights) of the World Trade Organization comes up for review in 1999. FAO has been a pioneer in the recognition of the contributions of farm families in genetic resources conservation and enhancement by promoting the concept of "Farmers Rights". Like WIPO, UPOV (Union for the Protection of New Varieties of Crops) should also undertake the task of preparing an integrated concept of breeders' and farmers' rights.

Precision farming and sustainable food security

Precision agriculture involves a system approach to experimental design and agronomic practices. It needs interdisciplinary research drawing on expertise in a range of subject areas such as agronomy, plant science, genetics, soil science, entomology, meteorology, weed science, plant physiology, plant pathology, ecology and economics (National Research Council, 1997). Agricultural extension workers, using information technology, will play an increasingly important role in crop production and natural resources management. The curricula of Agricultural Schools and Colleges need to be modified to make precision agriculture the road to an ever-green revolution. Precision agriculture is particularly valuable for increasing opportunities for skilled employment in the farm sector. For example, computer software development, equipment fabrication and sales, custom hiring of software and farm equipment, local production of biofertilisers, biopesticides and drip irrigation equipment and consultancy services can all provide new opportunities for unskilled workers to become skilled.

Precision farming methods have to be based on scientific land and water use planning. It will need concurrent attention to natural capital stocks and nature's services. Examples of stocks include: soils and soil nutrients, biodiversity, water, minerals, forests and oceans. Examples of natures' services include: water cycles, nutrient cycles, carbon sequestration and waste recycling, Agroforestry and other sustainable systems of land management need to be popularised in areas experiencing varying degrees of desertification. Costanza *et al.* (1997) have assessed the value of 17 ecosystem services and flows for 16 biomes (desert, tundra, ice/rock excluded) at $33 trillion per year. By comparison, global GNP is now $18 trillion per year. Food production services of croplands is $0.13 trillion per year. Thus, the need to conserve ecosystem services and flows under varying agro-ecological conditions through precision farming practices can hardly be over-emphasised.

In the emerging knowledge-intensive agriculture era, international cooperation is vital for taking the benefits of new technologies to those who have so far been bypassed by new knowledge and techniques. Women farmers and farm labour need particular attention in any agricultural extension and development programme designed to reach the unreached. The gender dimension needs to be internalised in all research, educational and development programmes intended to promote the conservation of biodiversity and natural resources (Swaminathan, 1998).

Substainable agriculture in the twenty-first century will be based on the appropriate use of biotechnology, information technology and ecotechnology. Practical achievements in bringing about the desired paradigm shift will depend upon public policy support and political action, Regulation through legislation, social mobilization through local level community organizations and education through the mass media and information shops will all be needed to meet the dual demands for food and ecological security.

The experience of the present century teaches us that unless technology and public policy are rooted in the principles of ecology, social and gender equity, employment generation and energy conservation, development will not be environmentally and socially sustainable. While technological progress, particularly in the areas of genomics, molecular breeding and information technology, is rapid, much of the new knowledge and technology is coming under monopolistic control through the system of intellectual property rights (IPR). Support for public-good research is shrinking, while investment in research for commercial projects is increasing at a fast pace. At the same time, it is known that knowledge is a continuum and that much of the progress in molecular breeding has become possible because of the genetic conservation traditions of rural and tribal women and men.

Also, a recent study in the USA showed that 73 per cent of private patents were based on knowledge generated by publicly funded institutions like Universities and government laboratories (Thurrow,

1997). The ethical aspects of genetic engineering research are also assuming significance (Reiss and Straughan, 1996). Thus, on the one hand we need new tools to solve the serious environmental and economic issues confronting humankind; on the other hand, there are important ecological, social and ethical issues relating to the use of such new tools, particularly recombinant DNA technology.

The Convention on Biological Diversity (CBD) shows the way to approach this dilemma. For the first time, an internationally built convention views the problems of conservation, sustainable use and equitable sharing of benefits in a holistic manner. Also, for the first time, international organizations like WIPO are looking at human creativity in an integrated manner. The edifice of both human food and health security has been built on the knowledge gained and material conserved by rural and tribal communities. Knowledge is a continuum and it is obvious that if its roots are starved, the tree of knowledge will not bear fruit for long. Traditional knowledge forms the backbone of modern discovery. Recognition and reward systems for traditional knowledge and genetic material should ensure that:

- commercialisation of knowledge does not result in inhibiting the sharing of knowledge and material for public good;
- the gender dimension is internalised in benefit-sharing mechanisms; and
- the reward system helps to preserve and revitalise the conservation of rural and tribal communities.

The emerging biological and knowledge millennium will need appropriate national, regional and global systems of knowledge creation and sharing based on Albert Einstein's dictum, 'concern for man himself and his fate must always form the chief interest of all technical endeavours in order that the creation of our minds shall be a blessing and not a curse.' As stressed earlier, effective mechanisms for biosafety, bioethics and biosurveillance will be needed to take advantage of genetic engineering without associated ecological and or social harm. There must be an internationally agreed protocol on biosafety as provided in Article

19 of the CBD in order to ensure the environmentally safe handling of genetically modified organisms.

The importance of information empowerment will be clear from the fact that in the referendum conducted on June 7, 1998 in Switzerland, contrary to general expectation, more than 66 per cent of the people and all 26 cantons voted against outlawing genetic alteration of animals and the release of genetically modified organisms into the environment. According to Prof. Heidi Diggelmann, President of the Swiss National Science Foundation, this was because of researchers talking to the people in the streets and clearing their concerns and fears (*New Scientist*, June 13, 1998).

Urgent tasks for the global community

(i) *Implementing the Convention on Biological Diversity*

The Convention on Biological Diversity (CBD), which has been in force since 1993, is a significant landmark among international agreements, since for the first time it incorporates the principle of equity and ethics in both access to genetic wealth and in sharing benefits. It aims to end the prevailing irony in which conservers and holders of genetic material and knowledge remain poor, while those who utilise their knowledge and germplasm get rich. Article 15(1) of the CBD recognises the sovereign rights of nations over the genetic resources occurring in their respective territories. Article 15(5) stipulates the need for prior informed consent in the use of genetic resources, while Article 15(4) recommends that access should be on mutually agreed terms. Several Articles of CBD stress the need to recognise and conserve traditional knowledge and wisdom. CBD also draws attention to the critical role of women in genetic resources conservation and enhancement.

CBD's triple goals are: conservation of biodiversity, sustainable use of biological wealth and equitable sharing of benefits. Any piece of legislation designed to empower governments and nations to give effect to the provisions of CBD obviously should strengthen ongoing efforts in the *in situ*, *ex situ* and *in situ*-on farm conservation of biodiversity as well as in their sustainable use for public good. Food, biotechnological and pharmaceutical industries have an economic stake in the use of biodiversity. The National Biodiversity Act should clearly establish that obtaining samples of biological material from areas within national jurisdiction for purpose of export, or commercial or industrial use, is prohibited without Prior Informed Consent (PIC) and reaching Mutually Agreed Terms.

Suitable Material and Information Transfer Agreements will have to be developed for this purpose. A system of recognition and reward will have to be developed that is capable of being implemented at the community level. A transparent mechanism will have to be developed for conferring financial benefits on the holders and conservers of knowledge and material. Clear guidelines for biodiversity access and benefit sharing will have to be developed, so that academic studies and research by University Ethnobiology and Anthropology Departments as well as by Agricultural, Medical and other Universities and industrial, agricultural and pharmaceutical R & D institutions, can proceed without interruption and harassment. It should not be forgotten that biological wealth is our most precious wealth and that it is the fundamental duty of our R & D institutions to convert biodiversity into economic wealth in an environmentally sustainable and socially equitable manner. Further, it should be our national goal to conserve all of our plant and animal species. Special efforts are needed to conserve and study micro-organisms.

Let me highlight a few features of the brief outline of a National Biological Diversity Act. *First*, any Act of this kind dealing with diversity should also take into account the diversity and pluralism characteristic of the socio-cultural and agro-ecological conditions prevailing in the country. *Second*, it should create an enabling environment where extensive public participation in the conservation and sustainable and equitable use of biodiversity is possible at the village, state and national levels. *Third*, the legislation should foster chronicling and documentation of biological wealth at the village/block level, using models like the Peoples' Biodiversiy Register developed by Professor Madhav Gadgil and his colleagues. This will help to safeguard the IPR rights of local communities. *Finally*, the legislation should help to address issues such as the following: (a) who owns biodiversity? (b) who is authorised to take decisions on PIC? (c) who benefits and how? and (d) has specific attention been given to the role and contributions of women?

It is obvious that to avoid frequent litigation, the Biodiversity Act should be simple, implementable and not overloaded with multiple objectives not directly related to the primary purpose. The Act should take into account the numerous existing pieces of legislation at the central and state levels relating to the protection of wildlife, forests and marine and terrestrial ecosystems. The Act should provide for structures and processes which can help to address, on a continuing basis, the complex issues relating to access and benefit sharing, rather than lay down rigid prescriptions and procedures. It is equally important that any piece of legislation, however good the intention may be, does not come in the way of the traditional uses of medicinal and food plants by tribal and rural families. Nor should the Act hinder scientific research in our Universities and Research Institutions. The experience gained in the Philippines in operating a Presidential decree on Biodiversity has indicated that legal measures for access, prior informed consent and benefit sharing should be simple and practical and should not lead to corruption and harassment.

(ii) *Plant Variety Protection Rights Act*

A system of recognition and reward of the work of plant breeders has been in existence in several European countries and the United States of America for over 50 years. The USA created, in 1930, a special form of protection termed "plant patent" for asexually reproduced varieties of plants like the potato and several fruit trees and ornamental plants.

An ordinance issued in the Netherlands in 1941, granted to breeders an exclusive right to market the first generation of certified seeds of agriculturally important species. Several other European countries took similar steps during the fifties.

In order to provide common guidelines and promote reciprocity in the field of plant variety protection, an International Convention for the Protection of New Varieties of Plants was adopted at a diplomatic conference held in Paris on December 2, 1961. This step helped to provide recognition of the rights of plant breeders on an international basis. The Member States of the Convention constitute the international Union for the Protection of New Varieties of Plants (UPOV). The UPOV convention has been revised in 1972, 1978 and 1991. Before the coming into force of the World Trade Agreement in 1994, UPOV consisted of 27 Member States. Now, it has nearly 40 Members. The 1991 UPOV Act came into force in April, 1998.

According to the TRIPS agreement, while "plants and animals other than micro-organisms" may be excluded from patentability, the member countries shall provide 'for the protection of plant varieties by an effective *sui generis* system or by any combination thereof.' Thus the choice of the system of protection of plant varieties is left to each country. If patent protection is not given, it is obligatory to provide an effective *sui generis* system (i.e. a system of its own). What is an effective *sui generis* system? This was not spelt out in the TRIPS agreement but it is commonly assumed that the provisions of the 1978 version of the UPOV convention will satisfy the criteria of effectiveness.

The Plant Variety Protection Acts in operation in industrialised countries have no provision for recognising and rewarding the contributions of the primary conservers of genetic variability in economic plants. Tribal and rural women and men have been saving and selecting seeds for cultivation under their respective agro-ecological conditions and socio-cultural needs. This is why we find enormous variability within a cultivated species. For example, over 100,000 strains occur in rice. Tribal and rural women in particular have not only conserved genetic diversity in species of value to the food and health security systems of the world, but have also added value to land races through selection for characters like yield, quality and tolerance to a wide range of biotic and abiotic stresses. Presently, their work goes unrewarded, while those who use their material and information to develop new varieties can seek varietal protection and royalties.

To end this unethical dichotomy, the concept of Farmers' Rights was developed in the forum of the Food and Agriculture Organization (FAO) of the UN over ten years ago. The goal of Farmers' Rights is to accord recognition to the past, present and ongoing contributions of farm families to genetic resources conservation and enhancement. Though the concept of Farmers' Rights has been accepted by Member Nations of FAO and its purpose is reflected in the access and benefit sharing provisions of CBD, no concrete measures have been taken so far to implement them. Methods of implementing, at field level, the concept of Farmers' Rights were developed at two interdisciplinary dialogues held at the M .S. Swaminathan Research Foundation, Chennai, in early 1994 and 1996. These dialogues led to the formulation of a draft Plant Variety Protection and Farmers' Rights Act, several features of which were incorporated by the Union Ministry of Agriculture in its draft legislation.

Farmers and breeders are partners in the struggle for sustainable food security and their contributions represent a continuum in the genetic innovation chain. Their respective rights should therefore become mutually reinforcing and should not be portrayed as antagonistic. So far, farm families have been conserving agrobiodiversity (i.e. biodiversity of agricultural value) for the public good at personal cost. It is time their contributions receive the same recognition as those of breeders.

The 1991 draft of the UPOV convention gives the option to member countries to restrict farmers' plant back rights. This draft strengthens breeders' rights and weakens farmers privileges. Since farmers can be cultivators of crops or conservers of genetic material or both, the MSSRF draft groups farmers' roles into two categories—cultivators and conservers—for the purpose of articulating their respective roles with clarity.

If India enacts a Plant Variety Protection and Farmers' Rights Act soon, it will be the first country in the world to recognise the invaluable role of tribal and farm families in the conservation and improvement of intra-specific genetic variability at the farm level. This method is now referred to as *in situ*-on farm conservation. Many of the critical genes which contribute to pest and disease resistance, grain quality and local adaptation characteristics of modern cultivars have come from this method of conservation and selection. The number of land races and folk varieties which go into the pedigrees of present day crop varieties is steadily increasing. Many modern rice varieties may have over 80 strains in their parentage, a quarter of which will be land races. By stipulating that breeders should provide, at the time of submitting their application for varietal protection, information on the approximate geographical area from which the genes critical to the unique features of their varieties came, it will be possible to recognise the work of farmer conservers.

An individual farmer who has developed a new strain can claim for recognition and reward in the same way as a breeder from a research institution or commercial company. The difficulty in recognition lies largely in the field of communities. Most valuable plant material has been selected and maintained by entire communities living in a specific ecological niche. Therefore, the 1994 MSSRF draft Act provided for creating a Community Gene Fund from which funds can be provided to the farmer-conservers to strengthen their *in situ*-on farm conservation traditions and for other purposes they may consider important. Such a step will not only help to accord social prestige and recognition to an activity of great importance to sustainable food and health security, but will also help to revitalise the conservation ethos of local communities. There will be a learning period in the implementation of the rights of farmer-conservers. However, only action can help to refine methods of community recognition and reward.

There is considerable experience in the world in implementing Breeders' Rights. In our legislation relating to the protection of new plant varieties based on criteria such as distinctiveness, novelty, homogeneity and stability, we should also address issues such as: how far back and for how long the recognition should go; unrestricted access to genetic material to breeders; the rights of farmers to retain seeds of protected varieties for their own use and for limited exchange in their neighborhood; and the extent of royalties and method of their collection. A provision has also to be made for automatic license of right, i.e. non exclusive Plant Variety Rights, if this is considered essential in the interests of national food and health security.

The pedigrees of modern plant varieties contain parents drawn from many countries. A national act can only deal with the material occurring within a country. A system of equitable benefit sharing as stipulated in CBD, can be developed only through agreements at the regional and global levels. About eight years ago, I suggested, for this purpose, the conversion of UPOV into an International Union for the Protection of Breeders' and Farmers' Rights. It is a matter for satisfaction that the World Intellectual Property Rights Organisation has started to consider questions relating to according recognition to traditional knowledge systems and informal innovations. ASEAN countries are now developing a common protocol on access and benefit sharing with reference to biodiversity. South-South collaboration in exchange of genetic material is even more important than North-North collaboration, since most of the centres of origin and diversity in economic plants occur in the South.

Developing countries should take the lead in adding the dimensions of ethics and equity in the TRIPS agreement, when it comes up for review next year. Only by ensuring that economic development has its roots in ecology, social and gender equity and ethics, can we help to realise Professor Mostafa K. Tolba's vision for global ecological security.

References

(1) Agrawal Anil and Sunita Narain (eds) (1997). *Dying Wisdom: Rise, fall and potential of India's traditional water harvesting systems*. Center for Science and Environment, New Delhi, 404 pp.

(2) Costanza Robert *et al.* (1997). The Value of the World's Ecosystem Services and Natural Capital. *Nature* Vol 387, 15 May 1997, pp. 253-260.

(3) Lester C Thurrow (1997). Needed a new system of Intellectual Property Rights. *Harvard Business Review* (September - October, 97). pp. 95-103.

(4) National Research Council, USA (1997). *Precision Agriculture in the 21st Century.* National Academy Press, Washington, DC. 149 pp.

(5) Oliver Klaffke, Basel. Altered minds, *New Scientist*, 13 June 1998 page 5.

(6) Reiss Michael J and Roger Straughan (1996). *Improving Nature? The Science and Ethics of Genetic Engineering.* Cambridge University Press.

(7) The Royal Society, London (1998). *Genetically modified plants for food use.* London, 16 pp.

(8) Swaminathan, M. S. (1968). The age of algney, genetic destruction of yield barriers and agricultural transformation. Presidential Address, Agricultural Science Section, Fifty fifth Indian Science Congress, January 1968. Proceedings Indian Science Congress, Varanasi.

(9) Swaminathan, M. S. (1974). Development and Spread of Integrated Nutrient Supply Systems. Proceedings of FAI-FAO seminar. pp. 61-63

(10) Swaminathan, M. S. (1975). I C A R Operational Research Projects, purpose and approach, *Indian Farming.* August 1975.

(11) Swaminathan, M. S. (1996). *Sustainable Agriculture. Towards an Evergreen Revolution*, Konark Publishers Pvt, Ltd, Delhi, 232 pp.

(12) Swaminathan, M. S. (ed.) (1998). *Gender dimensions in Biodiversity Management:* Konark Publishers Pvt, Ltd, Delhi, 229 pp.

Presidential Address

Genetic Conservation: Microbes to Man

M.S. Swaminathan

Interest in environmental assets began with the origin of settled agriculture. When the transition from gathering to growing food took place, there was immediate interest in the collection and utilization of useful plants and animals from the native flora and fauna. Thus, most of the important economic plants and animals were domesticated thousands of years ago. The need to conserve animal and forest resources was taught and decreed in parts of China and India as far back as 700 B.C. Genetic conservation, as we now understand and practise it, is however a relatively young venture, partly deriving its roots from the pioneering studies of De Candolle (1882) and Vavilov (1926).

At the VI International Congress of Genetics held in 1932 at Ithaca, New York, Vavilov pointed out that "the growing needs of civilized man and the development of industry make the introduction of new plants necessary. The vast resources of wild species, especially in the tropics, have been practically untouched by investigation." Even as Vavilov was explaining the importance of plant exploration and conservation, concern was expressed by several geneticists over the loss of genetic stocks due to the process of modernization of agriculture. It was pointed out that without coordinated efforts around the world valuable genes could be lost through neglect. The need for a clear national policy on. genetic stocks and genetic diversity has been stressed at numerous symposia on genetic conservation.

After the corn blight epidemic in the early 70s in the USA, the National Academy of Sciences set up a Committee on Genetic Vulnerability in Major Crops (1972) to review the situation relating to the diversity of the genetic material in widely cultivated major crops. The Committee found that the genetic diversity of many of the important crops in the USA was dangerously narrow. For example, 96 percent of the garden pea crop was planted to only two strains, and 95 percent of the groundnut crop to only nine varieties. Consumer preference and market forces were both held responsible for genetic uniformity and therefore to vulnerability. In recent years, many symposia and seminars have been organized by FAD, United Nations Environment Programme, and other agencies to review the situation relating to genetic diversity and vulnerability (Frankel and Bennett

1970, Frankel 1974, Frankel and Soule 1981). Today, only about 150 plant species with about a quarter million local strains are important in meeting the calorie needs of human populations. With the spread of high yielding varieties of crops, the existing variability is under threat of extinction. The UN Conference on Human Environment held UN Stockholm in 1972 drew attention to the serious consequences of losing the existing natural variability. The International Biological Programme also gave considerable attention to the conservation of habitats rich in genetic diversity. In 1974, the International Board for Plant Genetic Resources (IBPGR) was organized by the Consultative Group on International Agricultural Research (CGIAR) in order to provide structured support to the collection, conservation, and utilization of plant genetic resources.

THE TWO SCENARIOS

The year 1983 marks the 30th anniversary of the discovery of the chemical structure of DNA by Watson and Crick. This same year also marks the 20th anniversary of the UN-FAD World Food Programme (WFP). Seminars organized to commemorate the 30th anniversary of the discovery of the double helix structure of DNA have drawn attention to the numerous possibilities now open for recombining genetic material in different combinations. It is now possible to identify portions of the genetic molecule responsible for specific characteristics and then splice this portion onto genetic molecules from another strain. Research in cell biology, molecular genetics, recombinant DNA, tissue culture and related fields is opening up new possibilities for progress in agriculture, medicine, and industry.

While the vistas opened up by genetics are exciting, the real life scenario on the food front is depressing. The World Food Programme is the largest multilateral food aid programme currently in operation and has provided food to over 200 million people, both during emergencies and under normal conditions during the last 20 years. A seminar held to commemorate the 20th anniversary of WFP drew attention to the fact that over 500 million people may be going to bed hungry today. The dimensions of the world food problem are thus vast. In fact as recently as 1968 Paul Ehrlich, in his book "The Population Bomb," remarked: "The battle to feed all of humankind has been lost. In the seventies, the world will experience famines where hundreds of millions of people will starve to death." So far, the worldwide food catastrophe predicted by such experts has not materialized, thanks to recent progress in the genetic improvement of food crops. The rapid agricultural transformation witnessed in countries like Bangladesh, China, India, Indonesia, Pakistan, and the Philippines was triggered by the development and release of crop varieties with the capacity to respond to good management. Genes for semidwarf plant type, photoperiod insensitivity, and multiple resistance to pests and diseases as well as male sterility induced by gene-cytoplasm interaction provided the tools for developing high yielding varieties of hybrids in several cereals. Rapid population growth has however consumed many of the benefits of agricultural progress. It has been estimated that early in the year 2015, about 8 billion people will have to be fed and provided with jobs. If population stabilization programmes are successful, it is possible that the world population could be maintained at about 10 billion in another 50 years. The gains in production that appear possible from research and development in agriculture and aquaculture should meet much of the need for more food during the rest of the century.

For periods beyond, it is important that progress in fields like recombinant DNA technology, production of monoclonal antibodies, tissue culture, and protoplast fusion' should be fully harnessed. It is now feasible to construct entirely new genes and to modify the known ones. The principal practical results have been obtained with genetic manipulation of microorganisms, notably Escherichia coli. However, studies to alter the genomes of other prokaryotes and several eukaryotes are in progress using an increasing array of new vectors and techniques.

Human and other animal genes have been incorporated into plasmids of E. coli and expression of these genes has resulted in the production of human insulin and several other pharmaceutical products that are now undergoing clinical tests. By changing the sequences of nucleotides in known genes, new proteins have been obtained. In this way, interferons having enhanced effectiveness against viruses are being developed.

EVOLUTIONARY RESPONSIBILITY

Frankel (1974) described genetic conservation as an evolutionary responsibility at the XIII International Congress of Genetics held at Berkeley in the USA. Are we discharging this responsibility? To answer this question, it is important to classify spheres of responsibility and assign them to well-designed groups. We can discern three major groups of responsibilities based on the action needed, namely professional, political, and public. Professional responsibility has to be discharged by geneticists, plant and animal breeders, ecologists, conservationists, and a whole series of scientists and technologists concerned with the identification, collection, conservation, and utilization of genetic variability. The political aspect of the problem relates to the development of national policies which will help to accord priority to the protection of the environment conservation of genetic material, and appreciation of the dangers arising from genetic erosion and vulnerability and the consequent need to provide enough financial and technical resources to all work related to conservation. Above all, the political leaders must be committed to the cause of considering genetic material as common human heritage and hence should promote free exchange of material and coordinated action at the national, regional, and global levels on the utilization of germplasm for common good. Even if the necessary professional skill and political will are available, the cause of conservation will go by default if there is no widespread awareness among the general public of the need to promote development without destruction and of the pivotal role the people themselves can play in genetic conservation. Public awareness can be generated by mass media and through schools and colleges all over the country.

SCIENTIFIC ASPECTS OF CONSERVATION

I shall discuss briefly these three aspects of the problem of conservation starting with the purely scientific aspect. The scientific aspect of genetic conservation has been described and discussed in numerous books and symposia proceedings (see Hawkes 1983, Myers 1983). The first step in such work is to define the categories of material worthy of preservation and the major methods of preserving them. The following categories are usually regarded as important:

a) *Cultivated varieties (cultivars) in current use.* These are varieties of recognized value and performance released for cultivation after testing and approval. Successful cultivars are widely used in plant production in a given country or in several countries having similar cropping conditions. Uniformity of characteristics is a feature of this category.

b) *Obsolete cultivars.* These are varieties that were cultivated in, the past, but have now been replaced by the cultivars referred to in (a). Uniformity is also, to a certain degree, a feature of this category.

c) *Primitive cultivars or land races.* These are varieties that have been used for centuries in traditional agricultural systems. They were the products of selection by man, but have not undergone improvements by way of plant breeding. Variability in characteristics as observed in the field is a feature of this category.

d) *Wild species and weedy species closely related to cultivated varieties.* These ancestors of cultivars are species or relatives of crop plants that have not been cultivated varieties through plant breeding. This category also covers species of direct economic value, such as forest trees.
e) Wild species of potential value to man. These are species which are not cultivated and whose importance has not yet been assessed, but may be identified through exploration.

f) *Special genetic stocks.* This is material that has normally been developed by man and is or has been used in ongoing breeding programmes. It includes mutants, "breeders lines", lines with identified genes orgene combinations, gene marker stocks, and chromosomal variant types such as polyploids, aneuploids, translocations, and deficiencies. Material of this type is particularly useful because of the identification of genes for specific characteristics.

Conservation methodologies can take the following three forms:

a) Entire biomes. The preservation of vast tracts with the in situ conservation of animals and plants. This type of preservation will be extremely important in slowing the rate of species extinction.

b) In situ preservation of land races and wild relatives where genetic diversity exists and where wild/weedy forms are present, often hybridizing with the cultivated. These are evolutionary systems that are difficult for plant breeders to stimulate and should not be knowingly destroyed. Their disappearan<;¥ will give us more time to better understand how these systems evolved. Considerable potential for creative institutional arrangements exists for in situ preservation, especially in the developing countries.

c) Ex situ preservation of seed or in vitro cell lines stored in gene banks under appropriate conditions for long-term storage. This is the mode of preservation of most genetic resources. For doing this scientifically, considerable knowledge of the genetic structure of populations, sampling techniques, methods of regeneration and maintenance of varietal gene pool particularly in cross-pollinated plants will be necessary. Such a system to be useful requires documentation and evaluation so that a plant breeder will have enough information to know what to request. Information management will be as important as the physical arrangements of the gene bank. Ex situ preservation involves exploration, collection, storage, evaluation and documentation, and, lastly, breeding for the effective use of desirable genes.

Plucknett et at. (1983) have described the various steps taken by the International Board of Plant Genetic Resources since its inception in 1974 for identifying crop and geographical priorities in collection and conservation work. Considering the fact that the annual current budget of IBPGR is only about $4 million, the broad spectrum of work which it has been able to generate is indeed impressive.

Early in October 1983, 150 scientists from about 50 countries meeting in the USSR at the First International Congress of Biosphere Reserves jointly sponsored by UNESCO and UNEP drafted an Action Plan to protect the world's "wild genetic resources." The Action Plan calls for research on the use of biosphere reserves to conserve for posterity existing genetic variability. The global network of biosphere reserves promoted by UNESCO now consists of over 200 well defined areas characterized by variability in plants and animals.

Without green plants higher orders of life cannot exist since green plants are directly or indirectly the basis for nutrition for animals and man. This is why in situ conservation of the habitats where gene pools of wild plant species occur within their natural communities is so important. This aspect of conservation has hence received high priority in the

World Conservation Strategy launched in 1980. This strategy has recommended a three-pronged approach consisting of: (a) ex situ protection, as the chief and generally the only means of conserving cultivated material and as a back-up means of conserving wild material; (b) in situ protection, as the chief means of conserving wild material; and (c) a system of sound planning, allocation and management of land and water uses to achieve a sustainable combination of development, oil the one hand and of maintenance of the biological resources required for development, on the other. Robert and Christine Prescott-Allen (1982) have dealt with the various components of national and international action essential for promoting in situ conservation. In particular, they have suggested that the Threatened Plants Committee and the Commission on National Parks and Protected Areas of IUCN, in conjunction with UNEP, UNES¬CO/MAB and IBPGR should form a Crop Genetic Resources Group, the functions of which would be to promote action related to in situ conservation. There is need for setting up Genetic Resources Conservation Committees at national and in large countries at local levels in order to promote decentralized action and people's participation.

Gene banks with facilities for long-term storage now exist in several countries. According to IBPGR, the following countries have currently one or more base collections held for long-term conservation: Argentina, Belgium, Brazil, Canada, China, Colombia, Costa Rica, Ethiopia, Germany FR, Germany DR, Greece, India, Israel, Italy, Ivory Coast, Japan, Mexico, Netherlands, Nigeria, Peru, Philippines, Poland, Portugal, Spain, Syria, Thailand, United Kingdom, USA, and USSR; and at the Nordic Gene Bank which serves Denmark, Finland, Iceland, Norway, and Sweden.

These gene banks can be broadly grouped under three field categories: -base collection which consists of seed accessions dried to 5-7 percent moisture content, sealed and stored at between -10 and -20℃ for long term conservation. Samples are not generally distributed from a base collection.

active collection which complements the base collection, being one of the collections from which seed samples are drawn for distribution, exchange and other purposes such as multiplication and evaluation. Seeds are dried to about 7 percent moisture content and stored at about QOC, conditions suitable for medium-term storage.

-field collection which consists of living plants; crops such as temperate fruit trees, banana, cassava, ete., that are propagated vegetatively. Others with seeds that do not keep or are difficult to store are included, e.g. coconut, cocoa and oil palm. It is an all-purpose collection.

Wilkes (1983) has summarized the current status of crop plant germplasm conservation. It is generally recognized that a sound strategy for the conservation of genetic resources demands both ex situ and in situ conservation. Despite all recent advances in long-term in situ conservation, many scientists believe that gene banks-whether seed, clonal, or tissue culture-will never be adequate to ensure the availability of genetic varieties to the desired extent. For example, Hugh litis states, "I think germplasm banks are tremendously important for crop improvement and scientific research, but they will not save the genetic diversity of plants for the next 2,000 years. For one thing, such banks are vulnerable to accidents, power failures, even deliberate destruction." For example, germplasm conserved ex situ in several developing countries have been damaged as a result of power failures, breakdown of refrigeration equipment, strikes and civil disorders preventing personnel from entering the facilities, and drought and floods during the growing season when seed increases are made. Some experts will hence like to achieve some kind of freezing of the genetic landscape including hot spot locations for pests and pathogens where screening for resistance can be done in an effective manner under natural infection.

Robert and Christine Prescott-Allen have pointed out the following three major advantages of in situ conservation.

The first-continued coevolution-largely concerns those wild relatives that are sources of pest or disease resistance. In the wild these species can continue to coevolve with pests and pathogens, providing the breeder with a dynamic source of resistance that is lost in ex situ conservation. Second, maintenance of a crop relative in situ permits the breeder to study its autoecology and to obtain data that can assist in the selection of germplasm that might otherwise be overlooked. As Rick (1979) has stated: "... the capacity of a species or race to survive in a particular habitat implies the evolution of a genotype that determines that capacity... It follows that a study of the habitat should provide clues to useful genetic resources in such material." Rick gives as examples of characters revealed in this way tolerance of intense tropical moisture and temperatures in Lycopersicon esculentum, tolerance of saline soils in maritime ecotypes of L.

cheesmanii, insect resistance in L. hirsutum, and drought resistance in L. chilense and Solanum pennellii.

Third, natural parks and biosphere reserves may provide less expensive protection for the wild relatives of crops in ex situ measures. This is partly because certain kinds of maintenance (such as growing out) are not required; and partly because the costs of protection can be spread across other objectives besides genetic resources conservation-such as watershed protection, critical habitat protection, research, and recreation and tourism. Indeed the different reasons for protecting a given area can be mutually reinforcing, increasing the security of the area (by adding to the incentives for protecting it) at the same time as reducing the costs per benefit.

In addition there are several excellent reports on the role of tissue culture in preserving threatened and endangered plant species as well as current cultivars, particularly of vegetatively propagated material (Wochok 1981). IBPGR has recently reviewed the work on in vitro storage. In vitro work may involve many different culture systems ranging from protoplasts and disorganized cells and calluses, to organize shoots and somatic embryos (embryoids). Although plants can be propagated clonally using all types of culture, such studies as have been carried out for genetic conservation purposes have tended to concentrate upon nonadventitious shoot-tips (often erroneously referred to as "meristems", which in fact form only part of each explant or culture) and axillary buds. Whilst these may offer greater genetic stability (on present knowledge) and be entirely appropriate for some species (e.g. potato), their wide use may not be feasible for others which lack sufficient axillary means (e.g. Musa). Further, many basic studies on storage by cryopreservation (freezepreservation) have been carried out using disorganized systems which must regenerate by culture methods. Thus, only restricted use is being made of the full repertoire of available in vitro techniques, a limitation that is likely to become more apparent as in vitro work assumes greater importance in agriculture and plant breeding, particularly through the involvement of genetic engineering. A more balanced view should hence be taken on the relative merits of different types of culture; all may have a potential application in genetic conservation.

Taking rice as an example, I would like to illustrate the value of a properly planned and maintained gene bank. Chang has described in several articles the steps taken for the collection and conservation of genetic resources of rice (see Chang 1983). The genus Oryza has 20 wild species and two cultigens, O. sativa and O. glaberrima. The full spectrum of genetic resources in Oryza consists of: (1) wild species, natural hybrids. between the cultigen and wild relatives, and primitive cultivars of the cultigens in the areas of diversity; (2) commercial types, obsolete varieties, minor varieties, and special-purpose types in the centres of cultivation; and (3) pureline selection of farmers' varieties, elite varieties of hybrid origin, FI hybrids, breeding materials, mutants, polyploids, aneuploids, intergeneric and interspecific hybrids, composites and cytoplasmic sources from. Breeding and related research programs. It has been estimated that slightly more than 120,000 cultivars may occur in O. sativa and O. glaberrima in different parts of the world. The International Rice Germplasm Center at IRRI has about 67,000 Asian cultivars, 26,000 African rices, 1,100 wild rices, and 690 genetic testers. Between 1962 and 1982, the 'Germplasm' Center provided more than 91,000 seed samples of both cultigens and wild species to nearly 2,500 scientists. In addition, the Center provided IRRI scientists with nearly 350,000 seed samples.

Obviously having a large collection by itself has no significance unless the collection is carefully studied for its genetic representativeness and for minimizing genetic redundancy. The IRRI collection is hence carefully screened both to eliminate duplicates and assess the useful characters they possess. A multidisciplinary Genetic Evaluation and Utilization programme (GEU) helps to get maximum benefit from such an invaluable assemblage of genotypes. Recently in collaboration with IBPGR and national government agencies, a five-year plan (1983-87) has been drawn up to collect the remaining genetic material with the help of appropriate national organizations, so that we can preserve for posterity the fruits of thousands of years of natural evolution and human selection (IRRI-IBPGR 1983).

When the ecology of crop fields is changed in a direction that favors higher yield, conditions also become more favorable for pests and pathogens. Also, an increasing amount of land is being affected by soil constraints such as salinity, alkalinity, acid sulfate conditions and water logging as a result of unscientific irrigation. Fortunately, in nature there is considerable variability for many of these characteristics. Let me cite two examples from recent work in rice which underline the importance of genetic resources conservation. Resistance to the grassy stunt virus came from the wild species *Oryza nivara* collected from Uttar Pradesh in India. The single dominant gene for resistance to grassy stunt possessed by O. nivara has been incorporated into several varieties with multiple disease and insect resistance. They include IR36, IR42, and others now planted on about 20 million hectares of riceland. The famous hybrid rices of China, which now cover about 7 million hectares in that country, could be developed because of the discovery by Chinese scientists of cytoplasmic male sterility in a wild rice plant found on Hainan Island.

At present, the world's rice crop amounts to over 400 million tons a year, feeding a population of more than 2.3 billion people. Between 1968 and 1981, the total rice production in Asia rose by 42 percent. The average yields rose by about 30 percent during this period when unfortunately the population also increased by over 25 percent. Since there .is little difference between cultivable and cultivated land, further increases in production will have to come largely from productivity advances, For this purpose, we will need a wider range of high yielding varieties which can perform well both in ecologically handicapped regions and in farms cultivated by economically handicapped farmers. This emphasizes the urgency of collecting, classifying, and utilizing all available genetic variability. Considering the fact that the total expenditure on rice germplasm conservation in the world now may be about US $1.25 million every year, the returns from such an investment are obvious. The time available for saving the remaining genetic variability may hardly be a decade.

Forest genetic resources

Some estimates suggest that tree formations are being lost at the rate of 11 million hectares every year. For nearly 10 hectares of forests lost, hardly 1 hectare is being reforested in tropical America, Africa, and Asia. This has serious implications not only with regard to the in situ preservation of genetic variability but also for the quality of life of millions of forest dwellers. Some estimates indicate that so far only 1 in 6 of tropical plants and animals has been given a scientific name. Hence, we may not even know what we are losing due to the destruction of forest canopies.

Following the World Congress organized in 1971 by the International Union of Forestry Research Organizations (IUFRO), a Working Party of "Forest Gene Resources Conservation" was established by IUFRO. This Working Party has been collecting, synthesizing, and disseminating data on the genetic impoverishment of tree species. FAO also started in 1981 the publication of a data book on endangered forest tree species and provenances. An FAO/UNEP/IBPGR technical conference held in 1981 underlined the importance of greater attention to forest genetic resources, particularly species used in arid and semi-arid zones for fuel. Leguminous shrubs which can enrich soil fertility through biological nitrogen fixation and at the same time provide fodder, feed and fuel are particularly worthy of greater attention. Recent research by Dr. J.L. Brewbaker on Leucaena has shown how rewarding such research could be. Among other tree species which need greater scientific attention are neem and Casuarina. At an International Workshop on Casuarina held in Canberra, Australia, in August 1981, several useful suggestions were made regarding the collection, conservation, and utilization of Casuarina germplasm. Neem also is a unique plant with multiple uses particularly in pesticide and fertilizer industries. Papers presented at two international conferences on neem held in the Federal Republic of Germany in recent years suggest that the potential for improvement of the neem tree in yield and quality characteristics is considerable.

An FAO Newsletter entitled "Forest Genetic Resources" has become a useful medium for the exchange of information on genetic variability in tree species. In particular, there is need for a greater flow of information on emergency situations where delay in action could result in future generations not deriving the full benefit from our genetic heritage. The genuine needs of the poor for fodder and fuel, the greed of the rich who wish to exploit forests for short-term financial gains, and the spread of careless technology in areas like mining, road construction and industrial forestry are causing damage to natural forests in varying degrees. They require appropriate solutions if further damage to tropical forest ecosystems is to be prevented.

Microbial resources

The need for conservation has long been recognized by microbiologists. In the past, the greater scientific effort was directed primarily towards microorganisms causing diseases and spoilage. Recent genetical and biochemical advances, however, have led to a surge of interest in the use of miciuorganisms for the production of valuable foods, fuels, and a wide range of biochemicals. This has resulted in the establishment of microbiological gene banks particularly for the support of biotechnology (Kirsop 1983). Some countries favour a central facility on considerations of economy and efficiency, while others prefer a decentralized system with some arrangements for coordination. The Microbiological Resource Centers (MIRCENs) proposed by UNEP in 1974 aim to provide the infrastructure necessary for the management, utilization, and distribution of microbial gene pools. The MIRCEN network is now operating under the sponsorship of UNESCO, UNEP, and ICRO and promotes the maintenance and utilization of microorganisms important to agriculture, medicine, industry, and environment. The MIRCEN Newsletter is a good source of information on the activities of Microbiological Resource Centers.

Methods of preserving microorganisms are very diverse but fall generally into a few broad categories. These are serial subculture on solid or liquid medium, drying on a variety of substrates, freeze drying and cry preservation in a freezer at temperatures of -20℃, -70℃, -90℃ or in the vapor phase at -140℃ or liquid phase at -196℃ in liquid nitrogen.
The proper maintenance of cultures is a matter of continuing concern to culture collection specialists. While viability can be achieved readily, the stability of cultures presents problems and needs continuous monitoring. Culture collections are hence also appropriate locations for identification centers.

Developments in biotechnology have enlarged the scope of traditional culture collection centers. Collections of animal and plant cell lines, hybridomas, and plasmid-bearing strains are now getting organized. Some of these centers may become repositories for plasmids and gene fragments.

Increased use of microorganisms in industry, medicine, and agriculture has led to greater interest in patent rights. Patent laws and requirements vary among countries. The Budapest Treaty drawn up by the World Intellectual Property Organization has tried to rationalize and simplify the legal procedures.

From the early 70s when the cost of fossil fuels went up steeply, interest in biological nitrogen fixation has increased. It is now widely recognized that in breeding programmes relating to leguminous plants, the variability of *Rhizobium* bacteria needs to be considered in conjunction with the genetic variability of the host legume. Interest in the exploration, collection, documentation, and preservation of microorganisms of major importance to crop plants is now growing. The World Data Centre at the University of Queensland in Australia and the World Directory of Collections of Cultures of Microorganisms are useful sources of information. With rapid advances in recombinant DNA technology, complete libraries of nuclear, chloroplast, and mitochondrial DNA are also being developed. In view of the growing interest in bio-fertilizers, I would like to deal with blue-green algae and *Azolla* at some length.

Blue-green algae

Interest in blue-green algae (BGA) arises from:

1) The ability of a number of genera to fix atmospheric nitrogen, which has implications in the maintenance of soil fertility under both natural and cultivated ecosystems. In particular BGA are of importance in rice fields where their growth and Nz-fixing activity can be enhanced by inoculation and cultural practices, thus providing a cheap source of nitrogen for the crop.

2) The high nitrogen content of some genera, especially Spirulina,
and the possible usage of certain strains as food or feed (Venkataraman 1983).

3) The possible use of BGA as a source of useful biochemicals. 4) Academic curiosity for organisms that constitute the largest, the most diverse, and the most widely distributed group of photosynthetic prokaryotes. The collective properties of BGA (oxygenic photosynthesis, chromatic adaptation, Nz-fixation, cellular differentiation) make them attractive organisms for genetic research.

BGA are the only photosynthetic prokaryotes performing oxygenic photosynthesis. According to the classical definition of the term algae, which encompasses all oxygenic phototropns of simple general structure, BGA have traditionally been placed under the International Code of Botanical

Nomenclature. In 1977, Stanier and Cohen-Bazire, after an extensive study of the properties of BGA, emphasized their prokaryotic nature and stated that "the only logical treatment of the cyanobacteria (BGA) was to place them in the super kingdom Prokaryotae as a division, class, or order of bacteria" under the rules of the Bacteriological Code. At present there is no general agreement for placing BGA under a given code and most probably their future taxonomic treatment will take into account botanical and bacteriological criteria.

The ambivalent taxonomical position of BGA has been a hindrance to their genetic conservation. When placed only under the botanical code, the deposition of a "type culture" in a collection was not required. Therefore in 1968, among some 8,000 strains of microorganisms listed in the catalogue of the American Type Culture Collection, only one strain of BGA (*Plectonema boryanum*) was recorded.

Among the 349 collections recorded in the 1972 World Directory of Collections of Cultures of Microorganisms, only 9 comprised BGA. These 9 collections comprised 130 species of BGA in 41 genera.

In 1973 the first Botanical Monograph on BGA was published by Carr and Whitton. In the Appendices Komarek presented a list of 17 laboratories where BGA cultures were maintained. The records indicated that 968 strains (in 55 genera) were maintained and it was clearly stated that "by no means were all these cultures bacteria-free."

The availability of pure cultures is essential for effective experimental work on any microorganism. As pointed out by Stanier and Cohen-Bazire (1977) the' long prevalent myth that is exceptionally difficult, if not impossible, to isolate BGA in pure culture seems to have discouraged scientists from tackling the problem. Work conducted by Stanier and his co-workers at the Pasteur Institute in Paris has clearly demonstrated that production of axenic BGA is possible through classical bacteriological methods even if it is tedious and technically demanding. More than 200 strains representative of all BGA subgroups were made axenic in his laboratory. This work has led to a new taxonomic treatment of BGA epitomized by 3 publications in the Journal of General Microbiology. These comprise strain histories, properties and generic assignment of 176 pure cultures (Rippka et at. 1979) together with genome sizes and mean DNA base composition on the same collection (Herdman et at. 1979a, b).

However, the establishment of a collection of axenic strains initiated by Stanier needs to be continued by other laboratories. In the 15th edition of the ATCC catalogue (1982) only 119 strains of Cyanobacteria (BGA) belonging to 19 genera were listed. Among these, 103 originated from

Pasteur Institute collection. For comparison Desikachary (1959) reported five orders, 20 families, 160 genera, and 1,500 species of BGA.

At present, there is no organized approach for collecting BGA for genetic conservation purposes. In most cases strains recorded during ecological studies are not kept by the authors because the maintenance of a culture collection is expensive and time consuming. Also the acceptance of a strain in a collection such as A TCC requires purification of the strains and prerequisites which restrain the authors from depositing their material. In addition most of the laboratory studies on BGA are conducted with a limited number of easily available axenic strains because authors prefer to utilize a well known material and also are reluctant to purify new strains. This does not help in increasing the number of axenic strains deposited in official collections.

For short-term preservation, stock cultures incubated at laboratory temperature under continuous dim light, either on liquid medium or on agar slants, generally proved satisfactory with axenic and non-axenic strains. However, with axenic strains intervals between successive transfers are very variable: from several months to one to two weeks for more delicate strains (Rippka, personal communication). For long-term preservation of axenic and non-axenic strains deep freezing in liquid nitrogen has been successfully used. For medium-term preservation of non-axenic material, simple drying at room temperature (on filter paper if little material is available) is frequently successful. This method is used at IRRI for keeping Nz-fixing strains isolated from rice soils.

Among the possible uses of a BGA germplasm, the two major ones concern Nz-fixing strains and high protein producers.

The success of Rhizobium strain selection in increasing Nz-fixation in legumes has drawn attention to the possibility of selecting superior strains of other Nz-fixing microorganisms. However, requirements for a symbiotic prokaryote are different from those of a free-living BGA exposed to the harsh competitive world of the rice field. According to Venkataraman (1975), induction of mutations to enhance nitrogen-fixing capacity does not seem to be a profitable approach to improve strain efficiency as most of the mutants seem to have lost their ability to fix nitrogen. There is hence need for the collection and screening of BGA strains which can fix Nz well under diverse conditions of soil environment.

Spirulina, a non-Nz-fixing filamentous BGA particularly rich in protein (50-70 percent d. in.) has been traditionally used as food in Lake Chad area where it grows spontaneously. The Italian company "Montedison" is planning to install a large-scale Spirulina production plant along the boundaries of Lake Chad in Africa. This project is likely to be commercialized in 1985, and if it turns out to be real, it can capture 5 percent of protein market in 1990 (Venkataraman 1983). Many BGA strains are high protein producers and need to be collected and screened for their potentialities as food, feed, and source of useful biochemicals.

Azolia, an aquatic fern capable of fixing atmospheric nitrogen through a symbiotic blue-green alga living in cavities of its leaves, has long been used in China and Vietnam as green manure for rice. Its usage is now being tested and expanded in many rice growing countries of Southeast Asia.

In their recent book on Azalia, Lumpkin and Plucknett (1982) pointed out the need for:

-Establishment of a world Azolia germplasm collection, and continued collection and characterization activities.

-Breeding for properties and combination not found through mass selection.

-Development of techniques for the introduction of Anabaena azallae strains into Anabaena-free Azalia.

At present IRRI acts as an international center for Azalia germplasm collection, maintenance and distribution. Eighty-two strains in 6 species are maintained at IRRI: A. pin nata: 56; A. filiculoides: 8, A. mexicana: 4; A. caroliniana: 4; A. microphylia: 9; A. nilotica: 1. These strains have been or are being tested for growth under field conditions, response to different temperature, and phosphorus requirement. During the year 1983 more than 400 fresh Azolla samples ranging from a few grams to 0.5 kg have been given to laboratories or individuals.

A large collection of mainly Chinese strains has been established in the AzollaDepartment of the Ministry of Agriculture in Fuzhou. Besides these two large collections, smaller ones have been established at the Charles Kettering Laboratory, Ohio, U.S.A. and at the University of Hawaii, Honolulu.

At IRRI and in China, Azolla strains are kept as vegetative cultures on artificial nitrogen-free medium in small size containers in a greenhouse, some strains need to be grown on soil (A. filiculoides). Strains are grown also for safety in a phytotron on artificial medium at 21-29℃ under a light intensity of 30 klux (12 hours day-night cycle). Medium is changed every three weeks. Only one medium is used at IRRI.

The major problem for Azolla collection is that the material cannot be shipped by mail and has to be handcarried. Fronds deposited on a paper tissue impregnated with culture medium in a plastic Petri dish remain viable for eight to ten days if aerated at regular intervals and kept in the lower part of a refrigerator as often as possible.

Collecting and then disseminating and testing strains under a wide range of agroclimatic conditions will allow selection of more efficient strains. Azolla filiculoides and A. caroliniana from IRRI's collection are successfully utilized in Vietnam and China respectively. An A. Pinnata strain from India got successfully established in North Senegal whereas the indigenous strain (A. pinnata val. african a) was less productive.

Selection of strains resistant to pests is another way to improve Azolla use.

The symbiotic BGA has received the specific name A. azollae, no matter which Azolla species it accommodates. Knowledge of the symbiont is limited because attempts to cultivate it separately from the fern have been unsuccessful. Available results indicate homogeneity of the BGA samples extracted from different Azolla but it is possible that a detailed study may result in the identification of several species as it has happened in the case of Rhizobia in legumes. Possibility of replacing the existing Anabaena symbiont by a more efficient one may be considered but at present even the recombination of the two partners has not been performed.

Animal genetic resources

Although the number of animal species (excluding insects and fishes) which have been domesticated during the past several thousand years is scarcely more than a dozen, specific agro-ecological, economic, and cultural needs have resulted in numerous breeds and strains or races of each species. Consequently, we have now an enormous wealth of animal genetic resources in the world as a whole. While for a long time local communities were wedded to their own specific breeds of animals, there has been extensive replacement of indigenous animals with exotic strains in many developing countries in recent decades. Breed substitution, planned or unplanned crossbreeding or the creation of new composite breeds are all now important components of animal improvement programmes in cattle, sheep, goat, poultry, pigs, etc. Experience has shown that this process has two possible negative consequences-first, the new breed or hybrid may lack adaptation to local conditions, particularly with regard to pests and diseases and secondly, the animal genetic resources being replaced may be lost for all time.

Knowledge concerning the comparative advantages and disadvantages of indigenous and alternate breeds under different environments is lacking in many developing countries. Because of unhappy experience, there is the opposite trend among some animal geneticists of placing emphasis only on the improvement of indigenous breeds in isolation, without the use of strains which have proved their worth in other locations. More studies on the interaction between genotypes and environment particularly the management systems feasible of adoption under different socioeconomic and cultural conditions are needed. Meanwhile, breeds at present having little to offer a production system but which may have hidden or identifiable production strengths such as resistance to diseases, ticks, etc. should be retained in herds of optimal size or in cryogenic storage.

For developing a scientific strategy for conservation, there is need for more detailed information on breeds in danger of extinction in each country. The International Union for Conservation of Nature's Red Data Book does not contain information on the breeds threatened within a domestic species. The problem of genetic erosion in domesticated animals is not confined to developing countries alone. In developed nations, breeders keep only the best producers for reproduction. Thus, lines are formed whose genetic isolation becomes more and more strict. Breeders using the same lines often form a society whose chief purpose is to keep a herdbook in which only animals conforming to established standards of size, color, conformation, and performance will be registered. After a while, there is a tendency to close the herdbook and permit no more entries. A survey conducted under an FAO-UNEP pilot project on conservation of animal genetic resources revealed that in Europe and the Mediterranean region about 115 indigenous breeds of cattle were threatened with extinction in the early 70s (Lauvergne 1975). The reasons vary from breed to breed but include factors such as preference of specialized breeds for milk and meat production rather than dual purpose and the replacement of draught cattle with tractors. New reproduction techniques such as artificial insemination with foreign frozen emen also promote rapid gene replacement.

The proceedings of the FAO-UNEP Technical Consultation on Animal Genetic Resources Conservation and Management held in Rome in 1980 contain useful information and recommendations. Symposia organized by regional scientific organizations such as the Society for the Advancement of Breeding Researches in Asia and the Oceania (SABRAO) also contain valuable data (Barker 1980).

Following the FAO-UNEP Technical Consultation of June 1980, there has been considerable progress in conservation, documentation, and evaluation efforts. SABRAO, for example, is promoting standardization in the use of breed names, taking those of Mason's (1969) "World Dictionary of Livestock Breeds" as a basis. FAO and UNEP are now publishing a publication entitled "Animal Genetic Resources Information." The International Union for Conservation of Nature (IUCN) has set up a Genome Conservation Specialists Group as part of its Species Survival Commission. As a first task, the Group is compiling a list of existing collections of frozen germ cells of endangered species and their close relatives. The in situ conservation of animals suited to the production of food, fibre, and work is receiving greater attention at country level. India, for example, has set up a National Bureau of Animal Genetic Resources for promoting both in situ and ex situ conservation. Animal genetic resources data banks are being set up in different countries and training facilities in conservation and animal genetic resources are being expanded.

During the next few years, it is likely that the techniques of embryo transfer, long-term storage of embryos, and the cloning of embryos for even more rapid multiplication of desirable genotypes will be developed to commercially viable levels. These techniques will have applications both in the field of genetic conservation and in the rapid upgrading of animal productivity. It will be prudent for developing countries endowed with a large animal wealth to get a critical mass of scientists trained in the new techniques.

There are many symbiotic linkages between plant and animal genetic resources conservation. Where uncontrolled over-grazing takes place, the native flora soon takes the form of only plant species which are not eaten by animals. When plants that produce nectar, pollen, honey dew, propolis and are visited and pollinated by bees in general are destroyed, the bee population dwindles and disappears (Crane 1983). When pesticides are applied indiscriminately, the natural enemies of pests are destroyed and pest incidence could in many cases pose more serious problems. Conservation in crop fields hence involves the adoption of management procedures which will help to build up the population of natural enemies of pests. At IRRI, for example, it has been found that an intensive rice production system involving weekly planting and harvesting of rice with a total

production of about 18 to 20 tons of rice per hectare per year will be possible provided the following three steps are taken:

a) Protect and conserve the natural enemies of pests;

b) Cultivate varieties with multiple resistance to pests and diseases; and

c) Adopt timely applications of appropriate chemical pesticides based, on a careful monitoring of the buildup of pest populations. Thus, conservation will have to be a total approach, starting with micro-environments and extending up to macro-systems.

Marine and inland water genetic resources It has been estimated that about 200,000 species of plants and animals may have their homes in the oceans. This amounts to about 2-4 percent of earth's five to 10 million species. However, the oceans contain nearly two-thirds of all animal phyla. Because many marine species tend to be more uniformly distributed than terrestrial species, it has been assumed that the threat of extinction is less. Recent studies have indicated that many marine populations may be quite variable genetically and that this variability is particularly high in species-rich communities such as those of coral reefs (Polunin 1982 and Connell 1978).

The need for greater attention to the conservation of genetic resources of the sea arises from increasing damage to marine eco-systems from human activity. Sedimentation and siltation from land sources, pollution from inland and coastal sources and unscientific commercial eXploitation are all threats to marine genetic resources. The coastal zones also support some of the densest human populations on earth. Of the 50 or more cities that are projected to have populations exceeding 5 million in the year 2000, half are located on estuaries. In the tropics, many coral reefs and mangrove communities have been degraded or destroyed along extensive sectors of coasts of the Americas, Africa, India, Thailand, Malaysia, Vietnam, Indonesia, Philippines, and east and south Australia (UNESCO 1978). In recent years, some steps to promote conservation through the establishment of Marine National Parks have been taken but more coordinated action to protect both living inland water and ocean resources 'is necessary at the country level.

Even the small proportion of marine organisms examined to date has yielded numerous extracts and compounds of importance for medicine. The marine world is a valuable source of a wide range of biomedical compounds. Recent research on the development of cloning systems for producing commercially important chemicals, pharmacologically active compounds and metamorphosis-stimulating substances present in marine organisms, is one of promise. Attempts are underway to develop useful drugs from the sea, including antineoplastic, antibiotic, analgesic, and antispasmodic agents. Genetic engineering techniques are also being applied to improve fish, mollusks, and crustaceans in natural environments and hatchery systems (Colwell 1983).
In several parts of Europe and also in North America, "acid rain" affects freshwater fishes. For example, Gjedrem (1981') has reported that in southern Norway, there are hundreds of lakes without fish and that salmon is no longer to be found in many river systems. Both laboratory and field experiments indicate that acid water first affects fish eggs and fry. Massive kills of salmon and trout have been observed during snow melt and after heavy rain. Conservation of fish resources in such vulnerable areas will have to receive priority attention.

The FAO Technical Conference on Aquaculture held in Japan in 1976 recommended that appropriate methodologies should be developed for the establishment of gene pools in fish species. The Nordic Symposium on Gene Banks held in Helsinki in 1978 recommended that Gene Banks may be established in each country based on the following methods:

a) establishment of "Egg Banks" involving indefinite storage of eggs and milt at low temperature;

b) propagation in captivity of each strain; and

c) conservation in the natural environment with legal protection from fishing and other human interference.

Similar programmes are needed in every part of the world.

Human genetic resources

Any discussion on the conservation of human genetic resources cannot be divorced from ethical, social, racial, religious, and political issues.

Advances in the development of sperm and egg banks have opened up possibilities for the long-term preservation of human genotypes. The clinical technology of in vitro fertilization is now 5 years old. Grobstein et ai. (1983) have drawn attention to the various policy implications of external human fertilization. There are reports on cloning and on the establishment of sperm banks of "geniuses." All that can be said in favor of such research is that in addition to helping those who suffer from infertility or congenital disorders, safe storage of sperms and eggs of persons with valuable traits (particularly of those who promote cooperation) for future use might prove to be a step in the right direction, if the world is engulfed in a nuclear war.

FORWARD EDGE IN GENETIC CONSERVATION RESEARCH

In view of the enormous quantitative and qualitative dimensions of the problem of conservation, it is obvious that efforts will have to be selective. Developing countries will have to work on reliable and low cost ex situ conservation techniques, such as storage of seeds in permafrost areas and in cold desert regions. This will help to overcome some of the difficulties like power failure to which I had made a reference earlier.

With the advent of new techniques for genetic manipulation and analysis of biological functions, it may he possible in the future, as already demonstrated in some cases, to tailor-make genes for the hyper expression of functions in desired traits.

It can be visualized that cassettes of DNA clones carrying useful genes in forms for their maximal expression will be constructed. However, complete reconstruction of genetic background in which they should be functioning will be almost an impossible proposition for such vast numbers of organisms for which we have use. Clearly, the strategy would be to incorporate the genes from the cassette libraries into the genetic backgrounds of homologous or heterologous organisms already adapted to local field conditions or to the kinds of nutritions and substrates that can be made available for the exploitation of these organisms in new ways. This approach requires that organisms with the right kind of adaptive functions be conserved even if at this time they are of little economic value.

In other cases it would appear that specific characters would be transferred to organisms of economic value for their improvement or for the production of novel kinds of biochemicals for human welfare. In such cases it may not be essential to have very exhaustive conservation.

Nature of material to be conserved

Microorganisms. A variety of standard procedures are in vogue for the long time preservation of bacteria, fungi, and their parasites. These include storage of agar slants under oil and in cold, lyophilization, and deep freezing in glycerol. Since it has been noted that certain soil bacteria remain viable in soil samples for very long periods, soil storage should also be employed. In the developing countries where microbiology is not yet very advanced, soil samples and other kinds of primary samples from both typical and atypical habitats should be stored for isolation of the resident microorganisms in the future.

Man. Several human populations in the disease endemic areas should provide cellular material for preservation because of the adaptive traits they possess. In such cases germ line cells, specific kinds of B-cells and perhaps cells from some other specific tissues could be preserved in either of the following ways: cryogenically after culturing or as tissue and as hybridoa cell lines. Similar strategies could be adopted for preserving the genotypes of interesting proposition for analysis in the future. The sperm banks that are coming into being could complement this approach. It may perhaps be possible to store for long periods of time fertilized eggs or cells derived from very young embryos. These banks could serve a variety of purposes besides providing resource material to safeguard against extinction of races or genetic damage in vast populations of humans in the events of man-made and natural catastrophies.

New material for genetic conservation

The recombinant DNA procedures today allow cloning of any foreign DNA into E. coli. Current research is making this possible in yeast and other organisms. Further, the efficiency of the technology is bound to improve. Cloned DNA, DNA, and material having DNA in its native state should all be used for genetic conservation from now on.

Finally, nonviable material representing valuable genotypes stored in gene banks need not be discarded. They could become sources for DNA libraries from where a relevant gene or combination of genes can be recovered. Thus, today an integrated genetic conservation strategy can be planned at the level of

a) populations

b) individuals

c) tissues and organs d) cells

e) DNA libraries

A national conservation strategy could include an optimum mix of the above depending on needs and capabilities. Geneticists should help in preparing such an integrated strategy for action by each country as well as at regional and global levels.

POLITICAL CON'CFRNS AND ACTION

A recent evidence of political concern in the area of the conservation and free exchange of plant genetic resources is the adoption in November 1983 by member nations of F AO, with reservations by some countries, of a resolution on an International Undertaking on Plant Genetic Resources. The undertaking aims to achieve the following purposes:

I) to promote greater efforts in the exploration of plant genetic resources;

2) to facilitate the adoption of appropriate legislative and other measures for the preservation, evaluation, and documentation of genetic materials;

3) to allow' access to samples of genetic resources and to permit their export where the resources have been requested for the purpose of scientific research, plant breeding, or genetic resources conservation;

4) to strengthen international cooperation and to improve the capabilities of developing countries;

5) to foster a coordinated network of national, regional, and international network of base collections in gene banks under the auspices of FAO; .

6) to develop a global information system on available crop genetic resources; and

7) to promote greater financial support and security to conservation work.

To implement the recommendations relating to the international undertaking on plant genetic resources, FAO has also established a Commission on Plant Genetic Resources. This commission whose membership is open to all member nations and associate members of F AO will monitor the operation and arrangements for the implementation of the international undertaking and take or recommend measures that are necessary or desirable to ensure the comprehensiveness of the global system of conservation.

These initiatives on the part of member countries of FAO had their genesis in the growing apprehension among several developing countries that the introduction of plant breeders' rights and crop variety patenting may hinder the availability of useful genetic material to developing countries (Mooney 1983 and Barton 1982). Under a convention signed by several countries in Europe, the breeders of a new plant variety will be eligible for royalty for at least 18 years in the case of trees and vines, and at least 15 years for all other plants. This convention is administered by the International Union for the Protection of New Varieties of Plants (UPOV) which is an intergovernmental organization. The International Convention for the Protection of New' Varieties of Plants was signed in Paris on December 2, 1961, and amended by an additional Act signed in Geneva on November 10, 1978. The objective of the convention is the protection of new varieties of plants. A revised version of the convention was adopted on October 23, 1978 and signed by 16 states but has not yet entered into force.

A new plant variety for the purpose of patent rights must be clearly distinguishable by one or more important characteristics from any other varieties whose existence is known at the time protection is applied for. Distinctions from existing varieties, homogeneity, and stability are important for recognition.

The major aim of legislation relating Plant Breeders Rights (PBR) is to provide stimulus to private sector breeding and investment. Various views have been expressed on the positive and negative implications of PBR legislation (Jain 1982, Brown 1983, and IBPGR 1983). According to IBPGR, the issue of PBR in relation to genetic resources is of no great significance. What is urgently needed is greater effort in the collection and conservation of the basic sources of variation in wild and weedy species and primitive cultivars to prevent genetic erosion. However, IBPGR has also warned that "consent for the transfer of material under development may be more difficult to obtain than hitherto."

So far the UPOV Convention has been concerned only with patenting of genotypes. However, there are suggestions that the convention should also allow the patenting of specific genes. Mooney (1983) has expressed the fear that when plant breeding companies are purchased and managed by pesticide companies, there may be a temptation for not including genes for resistance to those pests for which chemical pesticides are available. If this happens, the cost of production of high yielding varieties could further increase. Today, some of the most widely grown varieties like the 11UJ6 variety of rice are preferred by farmers mainly because of the broad spectrum of resistance to important pests and diseases which such varieties possess. The breeding of such varieties is facilitated by international cooperation in getting segregating populations subjected to severe pest, pressures in hot spot locations.

It is true that patent rights could stimulate private enterprise in plant and animal breeding. At the same time, they provide scope for the emergence of monopolies in genetic material for specific traits unless a sample of existing variability is also maintained at a government controlled genetic resource center. There is hence need for examining whether adequate incentives could be provided for plant and animal breeding initiatives without in any way endangering the possibility of free exchange of genetic material among and within nations.

One immediate method of assisting developing countries in genetic conservation and utilization is the establishment of a global grid of genetic resource centers all committed to the cause of free exchange of material. For many developing countries, the availability of genetic variability itself will not provide any insurance against genetic vulnerability.

They need help in the short term in breeding and evaluation. This is one of the tasks undertaken by international agricultural research centers supported by the Consultative Group on International Agricultural Research (CGIAR) which distribute segregating populations and advanced breeding lines for selection under different growing conditions. Therefore a global grid of genetic resource centers will have to be linked to a complementary grid of national, regional, and international selection and hybridization centers. The national hybridization garden in sugatcane operated by the Indian Council of Agricultural Research at Coimbatore and Cannanore is a good example of how purposeful crossbreeding centers can be organized.

All these activities will require considerable financial resources. A World Gene Fund was suggested by the Government of The Netherlands in a letter to the FAO Secretariat in June 1983. Such a fund will help to accelerate collections in endangered habitats, train more geneticists and breeders from developing countries, and organize international selection and hybridization gardens to generate material of relevance to the countries needing them.

FAO has estimated that USA and Europe may hold about 340,000 and 750,000 accessions of seed crops respectively. The People's Republic of China has over 300,000 samples of seed crops. Some of the other major collections are the following:

		Accession (nearest thousand)
USSR	Wheat and wild relatives	70000
USSR	Barley, oats and rye	30000
IRRI	Rice	70000
ICRISAT	Sorghum	20000
ICRISAT	Millets	15000
CIMMYT	Maize	15000
USSR	Food legumes	26000
USSR	Vegetables	35000
USSR	Oil seeds	20000
USA	Wheat, oats, barley and rye	82000

It is necessary to have these collections replicated and conserved at different places to ensure their security. Also, there is need for an integrated computerized data bank from where anyone interested can get information on the availability of genetic material for specific breeding requirements. If a World Gene Fund comes into existence, a priority use of the additional resources should be on steps necessary for ensuring the security of germ plasm collections. In addition, the operation of Hybridization Gardens to provide a wide range of segregating populations for selection under different growing conditions and of an International Information System deserves support.

Apart from political commitment of the kind needed for fulfilling the obligations indicated in the International Undertaking on Plant Genetic Resources, there is need also for governments to consider mobilizing the help of defense services in work relating to the collection of genetic resources from endangered habitats. Some of the areas with considerable genetic wealth are not easily accessible due to political or defense considerations. Given the requisite training, army personnel will be easily able to collect material from remote areas which cannot readily be reached by civilians. It will be useful if a small corps of officers especially trained in genetic resources conservation is established within the defense services of each country. This corps could assist the concerned agencies of government both in germ plasm collection and in situ conservation work.

Political considerations become particularly important while developing public policies for the establishment of human sperm and egg banks. In vitro fertilization, for example, has now been in use for 5 years. In Australia and United Kingdom, policy governing its use is actively being formulated. The President's Commission for the Study of Ethical Problems in the USA in its 1982 report entitled "Splicing Life" has referred to a number of anticipated impacts of rapidly growing knowledge of molecular genetics, genetic screening and diagnosis, and the curing of genetic disorders. All governments and social scientists should pay attention to such issues.

PUBLIC PARTICIPATION

The first human form, referred to as hominid, is believed to have evolved in East Africa some 25 million or more years ago. Human beings may have existed as a species for more than 3 million years. Yet it was only about 10,000 years ago that human beings started growing food, rather than merely gathering it from the wild state and hunting wildlife. Thus, if the existence of human beings as an independent species is equated to a 24-hour day, we have been farmers for only about seven minutes. Even during these seven minutes, we have practised market-oriented agriculture only for a few seconds. Within these few seconds, we have been confronted with numerous problems including changing consumer preferences. We do not know what all new pests and soil and atmospheric (including temperature changes from higher CO2 content) constraints we will have to face in the future. We do not know all physiological and morphological traits that will- be needed for plants to perform well in a post-nuclear war era, if unfortunately such a calamity befalls our planet. Future generations of scientists and farmers will not have the tools with which they can solve such problems if we do not make genetic resources conservation, evaluation, and utilization a common cause and accord it the highest priority.

How can the above message become part of the daily life of every citizen in every country? How can we generate widespread awareness of the economic and biological necessity of conservation measures? There is no single or simple method of involving people in conserving their own environmental assets. The message has to start from the primary school and will have to become over a period of time a way of life.

In the past, rural communities replenished soil fertility through organic recycling. Women selected the best cobs or panicles for use as seed during the next growing season. They unconsciously selected in this process genotypes possessing resistance or tolerence for the major pests in the particular region. Similarly, selection was practised among animals with the result that adaptation to local conditions is the major characteristic of indigenous breeds of farm animals.

With the onset of modernization, conservation agriculture gave way to exploitative farming. With specialist agencies taking over the responsibility for seed production and fertilizer and pesticide distribution, the involvement of the local people, particularly illiterate women, has diminished or vanished. Consequently, conservation is now regarded as a government responsibility rather than a joint sector activity between the people and public agencies.

In modern mass media like television, radio, and the press, we have powerful tools for education and awareness generation. Where there is no awareness, there is little or no action. Hence, I hope the choice of the

theme "Genetic Conservation: Microbes to Man" for this Congress would help to create a sense of urgency in dealing with this problem. I would like to suggest that every T.V. and radio station and every newspaper should carry at least one small item on genetic conservation each day. Just as reporting on weather has become an integral part of daily news reporting, reporting on genetic conservation and eco-development should also become a daily feature.

Early in 1984, we will be observing the 100th death anniversary of Gregor Mendel. When I visited the village in Czechoslovakia where Mendel was born during the Mendel Centenary Symposium in 1965, the local villagers remembered Mendel for the fire station he had built in the village more than for the laws of heredity he had left behind. Let this heritage of Mendel be reflected in all our work and let us keep in mind what Albert Einstein once said: "Concern for man himself and his fate must always form the chief interest of all technical endeavors in order that the creation of our minds shall be a blessing and not a curse."

Let us use the power genetics has conferred upon us for advancing the frontiers of knowledge and of production and resist efforts to use it for refining the tools of destruction.

REFERENCES

Barton, J.H. 1982. The international breeders' rights system and crop plant innovation. Science 216: 1071-1075.

Brown, W.L. 1983. Genetic diversity and genetic vulnerability-an appraisal. Econ. Bot. 7(1): 4-12.

Chang, T.T. 1983. Genetic resources of rice. Outlook on agriculture 12(2): 57-62. Pergamon Press, UK.

Colwell, R.R. 1983. Biotechnology in the marine sciences. Science 222: 19-24.

Connell, J.H. 1978. Diversity in tropical rain forests and coral reefs. Science 199: 1302-1310.

Crane, E. 1983. Surveying the world's honey plants. 12.p. Central Association of Beekeepers, Essex, UK. Fisher, A. 1982. Preserving a diverse lineage. pp. 47-50. Mosaic May/June. Food and Agriculture Organization. 1983. Plant genetic resources report of the director-general. 45 p. Conference, FAO, Rome.

Food and Agriculture Organizatioll. 1983. Proposal for the establishment of an international gene bank and the preparation of a draft interna¬tional convention for plant genetic resources. 43 p. Conference, FAO, Rome.

Frankel, O.H. 1974. Genetic preservation: our evolutionary responsibility. XIII International Congress of Genetics. 78: 53-65.

Frankel, O.H. and Soule, M.E. 1981. Conservation and Evolution.
Cambridge University Press, Cambridge.

Gjedrem, T. 1981. Conservation of fish populations in Norway. In: Fish gene pools Ecol. Bull. (Stockholm), N. Ryman (ed.), 34: 33-36.

Grobstein, C., Flower, M. and Mendeloff, J. 1983. External human fertilization: an evaluation of policy. Science 222: 127-133.

Hawkes, J.G. 1983. The Diversity of Crop Plants. 184 p. Harvard University Press, Cambridge and London.

Herdman, M., Janvier, M., Rippka, R. andStanier, R.Y.1979. Deoxyribonucleic acid base of cyanobacteria. J. Gen. Microbiol. III: 63-71.

Herdman, M. 1982. Evolution and genetic properties of cyanobacterial genomes. pp. 263-305. In: The Biology of Cyanobacteria, N.G. Carr and B.A. Whitton (eds.), Blackwell Scientific Publication.

International Board for Plant Genetic Resources. 1983. A global network of genebanks. 7 p. IBPGR, Rome.

International Board for Plant Genetic Resources. 1983. IBPGR advisory committee on in vitro storage report of the first meeting. 11 p. IEPGR. Secretariat, Rome.

International Board for Plant Genetic Resources. 1983. Plant varieties rights and genetic resources 8 p. IBPGR, Rome.

International Rice Research Institute and International Board for Plant Genetic Resources. 1983. Proceedings of the 1983 Rice Germplasm Conservation Workshop. 109 p. IRRI, Los Banos Jain, H.K. 1982. Plant breeders' rights and genetic resources. Indian J. Genet. 42. 121-128. Kirsop, B. 1983. Culture collections: microbiological resource centres. Biologists 30(3): 139-143.

Lauvergne, J.J. 1975. Conservation of animal genetic resources. 60 p. FAO, Rome. Lumpkin, T.A. and D.L. Plucknett. 1982. Azolla as a green manure: use and management in crop production. 230 p. Westview Press, Boulder, Colorado.

Martin, S.M. and Skerman, V.B.D. 1972. World Directory of Collection of Cultures of Microorganisms. 560 p. Willey Interscience Publ.

Mooney, P.R. 1983. The law of the seed-another development and plant genetic resources. Development Dialogue 1983. 1-2.

Myers, N. 1983. A Wealth of Wild Species. 274 p. Westview Press, Boulder, Colorado. Plucknett, D.L., Smith, N.J.H., Williams, J.T. and Anishetty, N. Murthi. 1983. Crop germplasm conservation and developing countries. Science 220: 163-169.

Prescott-Allen, R.C. 1981. In situ conservation on crop genetic resources. 40 p. International Union for Conservation of Nature and Natural Resources. Switzerland.

Prescott-Allen, R.C. 1982. The case for in situ conservation of crop genetic resources. Nature and Resource 18(1): 15-20.

President's Commission for the Study of Ethical Problems in Mand Biomedical and Behavioral Research. 1982. Splicing Life. Government Printing Press. Washington, D.C. Rick, C.M. 1979. Potential improvement of tomatoes by controlled introgression of genes from wild species, pp. 167-173. In: Conference on Broadening Genetic Base of Crops. Wageningen.

Rippka, R., Deruelles, J., Watterbury, J.B., Herdman, M. and Stanier, R.Y. 1979. Generic assignments, strain histories and properties of pure cultures of cyanobacteria. J. Gen. Microbiol. Ill: 1-62.

Roger, P.A. and Kulasooriya, S.A. 1980. Blue-green algae and Rice. 112 p. The International Rice Research Institute.

Stanier, R.Y. and Cohen-Bazire. 1977. Phototrophic prokaryotes: the cyanobacteria. Ann. Rev. Microbiol. 31: 225-274.

Stewart, W.D.P., Rowell, P., Ladha, J.K. and Sampaio, M.J.A.M. 1979. Blue-green algae (Cyanobacteria) some aspects related to their role as source of fixed nitrogen in paddy soils. pp. 263-285. In: Nitrogen and Rice. The International Rice Research Institute. Los Banos, Philippines.

The American Type Culture Collection. 1968. Catalogue of strains. Eighth edition.

The American Type Culture Collection. 1982. Catalogue of strains. Fifteenth edition. Vavilov, N.!. 1932. The process of evolution of cultivated plants. pp. 331-342. In: Proceedings of the VI International Congress of Genetics. Brooklyn Botanic Garden, New York. Venkataraman, G.S. 1975. The role of blue-green algae in tropical rice cultivation. pp. 207-218. In: Nitrogen fixation by free-living micro organisms. Cambridge University Press. Venkataraman, L.V. 1983. Blue green alga Spirulina. 100 p. Department of Science and Technology, India.

Whitton, B.A. 1969. The taxonomy of blue-green algae. Br. Phycol. J. 4(1): 121-123.

Wilkes, G.H. 1983. Current status of crop plant germplasm. pp. 133-181. CRC Critical Reviews in Plant Science 1(2). CRC Press, Florida. W ochok, Z.S. 1981. The role of tissue culture in preserving threatened and endangered plant species. Biol. Cons. 20: 83-89.

Monsoon Management in an Era of Climate Change

M.S. Swaminathan*

Online edition of India's National Newspaper – The Hindu
Monday, Jul 13, 2009

Mastering the art and science of monsoon management holds the key to adaptation to adverse changes in temperature and precipitation caused by climate change leading to the more frequent occurrence of drought and floods. This year, we are experiencing an erratic behaviour of the South West Monsoon. Media accounts of the monsoon behaviour vary from day to day, from agony to ecstasy and then again to agony. Whether or not such deviations are related to climate change, the preparedness needed to enhance our coping capacity to meet the challenge of a very variable rainfall pattern is at present limited.

The South West Monsoon period (June to September), when we receive over 80% of annual rainfall, is not only important for our food and water security, but also for the work and income security of over 60 percent of our population engaged in farming. The health and survival of nearly 500 million farm animals, including cows, buffaloes, goats and sheep also depend upon the availability of water for direct consumption and for the production of feed and fodder. In addition, power generation and the recharge of the aquifer are affected. A water emergency is currently prevailing in several parts of the country. Climate models predict that sub-regional variation in more extreme rainfall pattern resulting in drought or floods may increase. Experience indicates that some of the areas experiencing drought at present may face floods in August-September, resulting in a further damage to crops. Heavy rainfall and floods in August-September will however help to fill up tanks and reservoirs and augment ground water supply, which will be very helpful for raising good **rabi** season crops.

So far, Punjab, Haryana and Western UP have had scanty rainfall. This is the region which supplies most of the wheat and rice for the Public Distribution System. Even if rains come now, the yield will be reduced since crop duration and yield are related. In the case of monsoon rains, it is not only the total rainfall which is important, but more importantly, the distribution of the rainfall which will determine the health of the crop. This is an important problem we face in our country, where most of the rainfall occurs in about 100 hours in a year. The loss occurring from a skewed distribution of rainfall is particularly high in soils with limited water holding capacity.

Another aspect of climate change is the possibility of a rise in sea level and more frequent occurance of coastal storms and cyclones. We should take anticipatory action to protect both the ecological security of coastal areas and the livelihood security of coastal communities. Preparing for climate change therefore should become an integral part of the planning process for sustainable food, water and livelihood security.

---

* Member of Parliament (Rajya Sabha)

What Should we do Immediately?

While the heartland of the green revolution is experiencing severe meteorological distress, other parts of our country are experiencing a reasonably satisfactory monsoon. This is particularly true of Eastern India comprising Bihar, Jharkhand, Chattisgarh, Orissa, West Bengal and Assam. Fortunately, the whole of Eastern India as well as parts of peninsular India have a very large untapped yield reservoir even with the currently available technologies. Since there is enough soil moisture for raising good crops in the Southern and Eastern parts of the country, we should launch a **Compensatory Production Programme** in this region. Such a programme should involve the intensification of the extension efforts, as well as supplying an additional dose of nutrients free of cost. It is important that the needed macro and micro-nutrients are supplied in the form of top dressing of standing crops. The addition of 20 to 25 kg of nitrogen per hectare will help to increase yield by a ton. The loss in yield sustained in the north west part of the country can to some extent be made good by increasing production in those parts of the country which are experiencing near normal monsoon conditions. If such a compensatory programme can be initiated immediately by mobilizing the requisite nutrients, it will be prudent to request the faculty and scholars of Agricultural and Animal Sciences Universities in the region to stop class room education for a month, and go to villages to work with farm families in improving water and nutrient use efficiency and in controlling pests and diseases. I am confident that if the Farm Universities do come to the rescue of the national food and livelihood security system by working with farm women and men for a month, they will also gain by learning more about the field problems than they will be able to do in classrooms. With a National Food Security Act on the anvil, it will be prudent to develop procedures which can help us to optimize production in the most favourable areas from the point of view of moisture availability.

Over 60 percent of our cultivated area is rainfed. Considerable research has been carried out on increasing the yield and stability of crops in dry farming areas. The gap between potential and actual yields in these areas is as high as 200 to 300 percent. There is need for launching a "**bridge the yield gap movement**" in all rainfed areas through attention to varietal choice, soil fertility, water conservation and management and plant protection. Fortunately, the Rashtriya Krishi Vikas Yojana involving an outlay of Rs.25,000 crores provides an opportunity for State Governments to help farmers to increase the production of dryland crops. Pulses and oilseeds are mostly cultivated with rain fall and we should increase the production of these crops to minimize imports and enhance the income of farmers.

There is also need for a decentralized grid of ware houses and grain storage structures based on recent technologies. This will help to prevent both distress sale and panic purchase. The production of vegetables and fruits is going up and it is important to strengthen facilities for the safe storage of perishable commodities.

Enhancing the coping Capacity

In the Dr Rajendra Prasad Memorial lecture I delivered under the auspices of Indian Society of Agricultural Statistics, in 1972 on the topic, "Can we face severe drought again without food imports?", I stressed that just as grain reserves are important for food security, seed reserves are important for crop security. This is because the implementation of contingency plans involving the cultivation of crops based on the rainfall pattern will be possible only if the seeds of the alternative crops are available. Later in 1973, delivering the Sardar Patel Memorial Lectures over the All India Radio, I stressed the need for preparing for each agro-climatic region in our country Drought, Flood and Good Weather Codes. The aim of such Codes, detailing the anticipatory measures that should be undertaken, is to minimize the adverse impact of aberrant weather and to maximize the benefits of good monsoons. Such Action Codes should become integral parts of our preparedness for insulating to the maximum possible extent our food, water and livelihood security systems from the impact of global warming.

Message of the G8 Meeting in L'Aquila

An agreement was reached at the recent G8 meeting, attended by our Prime Minister, that green house gas emissions should be reduced by 40% by 2020, so as to limit the maximum rise in temperature to 2 degrees Celsius over the pre-industrial period. Even such a 2 deg rise in mean temperature will have serious implications for our agriculture. Over 20 years ago, the late Dr S K Sinha and I showed that a 1 deg C rise in night temperature in the Punjab and Haryana will reduce the duration of the wheat crop by about a week. This in turn will reduce yield by 4 to 5 quintals per hectare, resulting in a considerable loss in wheat production in that region.

Unlike in industrialized countries, the average size of our farms is hardly 1 ha. The smaller the farm, the greater is the need for marketable surplus to get cash income. Therefore, the implications of reduced yield should not be considered just in terms of production of wheat or rice or other crops, but also in terms of the income of the small farmer. To keep farmers in farming, there will be need for proactive action in enlarging rural livelihood opportunities in the non-farm sector.

At the same time, we should intensify research on assembling genetic material for a warming India. Novel genetic combinations for tolerance to higher temperature and moisture stress can be developed through the tools of recombinant DNA technology. Similarly, flood and salinity tolerant strains of major crops can be developed. Farming has to become knowledge and technology intensive if we are to overcome successfully the emerging challenges. This year provides opportunities for initiating a well planned monsoon management strategy based on the best available scientific tools in rural India. The M S Swaminathan Research Foundation is developing training modules to build a cadre of Climate Risk Managers (atleast 1 woman and 1 male in every Panchayat), well versed in the science and art of monsoon management based on the integration of traditional wisdom with frontier technologies. A national cadre of grassroot Climate Risk Managers will help to initiate the proactive action needed to minimize human suffering and ensure both food and drinking water security in the emerging era of global warming and climate change.